AF792507

Trace Metals in the Environment - New Approaches and Recent Advances

Edited by Mario Alfonso Murillo-Tovar,
Hugo Saldarriaga-Noreña
and Agnieszka Saeid

Published in London, United Kingdom

IntechOpen

Supporting open minds since 2005

Trace Metals in the Environment - New Approaches and Recent Advances
http://dx.doi.org/10.5772/intechopen.83504
Edited by Mario Alfonso Murillo-Tovar, Hugo Saldarriaga-Noreña and Agnieszka Saeid

Contributors
Mpitloane Joseph Hato, Kwena Desmond Modibane, Gobeng Release Monama, Kabelo Edmond Ramohlola, Arjun Maity, Mogwasha Makhafola, Thabiso Carol Maponya, Lebogang Katata-Seru, Thabang Somo, Barbara Mueller, Eleazar Salinas, Juan Hernández, Eduardo Cerecedo, Ma. Isabel Reyes, Ventura Rodriguez, E. Omar Serrano-Mejía, Ma. Pilar Gutierrez, Cengiz Soykan, Luqman Ali Shah, Tanzil Ur Rehman, Noor Saeed Khattak, Abbas Khan, Noor Rehman, Dr Sultan Alam, Eman Noori Ali, Indu Sharma, Boukhlifi Fatima, Yao Shan, Jianjun Shi, María Luisa García Betancourt, Sandra Ramírez-Jiménez, Apsahara Nohemí González-Hodges, Zandra Nuñez-Salazar, Ismailia Leilani Escalante García, Jeannette Ramírez Aparicio, Uchenna Okereafor, Lukhanyo Mekuto, Vuyo Mavumengwana, Elizabeth Makhatha, Godwin Okereafor, Jumina Jumina, Harizal Harizal, Diana Linhares, Patricia Garcia, Armindo Rodrigues, Elizabeth Makhatha, Lukhanyo Mekuto, Vuyo Mavumengwana

© The Editor(s) and the Author(s) 2021
The rights of the editor(s) and the author(s) have been asserted in accordance with the Copyright, Designs and Patents Act 1988. All rights to the book as a whole are reserved by INTECHOPEN LIMITED. The book as a whole (compilation) cannot be reproduced, distributed or used for commercial or non-commercial purposes without INTECHOPEN LIMITED's written permission. Enquiries concerning the use of the book should be directed to INTECHOPEN LIMITED rights and permissions department (permissions@intechopen.com).
Violations are liable to prosecution under the governing Copyright Law.

Individual chapters of this publication are distributed under the terms of the Creative Commons Attribution 3.0 Unported License which permits commercial use, distribution and reproduction of the individual chapters, provided the original author(s) and source publication are appropriately acknowledged. If so indicated, certain images may not be included under the Creative Commons license. In such cases users will need to obtain permission from the license holder to reproduce the material. More details and guidelines concerning content reuse and adaptation can be found at http://www.intechopen.com/copyright-policy.html.

Notice
Statements and opinions expressed in the chapters are these of the individual contributors and not necessarily those of the editors or publisher. No responsibility is accepted for the accuracy of information contained in the published chapters. The publisher assumes no responsibility for any damage or injury to persons or property arising out of the use of any materials, instructions, methods or ideas contained in the book.

First published in London, United Kingdom, 2021 by IntechOpen
IntechOpen is the global imprint of INTECHOPEN LIMITED, registered in England and Wales, registration number: 11086078, 5 Princes Gate Court, London, SW7 2QJ, United Kingdom
Printed in Croatia

British Library Cataloguing-in-Publication Data
A catalogue record for this book is available from the British Library

Additional hard and PDF copies can be obtained from orders@intechopen.com

Trace Metals in the Environment - New Approaches and Recent Advances
Edited by Mario Alfonso Murillo-Tovar, Hugo Saldarriaga-Noreña and Agnieszka Saeid
p. cm.
Print ISBN 978-1-83880-331-5
Online ISBN 978-1-83880-332-2
eBook (PDF) ISBN 978-1-83880-617-0

For EU product safety concerns:
IN TECH d.o.o., Prolaz Marije Krucifikse Kozulić 3, 51000 Rijeka, Croatia, info@intechopen.com or visit our website at intechopen.com.

We are IntechOpen, the world's leading publisher of Open Access books

Built by scientists, for scientists

5,100+
Open access books available

127,000+
International authors and editors

145M+
Downloads

156
Countries delivered to

Our authors are among the
Top 1%
most cited scientists

12.2%
Contributors from top 500 universities

CLARIVATE ANALYTICS BOOK CITATION INDEX INDEXED

WEB OF SCIENCE™

Selection of our books indexed in the Book Citation Index in Web of Science™ Core Collection (BKCI)

Interested in publishing with us?
Contact book.department@intechopen.com

Numbers displayed above are based on latest data collected.
For more information visit www.intechopen.com

Meet the editors

Dr. Mario Alfonso Murillo Tovar is currently working at CIQ-IICBA Universidad Autónoma del Estado de Morelos, Cuernavaca, México as a Professor-Researcher and he is involved in teaching, research, management, and academic work. He received his undergraduate Bachelor of Chemistry degree from Universidad del Valle, Colombia and he obtained his Master´s degree and Doctorate in Chemical Sciences from Universidad Nacional Autónoma de México. His research has focused on the development and validation of analytical methods, chemical characterization of environmental samples, and treatment and removal methods. He has worked on many projects, including determination of trace metal, inorganic species and toxic organic compounds using ICP-MS, GC, and LC tandem MS techniques, identification of emission sources, chemical degradation of emerging compounds and risk assessment.

Since 1991, Hugo Saldarriaga-Noreña has worked in the environmental pollution field, specifically in water purification. Later in 1995, I joined the University of Antioquia (Colombia), combining research with teaching, until 2001. In 2007, I joined the Center for Research and Assistance in Technology and Design of the State of Jalisco, Mexico, as a leading researcher in the area of air quality. Since 2012, I have been a full-time professor at the Autonomous University of the State of Morelos, Mexico. My research area is environmental chemistry, specifically oriented to the characterization of environmental matrices, through the application of high-performance liquid and gas chromatography, ICP mass spectrometry, nuclear magnetic resonance, and IR among others.

Agnieszka Saeid graduated from the Department of Chemistry, Wrocław University of Science and Technology (WUST) in 2005. In 2010, she defended her doctoral thesis, and in 2019 she obtained habilitation and she was granted a Marie Sklodowska-Curie Actions (MSCA) to participate in the Postgraduate School of Industrial Ecology (PSIE) - IndEcol at the Norwegian University of Science and Technology in Trondheim. She has published over 77 papers, including 69 papers in world-known peer-reviewed scientific journals from the JCR list. Her works have been cited more than 300 times and she has an h-index: 15. She is the author of 5 patents. She cooperates with many universities in the field of multidisciplinary projects. She has edited books for CRC Press, Wiley, and NOVA Science and has also contributed to 10 book chapters published by Elsevier, Wiley, CRC Press, Nova Science, Studium Press LLC, and IntechOpen.

Contents

Preface

Environmental pollution with trace metals is one of the most severe ecological problems with a probable threat to human health and negative effects on nature and properties.

This book covers the current literature on the effect of trace elements on the environment and provides the readers with a more comprehensive overview of the main areas of concern to improve the understanding of the complexity associated with trace element research in the environment. This book is a comprehensive approach to those main aspects of metal pollution that should be considered when trying to reduce its emissions and diminish its impact on ecosystems and human health.

The reviews and research papers contained in this book summarize different methodologies for the identification and contribution by emissions sources and describe some of the most important processes involved in the mobility and transport of heavy metals through different compartments in the environment. This book also presents some information about toxic metal effects from direct exposure and environmental health risk assessment, and focuses on the potential use of innovative and technological alternatives for its removal and remediation with synthesized nanomaterials, waste materials, bioremediation techniques and natural adsorbents.

The editors wish to acknowledge the support of CONACYT (No. 2121). We also would like to thank each one of the authors for their contribution and involvement.

Mario Alfonso Murillo-Tovar and Hugo Albeiro Saldarriaga-Noreña
Universidad Autónoma del Estado de Morelos,
Mexico

Agnieszka Saeid
Wroclaw University of Science and Technology,
Poland

Section 1

Source, Mobility and Fate Processes

Chapter 1

Data Mining for Source Apportionment of Trace Elements in Water and Solid Matrix

Yao Shan and Jianjun Shi

Abstract

Trace elements migrate among different environment bodies with the natural geochemical reactions, and impacted by human industrial, agricultural, and civil activities. High load of trace elements in water, river and lake sediment, soil and air particle lead to potential to health of human being and ecological system. To control the impact on environment, source apportionment is a meaningful, and also a challenging task. Traditional methods to make source apportionment are usually based on geochemical techniques, or univariate analysis techniques. In recently years, the methods of multivariate analysis, and the related concepts data mining, machine learning, big data, are developing fast, which provide a novel route that combing the geochemical and data mining techniques together. These methods have been proved successful to deal with the source apportionment issue. In this chapter, the data mining methods used on this topic and implementations in recent years are reviewed. The basic method includes principal component analysis, factor analysis, clustering analysis, positive matrix fractionation, decision tree, Bayesian network, artificial neural network, etc. Source apportionment of trace elements in surface water, ground water, river and lake sediment, soil, air particles, dust are discussed.

Keywords: trace elements, data mining, source apportionment, water, sediment, soil, particles

1. Introduction

On the issue of trace element contamination of environment, the trace elements refer to the elements with lower concentrations than the major elements, O, H, Si, Al, Fe, Ca, Mg, Na, K, Ti, which are usually take no more than 1% in rocks and minerals. The trace elements have attracting wide research attentions for their high potential on environmental contamination and health impact. In some articles, the phase heavy metals are frequently used to represent elements that have high density or is toxic or poisonous at low concentrations. From the view of environmental impact, the phases trace elements and heavy metals refer to similar research objects, which are used as group name for metals and metalloids that have been associated with contamination of water, river sediment, soil and air particles and potential toxicity and ecotoxicity. In this chapter, the phase trace elements (TEs) are used to present the elements that may cause contamination and health problems, and

IntechOpen

address issue relating to the behavior and mechanism of them among environmental bodies.

The TEs are widely studies in the areas of water, rock, coal geochemistry, leaching and mobility potential, bioaccumulation and human health risk, survey technologies, and other related topics [1, 2]. The harm to human health of TEs are amount related. Some TEs are essential to human in a concentration scale, while become toxic along with the concentration elevation. Some toxic TEs may cause acute and chronic effect even in very low content. In light of the levels of toxicity, trace elements lead (Pb), zinc (Zn), copper (Cu), nickel (Ni), chromium (Cr), cadmium (Cd), arsenic (As), selenium (Se), mercury (Hg), are most investigated, studied and regulated. For example, small amounts of lead in the body can make it difficult for children to learn, pay attention and succeed in school. Lead accounts for most of the cases of pediatric heavy metal poisoning. Arsenic is the most common cause of acute heavy metal poisoning in adults and does not leave the body once it enters. Mercury exposure put newborns at risk of neurological deficits and increased cardiovascular risk in adults.

The TEs may be released from sources of lithogenic or anthropogenic [3, 4]. With the industrialization and urbanization process, TEs released from anthropogenic source are increasing, including discharge of industrial and municipal wastes, storms, run-offs, dry deposition, mine discharge, waste incineration, application of pesticides and fertilizers, sewage irrigation and transportation, and other diffused sources [1, 5–11]. The environmental medias, including water [10, 12–14], sediment, soil [3, 4, 15–23], air particles [15] can be contaminated.

In order to understand and control pollution of the trace element, source identification and quantification of TEs in water, sediment, soil, and particles are of great importance. The traditional techniques are mostly based on geochemical method. Statistical method based on univariate analysis are also used. However, the univariate analysis is cumbersome, and sometimes hard to explain. The multivariate analysis provides a new technique system for the TE source apportionment. Multivariate analysis, and related method, machine learning, data mining have been approved to be successful in a very wide aspects of human living and production. In the area of geochemistry, environmental engineering, applications of the method are also increasingly used.

In this chapter, two related topics are reviewed and discussed. First, the advances of multivariate analysis on the issue of source apportionment, especially several kinds of multivariate analytical method; second, understanding of the contaminating origin of TEs on important environmental media, ground and surface water, sediment in river and lake, soil, precipitate dust, suspended particle matters, PM 2.5 and PM 10.

2. Methods for data mining

To investigate trace element concentration in time series and spatial distribution, migration source, and reaction pathway, technology of data mining is used. In narrow sense, the data mining refers to using multivariate analysis and machine learning method to find distributing or changing pattern in big data sets. In a broader concept, the data mining may include more techniques, such as geochemical, isotopic, univariate analysis, etc. In this chapter, techniques of multivariate analysis and machine learning are emphasized for its increasingly application and effective in source apportionment and reaction path analysis.

Table 1 lists application of data mining methods and implementation on the trace element migration. In which, PCA stands for principal component analysis,

Environmental media	Country/region	Anthropogenic source TEs	Data mining method	References
Water (surface)	Ethiopia	Cu, K	PCA/FA/CA/DA	[24]
Water (surface)	Turkey	Not found	PCA/HCA	[25]
Water (surface)	Belgium	N	Bayesian network	[26]
Water (surface)	USA	Mg, Cl, Na	CA/DA	[27]
Water (surface ground)	Greece	EC, Cl, SO_4, Mg, Ca, Na, K, As, Fe, B, Br, Sr, V (sea water intrusion)	PCA/DA	[28–30]
Water (surface ground)	Spain	—	PCA/regression	[31]
Water (ground)	China	—	PCA/DA	[32]
Water (ground)	China	—	DA	[33]
Water (ground)	China	Se, As, Hg, Cr, Pb	PCA	[34]
Water (ground)	India	Pb, Cu, Cr	PCA/CA	[35]
Water (ground)	India	As, Cd, Co, Pb, V	PCA/CA	[36]
Water (ground)	Greece	N	Bayesian network	[29]
Water (ground)	USA	Cr, Br, Cl, N, S	Semi-supervised ML	[37]
Water (ground)	China	—	Decision tree/CA	[38]
Water/soil	Nigeria	Cd, Cr, Pb, Ni, V	PCA/CA	[12]
Sediment (river)	China	Ni, Hg, Cr, Cu, Cd, Pb, Zn, As	PCA/DA/Monte Carlo	[39]
Sediment (river)	China	Cd, Zn	PCA	[40]
Sediment (river)	China	Cr, Cd, Pb, Hg	PCA, EF	[41]
Sediment (lake)	China	As, Cd, Hg, Pb, Zn	PCA/EF	[42]
Sediment (lake)	China	Cr, Pb, Zn, Cu, Co	PCA	[43]
Soil	China	Hg, Cr, Ni, Ba	PCA/CA	[44]
Soil	China	Cu, Zn, Cd, Hg	PCA	[45]
Soil	China	Not found	PCA/CA	[46]
Soil (peat)	Spain	Cd, Pb, P, Zn	PCA/CA	[47]

Environmental media	Country/region	Anthropogenic source TEs	Data mining method	References
Soil (dust)	China	Zn, Mn, Ni, As, Cu, Pb, Cr, Co	PCA	[48]
Soil (dust)	China	Pb, Cd	PCA/ANN	[49]
Soil	India	Ni, Co	PCA	[50]
Soil (city topsoil)	Armenia	Pb, Zn, Cu, Mo	PCA/CA	[51]
Soil (atmospheric deposition)	China	As, Hg, Cu, Cd, Mo, S, Zn, Cr, Ni, Pb, Se	PCA/CA	[52]
Soil	Spain	Pb, Tl, As, Sb, Cd, Cr, Ni, Be, V, Co	FA	[53]
Soil	Pakistan	Ni, Cr, Zn, Cu, Pb, Cd, Co	PCA/FA/ CA	[54]
Soil (agriculture)	Greece	Cu, Pb, Zn, As, Cd, P, K	PCA/CA	[55]
Soil	Italy	Ni, Cr, Pb, Zn	PCA/FA-MLR	[56]
Soil	China	Cd, Hg, Pb, Zn	PCA/CA	[57]
Soil	Iran	—	Semi-supervised ML	[58]
Soil	USA	—	(Six models)	[59]
Particle	India	Ni, Cu, Pb, Cd, Cr	PCA/CA	[60]
Particle (PM 2.5)	Canada	—	PMF	[61]
Particle (PM 2.5)	China	Cr, Mn, Fe, Cu, Zn, As, Pb, Ba	Regression/Monte Carlo	[62]
Particle (PM 2.5)	China	—	PCA-MLR	[63]
Particle (PM 2.5)	China	Zn, Pb, Cd, Cu	PCA	[64]
Particle (PM 2.5)	USA	—	PCA	[65]
Particle (PM 2.5, PM 10)	Costa Rica	Fe, Ni, V	PMF	[66]
Particle (PM 2.5, PM 10)	Nigeria	Cl, K, V, Cr, Ni, Br, Pb, S, Na, S, Zn, As	PMF	[67]
Particle (PM 2.5, PM 10)	USA	Zn, Pb, P	PMF	[68]

Table 1.
A list of data mining method on trace element migration in recent years.

CA stands for clustering analysis, ANN stands for artificial neural network, FA stand for factor analysis, MLR stands for multi-linear regression, DA stands for discriminate analysis, EF stands for enrichment factor, PMF stands for positive matrix fractionation.

2.1 Geochemical methods

To analyze geochemical properties, and reaction mechanisms, mass balance, piper diagrams, Gibbs diagrams [28] are usually applied. The piper diagram shows the major element composition of water, which category water into different types. Several software, PHREEQC, MINTEQ, geochemists' workbench, can be used to calculate mass balance, saturated index, and model the reaction path, draw piper diagram, etc. By the process of water-rock interaction, major elements and TEs may be released and immigrate to other water bodies, therefore the major elements and trace element with distinguishing feature could be used as source apportionment [69, 70]. However, the TEs undergo geochemical process of adsorption, desorption, mineralization, dissolution to change concentration in water. Therefore, the univariate analysis is not credible and robust. Comparatively, isotopic analysis, both stable and radiogenic [30], may be used as a univariate analysis method or combing some other indexes. The widely used isotopic method are $\delta^{18}O$ and δD in water, $^{87}Sr/^{86}Sr$, $\delta^{34}S$ and $\delta^{18}O$ in sulfate [71, 72], $\delta^{15}N$ and $\delta^{18}O$ in nitrate [29], etc. The isotope $\delta^{18}O$ and δD in water are used to identify water relations between precipitation and surface/ground water. The ratio of strontium isotope of water is strictly controlled by water-rock interaction. For a unisource water, the $^{87}Sr/^{86}Sr$ reflect the mineralogy of the rocks with which the water has been contact and does not change along the water flow. It is highlighted that differences in the strontium isotope ratio and strontium concentration are caused mixing of water of various origins with specific chemical characteristics and isotopic values. Therefore, the strontium isotope is an ideal tracer for element resources, groundwater movements, and water-rock interaction [70, 73–75]. The use of $\delta^{34}S$ and $\delta^{18}O$ in sulfate is increasing because they have wide range of stable isotope composition and the $\delta^{34}S$ value is derived from multiple sources and very close to that of the precursor sulfide mineral. A common anthropogenic source of sulfate is the coal and metal mining which is rich in pyrite and other sulfide minerals. In activity and abandon coal mines, the sulfide mineral may be oxidized and dissolved, with the release of trace elements. It has been proved by the sulfate isotopes that the ground water could be contaminated by the water-rock interaction in coal mines. Besides of the natural isotope, some tracers such as isotopes and stable organic compound are injected into groundwater to find out the flow pathway [71, 72, 76, 77].

To evaluate contamination of TEs and find the source of pollution on water and solid matrix, some calculations are used. The enrichment factor (*EF*) is an enrichment level of a certain TEs in environment, with an equation as shown in Eq. (1):

$$EF = \left(c_i/c_{ref}\right)_{samples} / \left(B_i/B_{ref}\right)_{baseline} \tag{1}$$

where the c_i is the measured concentrations of TEs in samples, the c_{ref} is the measured concentration of the reference element, B_i and B_{ref} are the background level of the local region and reference element in the same region [41]. An *EF* value close to 1 suggests a weathering origin of trace element, while a higher than 1 value means TEs enrichment in soil which is probably caused by human activities. An *EF* value between 2 and 5 indicate a moderate contamination, and a higher than 5 value show a heavily polluted by TEs [49]. The *EF* factor is frequently used combing wit data mining method, or as a verification to trace the contaminating sources.

The geo-cumulative index method (I_{geo}) is defined using Eq. (2):

$$I_{geo} = \log_2[c_i/1.5B_i] \tag{2}$$

where the c_i is the measured concentration of TEs, and the B_i is the background concentration of the particular TEs [41].

The Hakanson potential ecology risk method (RI) was proposed by Hakanson and can be used to evaluate the potential ecological risk posed by TEs in water and solid matrix. This comprehensive method considers four factors: concentration, type of pollutant, toxicity level, and the sensitivity of the water body to metal contamination in water and solid matrix [78, 79].

2.2 Machine learning

Studies of environmental processes exhibit spatial variation within data sets. The ability to derive predictions of risk from field data is a critical path forward in understanding the data and applying the information to land and resource management. Multivariate analysis, or machine learning methods present advantages of precise, robust, and can look insight the phenomena to find mechanism. On the other hand, the environment data usually composed of matrices. Therefore, the machine learning methods is an ideal tool to deal with environmental and geochemical issues. However, the calculation of machine learning and multivariate analysis are complex, which may prohibit their implementation. Thanks to recent advances in predictive modeling, open source software (R, Python, SPSS, SAS, Minitab, etc.), and computing, the power to do this is within grasp.

Basic principle of ML is to train models for the specific data frame using the obtained data, then apply the models on the target problems. The ML methods can be divided into three types, namely supervised, unsupervised and semi-supervised learning. When the training data has labeled data, it is a supervised ML, while unsupervised ML have no labeled data. The semi-supervised ML add labels to data during model training. Generally, the supervised ML has higher precise and robust than others. However, the geochemical and environmental data are usually unlabeled. For example, when the researchers try to identify source of water, or TEs in water and solid matrices, the results are usually not assured. Therefore, the unsupervised ML, up to present, has more widely implemented than the supervised and semi-supervised ML.

2.2.1 Unsupervised ML

It is undoubted that the unsupervised ML is mostly used in this area. Common techniques of unsupervised ML include: principal component analysis (PCA), factor analysis (FA), clustering analysis (CA), positive matrix factorization (PMF), etc.

In the scope of machine learning algorithm, PCA is a tool to reduce high dimensional matrix to a lower, usually two to five, dimensional matrix. Dimensional reduction is accomplished by transforming the data to a new set of variables (principal components), which are derived from linear combinations of the original variables and classified in such a way that the first principal components are responsible for most of the variation in all of the original variables [80].

A matrix M, with m observations (row) and n variates (column), is calculated following five steps to form a new matrix with less variates.

Step 1: the raw data in the M is standardized;

Step 2: the covariance matrix of the standardized M is calculated;

Step 3: eigen value and eigen vectors of the covariance matrix are calculated;

Step 4: contributing ratio and accumulative contribution of the eigen value was calculated, then the principal components can be determined according to mathematical and project criterion;

Step 5: loading of every principal component and score of every observation can be calculated.

Theoretically, the number of new variates is equal to variates of the original matrix. On the other hand, the new variates contribute different ratio to explain variance of variates, then the principal components are selected based on the explanation ratio. Different criterion was used, some researchers use the eigen value larger than 1, and some others use the accumulative contribution of the eigen values, say 80%.

Given the M has 30 observations and 10 variates ($x1, x2, \ldots, x10$), and three principal components ($y1, y2, y3$) are selected, eigen vector of $y1$, $y2$, and $y3$ are $A1$, $A2$, and $A3$, respectively, then:

$y1 = A1 * x1$
$y2 = A2 * x2$, and
$y3 = A3 * x3$.

After the calculation of PCA, some variates have higher loadings on specific principal components, while some variates have higher loadings on other PCs. Then it is inferred that the variates have similar pattern in the matrix may have similar pattern in the real world, i.e., the source, migration behavior, and reaction pathway. Theoretically, the PCA is similar with clustering analysis, but the PCA is not constrained to two dimensions, which allow the researchers mine the inner relationships in the matrix and understand real world more precise.

The Factor analysis (FA) is based on PCA, have similar principle, and aim to obtain similar result with PCA, but the applications are less than PCA.

The data on every variate should be normal distributed. Kaiser-Meyer-Olkin (KMO) test and Bartlett's sphericity test are usually used to determine the distribution of data for analysis of PCA/FA.

Because of the advantages of PCA and FA in analysis of environmental and geochemical data mining, they are widely implemented [1–4, 11, 12, 17, 19, 23, 28–30, 34, 41, 49, 64, 78, 81–83].

On the issue of source apportionment of particle matter and trace elements for the suspended particles and trace elements inside, a method of positive matrix factorization (PMF) is usually used. When we have a matrix M, with f of observations, and n of variates, the M can be calculated as Eq. (3):

$$M_{(f*n)} = W_{f*k} * H_{k*n} \quad (3)$$

In a source apportionment problem, W stands for source contributions, H stands for source profiles, k is the number of possibly sources. The least loss function determines the proper k value and the matrix W and K, then the source quantity, contribution ratio can be inferred. The W_{f*k} need to be normalized by their average value across all samples as shown in Eq. (4):

$$\overline{w}_{fk} = \sum_{f=1}^{n} \frac{w_{fk}}{n} \quad (4)$$

where $\overline{w}_{fk}$ is elements in the matrix W. The PMF is usually used in source apportionment for particle, such as PM 10 and PM 2.5 [61, 66–68], but seldom used in other environmental medias.

2.2.2 Bayesian network

A model of Bayesian Network has been implemented to estimate TE source contribution, and evaluate the contaminating levels [26, 29, 84–86]. A R package SIAR (Stable Isotope Analysis in R) can be run to calculate the isotope mixing model base on the Bayesian Network. The mixing model can be elucidated as the equation set Eq. (5):

$$\begin{aligned} X_{ij} &= \sum_{k=1}^{k} p_k\left(S_{jk} + c_{jk}\right) + \varepsilon_{ij} \\ S_{jk} &\sim N\left(\mu_{jk}, \omega_{jk}^2\right) \\ c_{jk} &\sim N\left(\lambda_{jk}, \tau_{jk}^2\right) \\ \varepsilon_{jk} &\sim N\left(0, \sigma_j^2\right) \end{aligned} \quad (5)$$

where X_{ij} is the isotope value j of the mixture i, in which i = 1, 2, 3, ..., N and j = 1, 2, 3, ..., J; S_{jk} is the source value k on isotope j(k = 1, 2, 3, ..., K), c_{jk} is the isotope fractionation factor for isotope j on source k. p_k is the proportion of source k, which needs to be estimated by the SIAR model. The S_{jk} and c_{jk} are normally distributed with mean μ_{jk} and standard deviation ω_{jk}, mean λ_{jk} and standard deviation τ_{jk}, respectively. ε_{ij} is the residual error representing the additional unquantified variation between individual mixtures and is normally distributed with mean 0 and standard deviation σ_j. Algorithm of Monte Carlo is usually to solve the equation of Bayesian network.

2.2.3 Decision tree

The decision tree is a kind of supervised ML, including a series of machine learning techniques to divide samples into different categories, such as algorithms ID 3, C 4.5, C 5.0, CART, etc. Different algorithms follow the same principle: the observations are divided by breakpoints on a variate. The selection of variates and breakpoints provide basis of decision, and all of the decisions make a tree for users to make a project decision system. Take the algorithm CART for example, the sample space needs to be split by using the variate breakpoints. Different split strategies make different decision efficiency. As an index, the Gini Coefficients are use. The decision tree machine will calculate Gini index for every split, the split method with lower Gini Coefficients is used. Advantages of the decision tree are easy to carry out and easy to explain. Some researchers have introduced the method to trace source of nitrogen and TE contaminations [38].

Some decision tree series methods, including random forest, boosting method are widely used in the data mining [59]. However, the applications in the source apportionment are rare. The obstacle that prohibit the implementation of the decision tree methods may be the acquirement of data labeled.

2.2.4 Artificial neural network

The artificial neural network (ANN) has been recognized as a powerful supervised ML and applied in a wide scope of engineering and research. The ANN research is one of the most active research in the ML algorithm, which have a lot of branches. It is also the basis of deep learning, which is used as figure and voice identification.

In the area of environmental and geochemistry, the implementation of ANN is not as popular as unsupervised ML. The most important reason is that it is a kind of supervised data mining method. In the data preparation step, the observations need to be labeled, while the environmental data are usually cannot or difficult to label. However, some researches have used this to predict the contaminating potential, Mclean et al. reviewed the application of ANN on ambient air pollution [87]. The second obstacle for its implementation is that ANN usually need larger amount of data than PCA, decision tree, and some other methods.

A basic ANN model has an input layer, one or several hidden layer, and an output layer. Variates of one observation are input through the input layer, and the output layer are labeled data, the hidden layer are used to calculate the model from input to output. The relationship of the layers is trained by input the variate data and labeled data. A trained model is used to predict while input data are obtained. Once the labeled data can be obtained, the ANN model is useful to predict, discriminate, divide samples with trained model in the engineering and research of geochemical and environmental purpose.

2.2.5 Discrimination analysis

The discrimination analysis (DA) is a kind of supervised ML, because the data set is labeled in the model training step. The DA has a similar concept with the principal component analysis. While the PCA tries to find principal components (new axis) to stand for the most variates, the DA tries to find axis that stand for the least variates, so that the different variates and observations can be divided. Prior to the application of DA, all of the variables were standardized to ensure that scale differences between the variables are eliminated. Hence, the absolute discriminant weights ranked the variables in terms of their discriminating power, i.e., the variables with large weights are those that contribute most to differentiating the groups.

In the model training step, if the origin of samples can be identified, then the DA could be used to identify sample source. For example, water intrusion in coal mines may come from different ground water aquifer, the different aquifers have specific geochemical characteristics and hazard level. As a water hazard control work, discrimination models can be set up by train the labeled data. The labeled data means water collected from different aquifers. Once water intrusion happens, the water characteristics is used to identify source of water by comparing with the model [27, 33].

The criterion applied in the discrimination analysis are mainly distance based or Bayes rule base. When the distance rule is used, Manhattan distance of samples to different groups are calculated, a group with less distance to the samples is labeled to the samples. The distance has to be calculated in pair, which constrains efficiency of the model training and implantation. Another popular method is called the Bayes discrimination method. A reasonable way to discriminate the group of a characteristic sample is to compare the conditional probability of the characteristic sample falling in different category. The class with the highest conditional probability is the final category result of this characteristic sample. Theoretically, the Bayes DA has a higher coefficient and accurate than the distance-based DA.

2.2.6 Semi-supervised machine learning

The unsupervised ML are easy to carry out, but low in accuracy, robust, reliability and duplicate, while reverse for the supervised ML method. As an improved strategy, the semi-supervised ML is an option. When the labeled data is not easy to acquire, and need to do unsupervised ML at first, then the semi-supervised algorithm may apply to add labels for the data while the model is being trained. Vesselinov et al. used the non-negative matrix factorization method for blind source separation in the first step, then a semi-supervised clustering algorithm was used to predict the sources of contaminates [37]. Fatehi and Asadi used a hybrid method combining hieratical clustering and fuzzy c-means clustering to classify soil types [58]. At present, this method used in this topic at present is rare.

2.3 Regression

The regression is easy to use, explain, and understand. Also, the regression is a big too box in the machine learning workshop. The most popular method is the multivariate linear regression, sometime logistical regression, lasso regression, ridge regression, plastic net regression can also be used. The regression method can be combined into other machine learning techniques, such as decision tree [56], support vector machine, etc. However, the regression has very distinction shortages. In a regression process, the model of data is fitted to a linear or curve function, which may not accord with the real situation. Second, the regression is prone to overfitted, while the training mode performed well, disaster results may be gotten when applied in real environment. To solve this problem, lasso, ridge, and plastic net regression are applied. Besides the two issues, another problem may bother the application of the regression, the data are usually not easy, or cannot to label. In this situation, unsupervised techniques should be used. Once the labeled data are acquired, regression method are applied [31, 56, 59].

2.4 Artificial tracers

In order to find ground-surface, ground-ground water relationship, artificial tracers are also used. The chemical traces sodium chloride, eosine, uranine and pyranine were used to analyze spring-ground water relationship. Conductivity meter and thermometer was yet installed for electrical conductivity (EC) monitoring and field fluorimeter was equipped for tracer detection [31].

2.5 Other methods

In a research from Alaska America, six models were set up to predict soil contamination. The model includes random forests, generalized boosted regression, elastic net regression, multivariate adaptive regression splines, generalized linear model with stepwise selection using Akaike's information regression, and partial least squares regression. Although got similar explanatory power overall among the models, the machine learning models performed much better than the linear models on predictive accuracy and were better able to identify variables of interest and describe non-linear relationships. In order to understanding the mechanisms behind trace element pollutant fate and transport and were less vulnerable to errors of omission, the machine learning techniques have priorities than the linear models [59].

3. Implementation of the data mining of TE source apportionment

The environmental medias that may be contaminated by trace elements are grouped into four types, water, sediment, soil, and particles in this chapter. In every case, the probably sources of trace elements are listed in order of importance. The main method to be used are listed in **Table 1**.

3.1 TE apportionment in water

3.1.1 Contamination sources of surface water

The TEs migrate from rock and coal to water through water-rock interaction. Then the surface and ground water may be contaminated.

In Turkey, TEs source in a large reservoir was identified. The PCA showed that PC1, PC2, and PC3 includes Co/Cr/Fe, Cu/Pb/Zn, and As/Cd, respectively. Combing with correlation analysis, the three PCs were identified to natural source, bedrock weathering, and bedrock weathering, respectively [25]. Another research revealed by PCA that mineral pollution, nutrient pollution, and organic pollution are major latent factors which influence the water quality of Asi River [88].

Because of the vandalization of pipeline, soil and water may be contaminated. The PCA results showed that the first source was associated with anthropogenic source, such as vehicular emission, which was composited by Cd, Cr, Pb, and Mn. The second source, including Cu and Zn, was related to natural geological origin, and the Ni and V were released from natural source collaborating with the petroleum contamination [12].

In Ethiopia, water samples were divided into four categories by clustering analysis: natural cluster, mixed cluster, agriculture cluster and urban cluster. In the agriculture cluster, VF1 has strong loadings on TN, NO_3^-, salinity, Fe, NH_3, hardness, and Mn, which is cultivated originated, VF2 were associate with turbidity, Chl-α, and Cu, which may come from farming and excavation sites of quarrying activities. Mg, and K were mainly loading on VF3, and VF4, respectively. K is mainly spread while potash fertilizer is used [24].

Supervised ML technique, discriminant analysis, was applied with the clustering analysis to assort and find spatiotemporal distribution of trace element in surface water, in the USA. Sources of salt ions (magnesium, chloride, and sodium) vary from natural sources (oceans, atmospheric deposition, weathering of common rocks, minerals and soils, and salt deposits and brines) to anthropogenic sources (landfills, wastewater and water treatment, agriculture, and application of deicing salts) [27].

A Bayesian isotope mixing model was used to estimate proportional contributions of multiple nitrate sources in surface water in Belgium. The result showed that "manure and sewage" contributed highest, "soil N", "NO_3^- fertilizer" and "NH_4^+ fertilizer and rain" contributed middle, and "NO_3^- in precipitation" contributed least [26].

3.1.2 Contamination sources of ground water

In southern India, potential TE source of ground water was analyzed, it was concluded that Fe and Mn were natural origin, Cr, Cu, Pb and Ni may come from mixed sources, natural and flow contaminated with fertilizers and pesticide. In another study of northern India, the sources of ground water were identified to be anthropogenic source via agrochemical and industrial wastes (As, Cd, Co, Pb and V), parent material from an adjacent area (U and Sr), lithogenic origin (Fe, Mn, Zn), and background level elements (Mo and Se), respectively [36].

In Greece, Matiatos et al. [28–30] investigated surface water and ground water combing the method of geochemical, isotopic and multivariate statistical analysis, such as PCA and Bayesian isotope mixing model. By the PCA analytical result, EC,

Na, K, Cl, and Mg, were found to be seawater inrush origin, Fe/Zn, Ca/hardness were from water-silicate rocks interaction, and dissolution of limestone, respectively, NO_3 stand for nitrogen pollution, and SO_4^{2-} and Mn were from dedolomitization process and increased agricultural input.

A semi-supervised ML technique was used to trace contaminants' source in the USA. Vesselinov et al. [37] proposed a contaminant source identification approach that performed decomposition of the observation mixtures based on non-negative matrix factorization (NMF) method for blind source separation (BSS), coupled with a custom semi-supervised clustering algorithm. As a result, the mixing coefficients of all the groundwater types (contaminant sources) for each observation well (samples) were obtained.

As a supervised ML technique, decision tree is used combing with isotope method in a study to determine nitrogen source in groundwater. The decision tree has made 97.5% success in the water quality analysis. However, concentration data alone could not identify the dominant NO_3^- sources for groundwater contamination. It is suggested that an integrated approach should be setup by the combination of the N and O isotopes of NO_3^- with land-uses and physical-chemical properties, especially in areas with specific activities [38].

3.1.3 Contamination by coal mine water

One of the most focused issues of surface and ground water contamination is the acid mine drainage (AMD). The AMD is formed when pyrite and other sulfide minerals oxidized and dissolved during coal and metal mining, highway construction, and other large-scale excavation [13]. In an anaerobic environment, the sulfide minerals are stable, while exposure to water and oxygen, and with other accelerating factors such as bacteria, they are oxidized to form sulfuric acid, accompanying release of trace elements to surrounding water bodies [89]. In coal mine water, some of the drainage is alkaline, the leaching behavior and the TE composition in the leaching water are different [14].

Mobility of the TEs in AMD depends on several conditions. First, what is the TE occurrence and abundance in the potential AMD source; second, during the water-rock interaction process, where and how the adsorption-desorption, dissolution-precipitation take place; third, what are the main flow path, and the river and lake geochemistry where the TEs may be adsorbed or released again.

If the flow path is known, the source and reaction rates of specific trace elements can be estimated by mass balance calculation. The post-dissolution behavior of TEs is controlled by solution composition, pH, Eh of the water, temperature, and contact-time with mineral surfaces. For example, metal elements will have little attenuation in the solid phases, and high mobility potential into water. The versus behavior can be observed for the metalloid elements. Along with the flow path, water geochemical characteristics and pH and Eh of water changes, the TEs may undergo very complex reaction process, leading to redistribution of TEs in surface water, ground water, and sediment in the water bodies. Therefore, the source identification of TEs is an important and challenging work.

The pH of AMD ranges from 2 to 8. In an acid environment, metal element, Pb, Cd, Cu, Ni, have high mobility, while some metalloid element, As, Se, tend to migrate in an alkaline environment. Damaging effects of AMD are reported in Asia [34, 90–93], North America [72, 94–96], Europe [83, 97], South America [47, 98, 99]. When AMD enters surface water bodies, the effects include biotic impacts on stream and lake organisms through direct toxicity, habitat alteration by metal precipitates, visual changes from orange or yellow staining of stream sediments, nutrient cycle disruptions, or other mechanisms, and the water often becomes unsuitable for domestic, agricultural, and industrial uses. Gammons et al. have found the contamination of abandon coal mines on ground water using method of isotope analysis [71, 72].

The TEs in coal are not only migrate while mining and dumping of gangue and dust deposit [100], but also accompanying spread by smoke, fly ash, bottom ash when combustion [101, 102]. Trace elements, As, Cu, Se are found concentrated in the fly ash, which indicate impact on water and soil quality [103, 104].

3.1.4 Source apportionment of water inrush in coal mines

The TE source apportionment technology is used in coal mines to determine the source of water inrush [32]. In coal mine, water inrush constantly threatens the production, human health and cause financial losses. The water inrushes are cauterized to four sources: quaternary sand-gravel pore aquifer, Dyas sandstone aquifer, limestone aquifer from Ordovician and Carboniferous, and abandoned coal mine districts, respectively. Different sources show varies features and need different treatment strategies. The main purpose of the water inrush analysis is to find categories of source aquifers. Huang et al. [32] proposed a technology system, Piper-PCA-Bayes-LOOCV discrimination model to determine water inrush types in coal mines. The piper diagram is a geochemical technique to show the water characteristics, and abnormal samples/points were screened in this research. PCA was used to lower dimension of the sample matrix, to make less variates standing for all the original variates. Then the supervised ML model, Bayes DA, is used to train and implement a model for water source discriminant. LOOCV means leave-one-out cross-validation, to validate and improve quality of the model. Wang et al. used discriminant analysis to determine water bursting source in coal mines [33].

3.1.5 TE occurrence and reaction pathway

The PCA method has also used to investigate trace element occurrence in rock/coal, and reaction pathway, which may be the source of TEs that have contaminating potential on surrounding water bodies. Shan et al. [34] found that in coal host rock seam, Se/Cd/Hg/ As occurred in sulfide minerals, Be and V occurred in carbonate minerals, Cr and Pb occurred in clay minerals, respectively; while in coal seam, Se/Cr/Pb occurred in clay minerals, As and Hg occurred in sulfide minerals. Se, As and Hg immigrated through dissolution of sulfide minerals, Cr immigrated through transformation of clay minerals in coal host rock. In coal seam, As and Hg occurred in sulfide minerals. Se, Pb and Cr immigrated through transformation of clay minerals, As and Hg immigrated through dissolution of sulfide minerals, respectively. Pumure et al. [105] investigated occurrence of selenium and arsenic in coal by the method of two step PCA, founding that ultrasound leachable selenium concentrations were associated with 14 Å d-spacing phyllosilicate clays (chlorite, montmorillonite and vermiculite all 2:1 layered clays) whilst ultrasound leachable arsenic concentrations were closely related to the concentration of illite, another 2:1 phyllosilicate clay.

3.2 TE apportionment in sediment

Surface water and sediment compose a reaction system, trace elements in water may be adsorbed by sediment, meanwhile, trace elements in sediment are released. Therefore, the sediment may be a sink or origin of trace elements. Because of the complex reaction pathway, and environmental persistence and biological accumulation, the trace elements in the aquatic environments has drawn special attentions [106].

In southwest China, lake sediment was analyzed [42]. PCA result showed that Cd/Hg/Pb/Zn, and As (as PC2 and PC3, respectively) were mainly from non-point anthropogenic sources, especially with the atmospheric emission from non-ferrous metal smelting and coal consumption [107].

In Jiangxi China, river sediment was investigated. As the metal mines are excavating in the study area, metal element contamination was found. The PCA analytical result show probably coal and gold mining, copper mining and refining, Zn/Pb deposits and agricultural activities origin associated with PC1, PC2, and PC3, respectively. The PC1 were high loaded with Ni, Hg, Cr, the PC2 were high loaded with Cu, the PC3 were high loaded with Cd, Pb, Zn, As, respectively [39]. A research on lake sediment in Jiangxi China showed that Cr, Pb, and Zn may be mainly derived from both lithogenic and human activities, such as atmospheric and river inflow transportation, whereas Cu and Cd may be mainly contributed from anthropogenic sources, such as mining activities and fertilizer application [43]. In northern China, Cd and Zn are found originating from agriculture source and Cu, Cr, Ni were natural source origin [40]. In northwest China, Zn, Cu, Ni, and As were high loaded on the PC1, and natural originated, Cr and Cd/Pb/Hg are high loaded on the PC2 and PC3, which were township/silicon chemical factories, and agriculture/urban construction origin, respectively [41].

3.3 TE apportionment in soil

Researches have focused on distinguish TE source from natural and anthropogenic [108, 109] contaminates in soil. The major natural contribution of heavy metals comes from the parent materials from which the soils developed. The anthropogenic source of heavy metals in soils includes acid mine drainage [110], agricultural and industrial waste discharges [111], atmospheric deposition [112], fertilizers and pesticides [113], which has a significant contribution to the content levels of heavy elements in soils. PCA are now a popular technique to trace source of TEs in soil, then enrichment factors are usually used to verify the sources. In order to investigate source of TEs and spatial distribution, the combination method of geochemical, multivariate analysis, and geostatistical analysis. GIS and multivariate analysis of soil contamination has been detailed reviewed [114]. Understanding sources of heavy metals in surface soils is imperative for the decision to implement the strategies for protecting the food safety, human health and ecosystem sustainability.

3.3.1 TE apportionment in agricultural soil

In Greece, two main sources explained 74.8% of all the variance for the agriculture soil contamination analysis. The TEs Cu, Pb, Zn, As, Cd, P and K were identified to be anthropogenic influence, and TEs Ni, Co, Fe and Cr were recognized to be natural source origin [55].

In soil samples on hills in India, four principal components were determining by using the PCA method, high loading TEs on which are Mn/Zn, Cr, Ni, Co, respectively. The PC1 and PC2 were inferred to be natural sources, and PC3 represent fossil fuel burning origin, which contribute most of Ni in soil, and PC4 represent irrigation sources, respectively [50].

In Shanxi province China, soil samples were collected in an area of 25k km^2. The PCA analytical result showed that Co, Cr, Cu, Mn, Ni, Se, V, and Zn were mainly originated from natural source, and Cd and Pb were affected by anthropogenic pollution heavily. Associated with the spatial data, Pb were strongly associated with road traffic, and Cd were linked to industrial activities. In order to predict Pb and Cd concentration in the following years, an ANN model was applied [49].

In Beijing China, the TEs can be represented by two PCs. The TEs Co, Ni, Cr and V were probably released from parent material of the soil, Cd, Cu, and Zn were primarily from agricultural cultivation. Hg may be originated from coal combustion or mineral fertilizers [45].

In Jijin China, Al, Fe, Mn, Zn, Cr, Ni, As, Cu, and Pb was found accounting for 55.16% of the total variance, which was identified as natural source. N, OC, P, Cd, and Hg have high loadings on the PC1, accounting for 16.75% of the total variance. The PC2 also seemed as natural source, high relationship of Hg and Cd was explained to high organic affinity [46].

In Iran, a kind of semi-supervised ML method was applied. The study area was located around a Cu-Au porphyry deposit, so the soil may be associated. Initially eleven soil geochemical variables were selected by using hieratical clustering analysis and expert knowledge. Then, the semi-supervised fuzzy c-means clustering method (ssFCM) was used to separate multivariate soil geochemical anomalies from back-ground for further drilling [58].

3.3.2 TE apportionment in urban and industrial top soil

The impact of ore deposit on surrounding soil was investigated in Beijing China. Frequent mining activities produce dust, acidic drainage from the oxides and mill tailing. Cu, Co, Zn, Cd and V was found to mixed sources originated; Be, Pb and As came from natural sources and are mainly affected by the weathering and erosion of parent rock material; Cr, Ni and Ba were polluted by fine particle, industrial and mining activities; transportation and soil minerals were the common sources of Cr and Ni; Hg came from anthropogenic sources, mainly impacted by mining, beneficiation, smelting and acid mine drainage waste [44].

Large urban and industrial areas along the coastline in Italy was investigated. Pb and Zn due to heavy traffic and alloy production. Some Cr and Ni contamination were discerned through releases from tannery industry. Zn and Pb enrichment were mainly related to the large volcanic complexes. Cr and Ni were enriched in the siliciclastic deposits [56]. Another large-scale investigation was carried out in Yangtze river delta China, industrialization lead to high contamination potential on environment. Four PCs were selected to present the sources of trace element. As, Hg, Cu, Cd, Mo, S and Zn are recognized as traffic origin. Fe and Mn were from natural resources. The PC3, including Cr and Ni, pointed to pyrometallurgical processes, especially non-ferrous metal industries, etc. The PC4, composited by Pb and Se, was inferred to be coal combustion originated [52]. Another research carried out in this area showed that Cr, Ni, Co, Mn, Cu, and As were mainly came from natural sources. Cd/Hg and Pb/Zn originated from anthropogenic sources in two different groups [57].

In Shaanxi province China, roadway dust was analyzed. TEs Zn, Mn, Ni, As had the highest variance. Because Zn was released mainly from wear vehicle tire and corrosion of galvanized automobile part. Cu, Pb, and Cr was inferred to traffic origin. The third source was dominated by Co and Ni, and they were released from machine manufacturing plant [48]. In a research from Alaska America, soil contamination was found to be caused and controlled mainly by distance to road, traffic category, including highway and refuge road, land cover category, paved or not, land cover category, traffic loading, and other parameters, in descending order [59].

In Pakistan, four factors were identified using the factor analysis to trace surface soil contamination in industrial cite. VF1 contains Ni, Cr, Zn, and Cu, which originate from vehicular emission and industrial activities. VF2, compositing by Pb, Cd, and Co, originated from anthropogenic activities such as automobiles. Fe, Mn, standing for VF3, and VF4, were natural source origin [54]. In Armenia, Ti, V, Mn, Fe and Co, were identified to be natural originated. The PC2 include two distinguished negative groups, As/Hg, and Pb/Zn. The PC3 is composited mainly with Cu and Mo, and recognized as anthropogenic origin [51]. In Spain, the first source, including Pb, Tl, As, Sb, Cd, pointed to coal combustion. The second source was traffic air pollution origin, which released Cr, Ni, Be, V, Co. The third and fourth factors explained a very low proportion of variance and were considered secondary. These factors included TEs Cu, Zn and Sn, showing mixed behavior with regard to the first two factors [53].

In Nigeria, because of the vandalization of pipeline, the soil and water may be contaminated. The PCA gave 78.68% of accumulative contribution of the covariance from the first three PCs. PCA analysis result in soil was similar with that in water [12].

3.3.3 TE apportionment in soil to recall human activities

In Spain, core was obtained from peat bog, to evaluate trace element distribution and human activity impact in the past 8000 years. It was found that Al, Ba, Cr, Ga, K, Na, Sr, Ti, V, Y and Zr were lithogenic and supplied by atmospheric soil dust, while Cd, Pb, P, and Zn were recognized to anthropogenic, especially the ore exploration. The depth of samples depicted the influence degree of human activities yearly [47]. The EF profile showed that Pb, Zn, and Hg were at peak values in atmospheric in Roman age and nineteenth to twentieth centuries.

From the recent researches, it is concluded that Pb is an important anthropogenic originated element. Some reports argued that the vehicle emissions, brake lining, coal burning, plastics and rubber production, and car barriers are potential source of Pb. Meanwhile, Cu might come from vehicle brake lining, Zn from vehicle tires [51]. For the agriculture soil, Cu are usually cumulated by application of commercial fertilizers and Cu-based pesticides and fungicides [115]. Cd was related to the use of phosphate fertilizers [116]. Mineral fertilizers and animal manure may lead to elevation of Zn and Cu levels in soil.

3.4 TE apportionment in air and particles

The TEs spread through the air usually as particles. The particle smaller than 10 μm is called PM 10, while PM 2.5 stand for that smaller than 2.5 μm. It is obviously that the haze-day rate has increasing in the past decade, several researchers have reported characteristics, composition, and sources of PM 10 and PM 2.5 in some Chinese cities [17, 64, 117]. At the same time, PM 10 and PM 2.5 in megacities all around the world are investigated [118–121]. TEs, such as Cu, Zn, Pb, Cd, Cr, relating to the PM 2.5 and PM 10, show deleterious effects to human health. Based on the epidemiological and toxicological studies [122, 123], the TEs in ambient PM 2.5 influence the severity of allergic respiratory disease and have a high cancer risk to the exposed populations [81, 82].

In the source apportionment analysis, six types of main resource of ambient particular matter are commonly found: natural sources (including soil dust and sea salt), domestic fuel burning, industry, traffic, unspecified source of human origin pollution. Soil dust refer to the bare soils by local wind. Sea salt particles can be found close to the coast. Domestic fuel burning includes coal, gas fuel and wood for cooking and heating. Traffic is a complex source of PM and TEs. All the burning of fuel and diesel, wear of brake linings, clutch, and tires are source of TEs [124]. The

"Unspecified sources of human origin" category mainly includes secondary particles formed from unspecified pollution sources of human origin. The reasons of the second outbreak of PM 2.5 are complex, including some chemical reaction. In fact, the reasons of the fog and haze are: (1) the accumulation of the fog and haze, namely the results of combustion, automobile exhaust, and dust effects; (2) the fog and haze particles' upward momentum—hot-air upward movement and wireless communication, namely the electromagnetic wave net sports; (3) no sustained wind. These three conditions indispensable lead to persistent fog and haze weather, and the second outbreak of PM 2.5 results from the above three conditions together. In a review for the source apportionment study, 87% of the record have traffic origin, 66% have industry origin, 45, 100, and 89% have domestic fuel burning, unspecified source of human origin, and natural sources origin, respectively [125].

In southern China, TE source in the PM 2.5 was identified using PCA technique. Three sampling sites were analyzed separately. In YL sampling site, the PC1, with Zn and Pb, were identified as the traffic source, Cu and Cd, high loading on the PC2, originated from coal and other kind of fossil fuel. In KF sampling site, Zn, Cd, and Pb were from vehicle emission and abrasion of automobile tire. Cu, high loaded on the PC2, is a tracer of fossil and other fuel combustion. In the YH site, Cu, Cd, and Pb were associated with domestic fossil fuel burning, and Zn represent brake and tire wear and other transportation processes [17]. During Chinese Spring Festival, haze may occur more frequently, and the PM 2.5 level can be elevated. In Henan province China, sources of PM 2.5 were identified by using PCA, and a model to predict PM 2.5 concentrations using multivariate linear regression was set up. The most important source was burning source, including coal combustion, fireworks, fire crackers and biomass burning, contributing 61% of all the PM 2.5. The second, third, and fourth sources were vehicle emission (27%), soil (8%), and road dust (3.28%), respectively [63].

In Costa Rica, by using the method of PMF, eight important sources of PM 2.5 and PM 10 and TEs were identified. Vehicle exhaust, containing EC, OC, SO_4^{2-} and certain amount of Fe, residual oil combustion, bringing Ni and V, fresh sea salt, including Cl^-, Na and Mg, were the first three source. The others are crustal, or dust aerosols originated, organic carbon and sulfate, secondary sulfate, secondary nitrate, and heavy fuels [66].

In the USA, sources of PM 2.5 were determined, variance of meat, secondary aerosols, motor oil/brake dust/other outdoor, dust, cigarette, gasoline, biomass burning, and retene were explained with 23, 14, 10, 9, 6, 6, 5, and 5%, respectively. For the chemicals, meat released cholesterol, the alkanoic acids, OC, and light n-alkanes. Ammonium, sulfur, and nitrate were mainly released by secondary aerosol. The PC3 includes cholestanes, hopanes, Ba, and nitrate, which was related to motor oil/brake dust/other outdoor [65].

A 6-year investigation of PM 2.5 levels, source and potential human risk was investigated in Canada. Secondary organic aerosol, secondary nitrate, secondary sulfate, transportation and biomass burning, contributed more than 85% to PM 2.5, the importance of which was in descent order [61].

In Nigeria, source of PM 2.5 was identified to be soil (44%), savannah burning (26%), scrap processing (18%) and vehicular emissions (12%), and soil plus biomass burning (71%), sea salt (22%), scrap processing (5%) and vehicle emissions (tire wear) (2%) for the PM 10. Elements Al, Si, Ca, Ti, Mg, Fe and Na were spread through fine particle, and crustal elements Al, Si, Ca, Ti, Mn, Fe and anthropogenic elements Cl, K, V, Cr, Ni, Br, Pb, and black carbon were spread through coarse particles. Savannah burning release Br, BC and Pb through fine particles. The vehicle emit Na, S, Zn, As, Br, and Pb, and spread through fine particles [67].

Coal mining impact of air pollution, including suspended particles was investigated in India. The PCA and CA results suggested PC1 represent PM 10, SO_2, PM 2.5, PM 1.0, Ni and Cu, which are originate from coal burning and active mine fire. PC2 was high loaded with NO_2, Pb, Cd and Cr, and originated from crude oil combustion and vehicular emission. The PC3, including Fe and Mn, was mainly contributed by earth crust, wind-blown soil, and coal fly ash [60].

In the USA, brake wear, tire wear, fertilized soil, and resuspended soil were found to be important sources of copper, zinc, phosphorus, and silicon, respectively, using the method of positive matrix factorization. Zn was found strongly related to tire wear but also contributed to the Pb-rich features and soil. At the same time, the Pb-rich contributions are highly correlated with the tire wear, elevated P contributions within the fertilized soil as well as the Pb-rich feature [68].

Brinkman et al. compared the performance of PCA and PMF on the source apportionment for the particle matters. It was found that most of the PCA factors were easily distinguishable from others by sharp differences in the factor loadings. For many individual compounds, the variance was explained primarily by a single factor. In contrast, the factors obtained with PMF were more difficult to distinguish because anticipated tracer compounds for certain sources appeared in multiple PMF factors [65].

3.5 Summary of method used to identify source of contaminates

Applications and implementations of multivariate analysis/data mining, combining with geochemical method, on source apportionment of trace element as contaminates in environmental medias are increasing, with the development of techniques of big data, machine learning, and computer software. Four environmental medias, water, sediment, soil, and particles are discussed.

Four types of application can be identified for water contamination: trace the source of TEs, evaluate water quality of surface water and ground water, identify intrusion in coal mines and other scenario, and find and quantify water relationship between different bodies, such as surface-ground water relationship. The sediment and water composite a reaction system, i.e., the sediment could be origin, sink of trace elements in water, or be sink at first step, then origin again. Therefore, the system should be analyzed together. The researches on sediment are less than water, and most of articles on this topic are from China.

The most used method for the source apportionment of TEs in water and sediment is principal component analysis (PCA), probably for it's easy to use and explain. With the developing of data mining algorithm and calculation software, the application of PCA become easier and more efficient. The similar method, factor analysis (FA) is also used. The PCA and FA are both unsupervised ML method. Although having less accuracy than the supervised method, these methods are suitable for this topic.

Supervised ML methods are also used in this area, though much less than the unsupervised ML methods, and its scope of application is different. For example, decision tree is used to classify the sample types [38]. Discriminant analysis is also a supervised method, its implementation can be found, especially on the identifying water inrush source in coal mines, as the labeled data can be obtained [32, 33]. In this sense, other supervised machine learning method, ANN, support vector machine, decision tree, can also be used to identify water inrush source. Usually, ANN need more data to improve predicting quality, than SVM and decision tree.

In order to combing the advantages of unsupervised and supervised machine learning methods, semi-supervised method has been introduced and implemented on this topic [52]. At present, related researches are rare, but promising reports are expected.

From the reviewed reports, it is concluded that the surface water is more contaminated by major elements, and nitrogen, which may stand for the organic contamination. The ground water is more contaminated by trace elements, As, Cr, Cd, Pb, Hg, Se, etc. The surface water may be impacted by civil and industrial activities, and the ground water may be impacted by water-rock interaction in the rock seam. The most important anthropogenic source of trace elements in the ground water is the coal and metal mines. These mines contain high content of toxic trace elements, which is stable in an anaerobic environment. Once the rock and coal are excavated, trace elements are released. Less contaminated by trace elements in the investigated surface water is not proving of safety of the surface water. Researches of sediment in rivers and lakes have found high content of anthropogenic source trace element, including As, Cr, Cd, Pb, Hg, Se, Cu, Zn, Ni, etc. The sediment and water in river and lake composite a reactive system, in which the sediment is both sink and source of the trace elements. Therefore, the source, reaction pathway in this system need thoroughly researches and regulations.

Researches on soil can roughly be divided into two large group, agriculture soil, and urban/industrial soil. Unsurprisingly, first TE source of agriculture soil is natural, and first TE source of urban/industrial soil is anthropogenic. As the impact of industrial development on environment, researches on urban/industrial soil are increasing, and carried out in a wider scale. Researches on particles have become popular because the air is easily impacted by human activities. In some countries, haze has become an important problem. As the main composition, suspended particles in air, especially PM 2.5, are the important media to transport and spread contaminates. The researches on PM 2.5 are carrying out all around the world, both developed and developing countries.

The most popular method used are PCA, FA, and positive matrix fractionation (PMF). The PMF is frequently used in the particle researches, but less in water and soil researches. In the study of soil and particle, semi-supervised ML techniques are also implemented [38]. Some researches combine the machine learning method with geochemical method, or two or more machine learning method together. For example, Petrik et al. [56] combined factor analysis and multivariate linear regression. The ANN is a tool to predict air quality based on history data, relative researches are abundant, Mclean et al. have made a thorough reviewed on this topic [87]. However, very little work has been carried out to identify TE source using ANN method.

From the reviewed reports, anthropogenic source of trace elements in soil and particle includes mainly metal element, Zn, Mn, Ni, Cu, and some other toxic elements, As, Cd, Cr, Hg, Pb, etc. The soil and particle have similar TE composite. More metal TEs are found in soil and particle than that in ground water.

4. Conclusions

The techniques of data mining are widely used to trace sources of TEs in water and solid matrix.

In water environment, ground water and surface water have relation in the flow network. Human activities, especially for the mining, change the natural reaction environment, releasing trace element into ground water and surface water. Then the sediment in river and lake may be contaminated and be a source to water that may release trace element again. Soil, dust, and air particles may be influenced by varies of human activities, especially in the urban and industrial area. The TE composition is different depending on the environmental media type, human activities, land use type, etc. However, some environmental concern element, As,

Pb, Cd, Hg, Cr, are frequently found in water, sediment, soil, and particle, showing high mobility and contaminating potential on environment.

The unsupervised machine learning algorithm, including principal component analysis, factor analysis, positive matrix fractionation is mostly used. The PCA is used in water is to find contamination source of trace element, and sometimes water inrush in coal mines. In the air particle researches, PCA and PMF are frequently used to trace the source of PM 2.5 and PM 10, and the TEs source in the particle sources. Some supervised algorithm, including discrimination analysis, Bayesian network, artificial neural network, decision tree is used when the data are labeled.

Generally speaking, the most popular methods used to apportion the source of trace elements as contaminants are unsupervised ML techniques, especially the principal component analysis. In a wider scope, supervised ML is a big tool box for investigations and researches, which is frequently applied and implemented in the areas of science and society. The supervised ML usually gives more accuracy and robust result than the unsupervised ML. In the area of trace element apportionment, some factors constrain the implementation of supervised ML techniques, as the sources are usually not known. However, some techniques are promising to treat the issues of trace element apportionment. First, the supervised ML methods could be implemented more frequently. The unsupervised ML methods are used in the first step. With the intensive research, as some sources have been identified, the supervised ML methods could be used. For example, water inrush is sometimes a threaten in some Chinese coal mines. As the potential source of inrush can be identified, supervised ML method, discriminant analysis is used to determine the water type of inrush, then the corresponding technologies to deal with the threaten or accidents could be implemented. At this stage, some other supervised ML method could also be used. However, the discriminant analysis was mostly used. Second, semi supervised ML may be used implemented more. This method is a series of relative novel techniques. Once more data is obtained in an investigation or research, the semi-supervised ML may be used. In a sense, this method combines the unsupervised and supervised techniques in one implementation. Third, the machine learning method could be combined with geochemical method together. Two technique system have their advantages and disadvantages, the combination may achieve its maximum consequences and efficiency.

Acknowledgements

Our study is funded by the Fundamental Research Funds for the Central Universities (3142014005), the Colleges and Universities in Hebei Province Science and Technology Research Project (ZD2016204), and the National Natural Science Foundation of China (51674119).

Conflict of interest

The authors declare no conflict of interest.

Author details

Yao Shan* and Jianjun Shi
School of Safety Engineering, North China Institute of Science and Technology, Yanjiao, China

*Address all correspondence to: 9106350@qq.com

IntechOpen

© 2019 The Author(s). Licensee IntechOpen. This chapter is distributed under the terms of the Creative Commons Attribution License (http://creativecommons.org/licenses/by/3.0), which permits unrestricted use, distribution, and reproduction in any medium, provided the original work is properly cited.

References

[1] Iqbal J, Shah MH. Distribution, correlation and risk assessment of selected metals in urban soils from Islamabad, Pakistan. Journal of Hazardous Materials. 2011;**192**(2): 887-898. DOI: 10.1016/j.jhazmat.2011.05.105

[2] Manta DS, Angelone M, Bellanca A, Neri R, Sprovieri M. Heavy metals in urban soils: A case study from the city of Palermo (Sicily), Italy. Science of the Total Environment. 2002;**300**(1–3): 229-243

[3] Qu MK, Li WD, Zhang CR, Wang SQ, Yang Y, He LY. Source apportionment of heavy metals in soils using multivariate statistics and geostatistics. Pedosphere. 2013;**23**(4): 437-444. DOI: 10.1016/S1002-0160(13)60036-3

[4] Xu X, Zhao Y, Zhao X, Wang Y, Deng W. Sources of heavy metal pollution in agricultural soils of a rapidly industrializing area in the Yangtze Delta of China. Ecotoxicology and Environmental Safety. 2014;**108**: 161-167. DOI: 10.1016/j.ecoenv.2014.07.001

[5] Dai S, Li W, Tang Y, Zhang Y, Feng P. The sources, pathway, and preventive measures for fluorosis in Zhijin County, Guizhou, China. Applied Geochemistry. 2007;**22**(5):1017-1024

[6] Hosono T, Su CC, Okamura K, Taniguchi M. Historical record of heavy metal pollution deduced by lead isotope ratios in core sediments from the Osaka Bay, Japan. Journal of Geochemical Exploration. 2010;**107**(1):1-8. DOI: 10.1016/j.gexplo.2010.05.003

[7] Ding F, He Z, Liu S, Zhang S, Zhao F, Li Q, et al. Heavy metals in composts of China: Historical changes, regional variation, and potential impact on soil quality. Environmental Science and Pollution Research. 2017;**24**(3): 3194-3209

[8] Tedoldi D, Chebbo G, Pierlot D, Branchu P, Kovacs Y, Gromaire MC. Spatial distribution of heavy metals in the surface soil of source-control stormwater infiltration devices—Inter-site comparison. Science of the Total Environment. 2017;**579**:881-892. DOI: 10.1016/j.scitotenv.2016.10.226

[9] Sridhara Chary N, Kamala CT, Samuel Suman Raj D. Assessing risk of heavy metals from consuming food grown on sewage irrigated soils and food chain transfer. Ecotoxicology and Environmental Safety. 2008;**69**(3): 513-524

[10] Khan S, Rehman S, Zeb Khan A, Amjad Khan M, Tahir Shah M. Soil and vegetables enrichment with heavy metals from geological sources in Gilgit, northern Pakistan. Ecotoxicology and Environmental Safety. 2010;**73**(7): 1820-1827. DOI: 10.1016/j.ecoenv.2010.08.016

[11] Chabukdhara M, Nema AK. Heavy metals assessment in urban soil around industrial clusters in Ghaziabad, India: Probabilistic health risk approach. Ecotoxicology and Environmental Safety. 2013;**87**:57-64. DOI: 10.1016/j.ecoenv.2012.08.032

[12] Ogunlaja A, Ogunlaja OO, Okewole DM, Morenikeji OA. Risk assessment and source identification of heavy metal contamination by multivariate and hazard index analyses of a pipeline vandalised area in Lagos State, Nigeria. Science of the Total Environment. 2019;**651**:2943-2952. DOI: 10.1016/j.scitotenv.2018.09.386

[13] Yue M, Zhao F. Leaching experiments to study the release of trace elements from mineral separates from

Chinese coals. International Journal of Coal Geology. 2008;**73**(1):43-51

[14] Shan Y, Qin Y, Wang W. Chromium leaching mechanism of coal mine water —A modeling study based on Xuzhou-Datun coal mine district. Mining Science and Technology. 2010;**20**(1):97-102. DOI: 10.1016/S1674-5264(09)60168-X

[15] Khuzestani RB, Souri B. Evaluation of heavy metal contamination hazards in nuisance dust particles, in kurdistan province, western Iran. Journal of Environmental Sciences. 2013;**25**(7): 1346-1354. DOI: 10.1016/S1001-0742 (12)60147-8

[16] Liu J, Yang T, Chen Q, Liu F, Wang B. Distribution and potential ecological risk of heavy metals in the typical eco-units of Haihe River Basin. Frontiers of Environmental Science & Engineering. 2016;**10**(1):103-113. DOI: 10.1007/s11783-014-0686-5

[17] Ma L, Yang Z, Li L, Wang L. Source identification and risk assessment of heavy metal contaminations in urban soils of Changsha, a mine-impacted city in Southern China. Environmental Science and Pollution Research. 2016; **23**(17):17058-17066

[18] Chen T, Liu X, Zhu M, Zhao K, Wu J, Xu J, et al. Identification of trace element sources and associated risk assessment in vegetable soils of the urban-rural transitional area of Hangzhou, China. Environmental Pollution. 2008;**151**(1):67-78

[19] Xia X, Chen X, Liu R, Liu H. Heavy metals in urban soils with various types of land use in Beijing, China. Journal of Hazardous Materials. 2011;**186**(2–3): 2043-2050. DOI: 10.1016/j. jhazmat.2010.12.104

[20] Huang J, Peng S, Mao X, Li F, Guo S, Shi L, et al. Source apportionment and spatial and quantitative ecological risk assessment of heavy metals in soils from a typical Chinese agricultural county. Process Safety and Environment Protection. 2019;**126**:339-347. DOI: 10.1016/j. psep.2019.04.023

[21] Mazurek R, Kowalska J, Gąsiorek M, Zadrożny P, Józefowska A, Zaleski T, et al. Assessment of heavy metals contamination in surface layers of Roztocze National Park forest soils (SE Poland) by indices of pollution. Chemosphere. 2017;**168**:839-850

[22] Chuncai Z, Guijian L, Ting F, Ruoyu S, Dun W. Leaching characteristic and environmental implication of rejection rocks from Huainan Coalfield, Anhui Province, China. Journal of Geochemical Exploration. 2014;**143**:54-61. DOI: 10.1016/j.gexplo.2014.03.010

[23] Li S, Jia Z. Heavy metals in soils from a representative rapidly developing megacity (SW China): Levels, source identification and apportionment. Catena. 2018;**163** (December):414-423. DOI: 10.1016/j. catena.2017.12.035

[24] Anteneh Y, Zeleke G, Gebremariam E. Assessment of surface water quality in legedadie and dire catchments, Central Ethiopia, using multivariate statistical analysis. Acta Ecologica Sinica. 2018;**38**(2):81-95 Available from: https://doi.org/10.1016/ j.chnaes.2017.05.005

[25] Varol M. Arsenic and trace metals in a large reservoir: Seasonal and spatial variations, source identification and risk assessment for both residential and recreational users. Chemosphere. 2019; **228**:1-8. DOI: 10.1016/j. chemosphere.2019.04.126

[26] Xue D, De Baets B, Van Cleemput O, Hennessy C, Berglund M, Boeckx P. Use of a Bayesian isotope mixing model to estimate proportional contributions of multiple nitrate sources

in surface water. Environmental Pollution. 2012;**161**:43-49. DOI: 10.1016/j.envpol.2011.09.033

[27] Hajigholizadeh M, Melesse AM. Assortment and spatiotemporal analysis of surface water quality using cluster and discriminant analyses. Catena. 2017; **151**:247-258. DOI: 10.1016/j.catena.2016.12.018

[28] Matiatos I, Paraskevopoulou V, Lazogiannis K, Botsou F, Dassenakis M, Ghionis G, et al. Surface–ground water interactions and hydrogeochemical evolution in a fluvio-deltaic setting: The case study of the Pinios River delta. Journal of Hydrology. 2018;**561**(April):236-249

[29] Matiatos I. Nitrate source identification in groundwater of multiple land-use areas by combining isotopes and multivariate statistical analysis: A case study of Asopos basin (Central Greece). Science of the Total Environment. 2016;**541**:802-814. DOI: 10.1016/j.scitotenv.2015.09.134

[30] Matiatos I, Alexopoulos A, Godelitsas A. Multivariate statistical analysis of the hydrogeochemical and isotopic composition of the groundwater resources in northeastern Peloponnesus (Greece). Science of the Total Environment. 2014;**476–477**:577-590. DOI: 10.1016/j.scitotenv.2014.01.042

[31] Barberá JA, Andreo B. River-spring connectivity and hydrogeochemical interactions in a shallow fractured rock formation. The case study of Fuensanta river valley (Southern Spain). Journal of Hydrology. 2017;**547**:253-268

[32] Huang P, Yang Z, Wang X, Ding F. Research on Piper-PCA-Bayes-LOOCV discrimination model of water inrush source in mines. Arabian Journal of Geosciences. 2019;**12**:334. DOI: 10.1007/s12517-019-4500-3

[33] Wang J, Li X, Cui T, Yang J. Application of distance discriminant analysis method to headstream recognition of water-bursting source. Procedia Engineering. 2011;**26**:374-381. DOI: 10.1016/j.proeng.2011.11.2181

[34] Shan Y, Wang W, Qin Y, Gao L. Multivariate analysis of trace elements leaching from coal and host rock. Groundwater for Sustainable Development. 2019;**8**(November): 402-412. DOI: 10.1016/j.gsd.2019.01.001

[35] Magesh NS, Chandrasekar N, Elango L. Chemosphere trace element concentrations in the groundwater of the Tamiraparani river basin, South India: Insights from human health risk and multivariate statistical techniques. Chemosphere. 2017;**185**:468-479. DOI: 10.1016/j.chemosphere.2017.07.044

[36] Kumar M, Ramanatahn AL, Tripathi R, Farswan S, Kumar D, Bhattacharya P. A study of trace element contamination using multivariate statistical techniques and health risk assessment in groundwater of Chhaprola Industrial Area, Gautam Buddha Nagar, Uttar Pradesh, India. Chemosphere. 2017;**166**:135-145. DOI: 10.1016/j.chemosphere.2016.09.086

[37] Vesselinov VV, Alexandrov BS, Malley DO. Contaminant source identi fi cation using semi-supervised machine learning. Journal of Contaminant Hydrology. 2018;**212**(November): 134-142. DOI: 10.1016/j.jconhyd.2017.11.002

[38] Xue D, Pang F, Meng F, Wang Z, Wu W. Decision-tree-model identification of nitrate pollution activities in groundwater: A combination of a dual isotope approach and chemical ions. Journal of Contaminant Hydrology. 2015;**180**: 25-33. DOI: 10.1016/j.jconhyd.2015.07.003

[39] Chen H, Chen R, Teng Y, Wu J. Contamination characteristics,

ecological risk and source identification of trace metals in sediments of the Le'an River (China). Ecotoxicology and Environmental Safety. 2016;**125**:85-92. DOI: 10.1016/j.ecoenv.2015.11.042

[40] Ke X, Gui S, Huang H, Zhang H, Wang C, Guo W. Ecological risk assessment and source identification for heavy metals in surface sediment from the Liaohe River protected area, China. Chemosphere. 2017;**175**:473-481. DOI: 10.1016/j.chemosphere.2017.02.029

[41] Zhang Z, Juying L, Mamat Z. Sources identification and pollution evaluation of heavy metals in the surface sediments of Bortala River, Northwest China. Ecotoxicology and Environmental Safety. 2016;**126**:94-101. DOI: 10.1016/j.ecoenv.2015.12.025

[42] Lin Q, Liu E, Zhang E, Li K, Shen J. Spatial distribution, contamination and ecological risk assessment of heavy metals in surface sediments of Erhai Lake, a large eutrophic plateau lake in Southwest China. Catena. 2016;**145**: 193-203. DOI: 10.1016/j.catena.2016.06.003

[43] Dai L, Wang L, Li L, Liang T, Zhang Y, Ma C, et al. Multivariate geostatistical analysis and source identification of heavy metals in the sediment of Poyang Lake in China. Science of the Total Environment. 2018; **621**:1433-1444. DOI: 10.1016/j.scitotenv.2017.10.085

[44] Qin F, Ji H, Li Q, Guo X, Tang L, Feng J. Evaluation of trace elements and identification of pollution sources in particle size fractions of soil from iron ore areas along the Chao River. Journal of Geochemical Exploration. 2014;**138**: 33-49. DOI: 10.1016/j.gexplo.2013.12.005

[45] Lin Y, Han P, Huang Y, Yuan GL, Guo JX, Li J. Source identification of potentially hazardous elements and their relationships with soil properties in agricultural soil of the Pinggu district of Beijing, China: Multivariate statistical analysis and redundancy analysis. Journal of Geochemical Exploration. 2017;**173**:110-118. DOI: 10.1016/j.gexplo.2016.12.006

[46] Chai Y, Guo J, Chai S, Cai J, Xue L, Zhang Q. Source identification of eight heavy metals in grassland soils by multivariate analysis from the Baicheng-Songyuan area, Jilin Province, Northeast China. Chemosphere. 2015;**134**:67-75. DOI: 10.1016/j.chemosphere.2015.04.008

[47] Silva LFO, Fdez-Ortiz de Vallejuelo S, Martinez-Arkarazo I, Castro K, Oliveira MLS, Sampaio CH, et al. Study of environmental pollution and mineralogical characterization of sediment rivers from Brazilian coal mining acid drainage. Science of the Total Environment. 2013;**447**:169-178. DOI: 10.1016/j.scitotenv.2012.12.013

[48] Lu X, Pan H, Wang Y. Pollution evaluation and source analysis of heavy metal in roadway dust from a resource-typed industrial city in Northwest China. Atmospheric Pollution Research. 2017;**8**(3):587-595. DOI: 10.1016/j.apr.2016.12.019

[49] Shangguan Y, Cheng B, Zhao L, Hou H, Ma J, Sun Z, et al. Distribution assessment and source identification using multivariate statistical analyses and artificial neutral networks for trace elements in agricultural soils in Xinzhou of Shanxi Province, China. Pedosphere. 2018;**28**(3):542-554. DOI: 10.1016/S1002-0160(17)60304-7

[50] Chandrasekaran A, Ravisankar R, Harikrishnan N, Satapathy KK, Prasad MVR, Kanagasabapathy KV. Multivariate statistical analysis of heavy metal concentration in soils of Yelagiri Hills, Tamilnadu, India—Spectroscopical approach. Spectrochimica Acta, Part A: Molecular and Biomolecular Spectroscopy. 2015;

137:589-600. DOI: 10.1016/j.saa.2014.08.093

[51] Tepanosyan G, Sahakyan L, Belyaeva O, Saghatelyan A. Origin identification and potential ecological risk assessment of potentially toxic inorganic elements in the topsoil of the city of Yerevan, Armenia. Journal of Geochemical Exploration. 2016;**167**:1-11. DOI: 10.1016/j.gexplo.2016.04.006

[52] Huang S, Tu J, Liu H, Hua M, Liao Q, Feng J, et al. Multivariate analysis of trace element concentrations in atmospheric deposition in the Yangtze River Delta, East China. Atmospheric Environment. 2009; **43**(36):5781-5790. DOI: 10.1016/j.atmosenv.2009.07.055

[53] Boente C, Matanzas N, García-González N, Rodríguez-Valdés E, Gallego JR. Trace elements of concern affecting urban agriculture in industrialized areas: A multivariate approach. Chemosphere. 2017;**183**: 546-556

[54] Malik RN, Jadoon WA, Husain SZ. Metal contamination of surface soils of industrial city Sialkot, Pakistan: A multivariate and GIS approach. Environmental Geochemistry and Health. 2010;**32**(3):179-191

[55] Kelepertzis E. Accumulation of heavy metals in agricultural soils of Mediterranean: Insights from Argolida basin, Peloponnese, Greece. Geoderma. 2014;**221–222**:82-90. DOI: 10.1016/j.geoderma.2014.01.007

[56] Petrik A, Thiombane M, Albanese S, Lima A, De Vivo B. Source patterns of Zn, Pb, Cr and Ni potentially toxic elements (PTEs) through a compositional discrimination analysis: A case study on the Campanian topsoil data. Geoderma. 2018;**331**(December): 87-99. DOI: 10.1016/j.geoderma.2018.06.019

[57] Liu Y, Ma Z, Lv J, Bi J. Identifying sources and hazardous risks of heavy metals in topsoils of rapidly urbanizing East China. Journal of Geographical Sciences. 2016;**26**(6):735-749

[58] Fatehi M, Asadi HH. Application of semi-supervised fuzzy c-means method in clustering multivariate geochemical data , a case study from the Dalli Cu-Au porphyry deposit in Central Iran. Ore Geology Reviews. 2017;**81**:245-255. DOI: 10.1016/j.oregeorev.2016.10.002

[59] Reeves MK, Perdue M, Munk LA, Hagedorn B. Predicting risk of trace element pollution from municipal roads using site-specific soil samples and remotely sensed data. Science of the Total Environment. 2018;**630**:578-586. DOI: 10.1016/j.scitotenv.2018.02.171

[60] Pandey B, Agrawal M, Singh S. Assessment of air pollution around coal mining area: Emphasizing on spatial distributions, seasonal variations and heavy metals, using cluster and principal component analysis. Atmospheric Pollution Research. 2013;**5**(1):79-86. DOI: 10.5094/APR.2014.010

[61] Bari MA, Kindzierski WB. Fine particulate matter (PM2.5) in Edmonton, Canada: Source apportionment and potential risk for human health. Environmental Pollution. 2016;**218**:219-229. DOI: 10.1016/j.envpol.2016.06.014

[62] Ying Q, Feng M, Song D, Wu L, Hu J, Zhang H, et al. Improve regional distribution and source apportionment of PM 2.5 trace elements in China using inventory-observation constrained emission factors. Science of the Total Environment. 2018;**624**:355-365

[63] Feng J, Yu H, Su X, Liu S, Li Y, Pan Y, et al. Chemical composition and source apportionment of PM 2.5 during Chinese Spring Festival at Xinxiang, a heavily polluted city in North China:

Fireworks and health risks. Atmospheric Research. 2016;**182**:176-188

[64] Zhai Y, Liu X, Chen H, Xu B, Zhu L, Li C, et al. Source identification and potential ecological risk assessment of heavy metals in PM2.5 from Changsha. Science of the Total Environment. 2014; **493**:109-115. DOI: 10.1016/j.scitotenv.2014.05.106

[65] Brinkman GL, Milford JB, Schauer JJ, Shafer MM, Hannigan MP. Source identification of personal exposure to fine particulate matter using organic tracers. Atmospheric Environment. 2009;**43**(12):1972-1981. DOI: 10.1016/j.atmosenv.2009.01.023

[66] Murillo JH, Roman SR, Rojas Marin JF, Ramos AC, Jimenez SB, Gonzalez BC, et al. Chemical characterization and source apportionment of PM10 and PM2.5 in the metropolitan area of Costa Rica, Central America. Atmospheric Pollution Research. 2013;**4**(2):181-190. Available from: https://linkinghub.elsevier.com/retrieve/pii/S1309104215303871

[67] Owoade KO, Hopke PK, Olise FS, Adewole OO, Ogundele LT, Fawole OG. Source apportionment analyses for fine (PM 2.5) and coarse (PM 2.5–10) mode particulate matter (PM) measured in an urban area in southwestern Nigeria. Atmospheric Pollution Research. 2016; **7**(5):843-857. DOI: 10.1016/j.apr.2016.04.006

[68] Sturtz TM, Adar SD, Gould T, Larson TV. Constrained source apportionment of coarse particulate matter and selected trace elements in three cities from the multi-ethnic study of atherosclerosis. Atmospheric Environment. 2014;**84**:65-77. DOI: 10.1016/j.atmosenv.2013.11.031

[69] Zongjie LI, Fei LIU, Yong S, Lingling S, Qing T, Bing JIA, et al. Chemical characteristics of precipitation and the indicative significance for sand dust events in the northern and southern slopes of Wushaoling Mountain , northwestern China. Journal of Arid Land. 2017;**9**:911-923

[70] Bo Y, Liu C, Zhao Y, Wang L. Chemical and isotopic characteristics and origin of spring waters in the Lanping-Simao Basin, Yunnan, Southwestern China. Chemie der Erde—Geochemistry. 2015;**75**(3):287-300. DOI: 10.1016/j.chemer.2015.04.002

[71] Gammons CH, Brown A, Poulson SR, Henderson TH. Using stable isotopes (S, O) of sulfate to track local contamination of the Madison karst aquifer, Montana, from abandoned coal mine drainage. Applied Geochemistry. 2013;**31**:228-238. DOI: 10.1016/j.apgeochem.2013.01.008

[72] Gammons CH, Duaime TE, Parker SR, Poulson SR, Kennelly P. Geochemistry and stable isotope investigation of acid mine drainage associated with abandoned coal mines in Central Montana, USA. Chemical Geology. 2010;**269**(1–2):100-112. DOI: 10.1016/j.chemgeo.2009.05.026

[73] Rao W, Han G, Tan H, Jiang S. Chemical and Sr isotopic compositions of rainwater on the Ordos Desert Plateau, Northwest China. Environment and Earth Science. 2015;**74**(7):5759-5771

[74] Fan Q, Ma H, Lai Z, Tan H, Li T. Origin and evolution of oilfield brines from Tertiary strata in western Qaidam Basin: Constraints from 87Sr/86Sr, δD, δ18O, δ34S and water chemistry. Chinese Journal of Geochemistry. 2010; **29**(4):446-454

[75] Soler A, Canals A, Goldstein SL, Otero N, Antich N, Spangenberg J. Sulfur and strontium isotope composition of the Llobregat river (Ne Spain): Tracers of natural and anthropogenic chemicals in stream waters. Water, Air, and Soil Pollution. 2001;**136**:207-224

[76] Samborska K, Halas S. 34S and 18O in dissolved sulfate as tracers of hydrogeochemical evolution of the Triassic carbonate aquifer exposed to intense groundwater exploitation (Olkusz-Zawiercie region, southern Poland). Applied Geochemistry. 2010; **25**(9):1397-1414. DOI: 10.1016/j.apgeochem.2010.06.010

[77] Bottrell S, Tellam J, Bartlett R, Hughes A. Isotopic composition of sulfate as a tracer of natural and anthropogenic influences on groundwater geochemistry in an urban sandstone aquifer, Birmingham, UK. Applied Geochemistry. 2008;**23**(8): 2382-2394

[78] Cui J, Zang S, Zhai D, Wu B. Potential ecological risk of heavy metals and metalloid in the sediments of Wuyuer River basin, Heilongjiang Province, China. Ecotoxicology. 2014; **23**(4):589-600

[79] Maanan M, Saddik M, Maanan M, Chaibi M, Assobhei O, Zourarah B. Environmental and ecological risk assessment of heavy metals in sediments of Nador lagoon, Morocco. Ecological Indicators. 2015;**48**:616-626. DOI: 10.1016/j.ecolind.2014.09.034

[80] Ujević Bošnjak M, Capak K, Jazbec A, Casiot C, Sipos L, Poljak V, et al. Hydrochemical characterization of arsenic contaminated alluvial aquifers in Eastern Croatia using multivariate statistical techniques and arsenic risk assessment. Science of the Total Environment. 2012;**420**:100-110

[81] Taner S, Pekey B, Pekey H. Fine particulate matter in the indoor air of barbeque restaurants: Elemental compositions, sources and health risks. Science of the Total Environment. 2013; **454–455**:79-87. DOI: 10.1016/j.scitotenv.2013.03.018

[82] Massey DD, Kulshrestha A, Taneja A. Particulate matter concentrations and their related metal toxicity in rural residential environment of semi-arid region of India. Atmospheric Environment. 2013;**67**:278-286. DOI: 10.1016/j.atmosenv.2012.11.002

[83] Cánovas CR, Olías M, Macias F, Torres E, San Miguel EG, Galván L, et al. Water acidification trends in a reservoir of the Iberian Pyrite Belt (SW Spain). Science of the Total Environment. 2016;**541**:400-411. DOI: 10.1016/j.scitotenv.2015.09.070

[84] Erhardt EB, Bedrick EJ. A Bayesian framework for stable isotope mixing models. Environmental and Ecological Statistics. 2013;**20**(3):377-397

[85] Qian SS, Craig JK, Baustian MM, Rabalais NN. A Bayesian hierarchical modeling approach for analyzing observational data from marine ecological studies. Marine Pollution Bulletin. 2009;**58**(12):1916-1921. DOI: 10.1016/j.marpolbul.2009.09.029

[86] Qian SS, Miltner RJ. A continuous variable Bayesian networks model for water quality modeling: A case study of setting nitrogen criterion for small rivers and streams in Ohio, USA. Environmental Modelling & Software. 2015;**69**:14-22. DOI: 10.1016/j.envsoft.2015.03.001

[87] Mclean S, Kaiser J, Richard B. Environmental modelling & software a review of artificial neural network models for ambient air pollution prediction. Environmental Modelling & Software, June. 2019;**119**:285-304. DOI: 10.1016/j.envsoft.2019.06.014

[88] Kilic E, Yucel N. Determination of spatial and temporal changes in water quality at Asi River using multivariate statistical techniques. Turkish Journal of Fisheries and Aquatic Sciences. 2018; **19**(9):727-737

[89] Seal RR, Shanks WC. Sulfide oxidation: Insights from experimental,

theoretical, stable isotope, and predictive studies in the field and laboratory. Applied Geochemistry. 2008;**23**(2):101-102. DOI: 10.1016/j.apgeochem.2007.10.006

[90] Liu P, Hoth N, Drebenstedt C, Sun Y, Xu Z. Hydro-geochemical paths of multi-layer groundwater system in coal mining regions—Using multivariate statistics and geochemical modeling approaches. Science of the Total Environment. 2017;**601–602**:1-14. DOI: 10.1016/j.scitotenv.2017.05.146

[91] Liu B, Tang Z, Dong S, Wang L, Liu D. Vegetation recovery and groundwater pollution control of coal gangue field in a semi-arid area for a field application. International Biodeterioration and Biodegradation. 2018;**128**:134-140. DOI: 10.1016/j.ibiod.2017.01.032

[92] Sahoo PK, Tripathy S, Panigrahi MK, Equeenuddin SM. Geochemical characterization of coal and waste rocks from a high sulfur bearing coalfield, India: Implication for acid and metal generation. Journal of Geochemical Exploration. 2014;**145**:135-147. DOI: 10.1016/j.gexplo.2014.05.024

[93] Park JH, Edraki M, Baumgartl T. A practical testing approach to predict the geochemical hazards of in-pit coal mine tailings and rejects. Catena. 2017;**148**:3-10. DOI: 10.1016/j.catena.2015.10.027

[94] Hendryx M. The extractive industries and society the public health impacts of surface coal mining. Extractive Industries and Society. 2015;**2**(4):820-826. DOI: 10.1016/j.exis.2015.08.006

[95] Cravotta CA. Monitoring, field experiments, and geochemical modeling of Fe(II) oxidation kinetics in a stream dominated by net-alkaline coal-mine drainage, Pennsylvania, USA. Applied Geochemistry. 2015;**62**:96-107. DOI: 10.1016/j.apgeochem.2015.02.009

[96] Nordstrom DK. Hydrogeochemical processes governing the origin, transport and fate of major and trace elements from mine wastes and mineralized rock to surface waters. Applied Geochemistry. 2011;**26**(11):1777-1791. DOI: 10.1016/j.apgeochem.2011.06.002

[97] Tozsin G. Hazardous elements in soil and coal from the Oltu coal mine district, Turkey. International Journal of Coal Geology. 2014;**131**:1-6. DOI: 10.1016/j.coal.2014.05.011

[98] Martins-Ferreira MAC, Campos JEG, Pires ACB. Near-mine exploration via soil geochemistry multivariate analysis at the almas gold province, Central Brazil: A study case. Journal of Geochemical Exploration. 2017;**173**:52-63. DOI: 10.1016/j.gexplo.2016.11.011

[99] Oliveira MLS, Ward CR, French D, Hower JC, Querol X, Silva LFO. Mineralogy and leaching characteristics of beneficiated coal products from Santa Catarina, Brazil. International Journal of Coal Geology. 2012;**94**:314-325. DOI: 10.1016/j.coal.2011.10.004

[100] Raja R, Nayak AK, Shukla AK, Rao KS, Gautam P, Lal B, et al. Impairment of soil health due to fly ash-fugitive dust deposition from coal-fired thermal power plants. Environmental Monitoring and Assessment. 2015;**187**:679. DOI: 10.1007/s10661-015-4902-y

[101] Munawer ME. Human health and environmental impacts of coal combustion and post-combustion wastes. Journal of Sustainable Mining. 2018;**17**(2):87-96. DOI: 10.1016/j.jsm.2017.12.007

[102] Schwartz GE, Rivera N, Lee SW, Harrington JM, Hower JC, Levine KE, et al. Leaching potential and redox transformations of arsenic and selenium in sediment microcosms with fly ash. Applied Geochemistry. 2016;**67**:177-185. DOI: 10.1016/j.apgeochem.2016.02.013

[103] Pandey VC, Singh JS, Singh RP, Singh N, Yunus M. Arsenic hazards in coal fly ash and its fate in Indian scenario. Resources, Conservation and Recycling. 2011;**55**(9–10):819-835. DOI: 10.1016/j.resconrec.2011.04.005

[104] Lokeshappa B, Dikshit AK. Fate of metals in coal fly ash ponds. International Journal of Environmental Science and Development. 2013;**1** (January):43-48. DOI: 10.1016/j.apcbee.2012.03.007

[105] Pumure I, Renton JJ, Smart RB. The interstitial location of selenium and arsenic in rocks associated with coal mining using ultrasound extractions and principal component analysis (PCA). Journal of Hazardous Materials. 2011; **198**:151-158. DOI: 10.1016/j.jhazmat.2011.10.032

[106] Yi Y, Yang Z, Zhang S. Ecological risk assessment of heavy metals in sediment and human health risk assessment of heavy metals in fishes in the middle and lower reaches of the Yangtze River basin. Environmental Pollution. 2011;**159**(10):2575-2585. DOI: 10.1016/j.envpol.2011.06.011

[107] Tian HZ, Zhu CY, Gao JJ, Cheng K, Hao JM, Wang K, et al. Quantitative assessment of atmospheric emissions of toxic heavy metals from anthropogenic sources in China: Historical trend, spatial distribution, uncertainties, and control policies. Atmospheric Chemistry and Physics. 2015;**15**(17):10127-10147

[108] Lu Y, Zhu F, Chen J, Gan H, Guo Y. Chemical fractionation of heavy metals in urban soils of Guangzhou, China. Environmental Monitoring and Assessment. 2007;**134**(1–3):429-439

[109] Sheng J, Wang X, Gong P, Tian L, Yao T. Heavy metals of the Tibetan top soils: Level, source, spatial distribution, temporal variation and risk assessment. Environmental Science and Pollution Research. 2012;**19**(8):3362-3370

[110] Skousen JG, Ziemkiewicz PF, McDonald LM. Acid mine drainage formation, control and treatment: Approaches and strategies. Extractive Industries and Society. 2019;**6**(1):241-249

[111] Solgi E, Abbas ES, Alireza RB, Hadipour M. Soil contamination of metals in the three industrial estates, Arak, Iran. Bulletin of Environmental Contamination and Toxicology. 2012; **88**(4):634-638

[112] Hovmand MF, Kemp K, Kystol J, Johnsen I, Riis-Nielsen T, Pacyna JM. Atmospheric heavy metal deposition accumulated in rural forest soils of southern Scandinavia. Environmental Pollution. 2008;**155**(3):537-541

[113] Nziguheba G, Smolders E. Inputs of trace elements in agricultural soils via phosphate fertilizers in European countries. Science of the Total Environment. 2008;**390**(1):53-57

[114] Hou D, O'Connor D, Nathanail P, Tian L, Ma Y. Integrated GIS and multivariate statistical analysis for regional scale assessment of heavy metal soil contamination: A critical review. Environmental Pollution. 2017;**231**: 1188-1200. DOI: 10.1016/j.envpol.2017.07.021

[115] Acosta JA, Faz A, Martínez-Martínez S, Arocena JM. Enrichment of metals in soils subjected to different land uses in a typical Mediterranean environment (Murcia City, Southeast Spain). Applied Geochemistry. 2011; **26**(3):405-414. DOI: 10.1016/j.apgeochem.2011.01.023

[116] Atafar Z, Mesdaghinia A, Nouri J, Homaee M, Yunesian M, Ahmadimoghaddam M, et al. Effect of fertilizer application on soil heavy metal concentration. Environmental Monitoring and Assessment. 2010;**160**(1–4):83-89

[117] Fu Z, Zhai Y, Wang L, Zeng G, Li C, Peng W, et al. Morphological,

geochemical composition and origins of near-surface atmospheric dust in Changsha city of China. Environment and Earth Science. 2012;**66**(8): 2207-2216

[118] Liacos JW, Kam W, Delfino RJ, Schauer JJ, Sioutas C. Characterization of organic, metal and trace element PM2.5 species and derivation of freeway-based emission rates in Los Angeles, CA. Science of the Total Environment. 2012;**435–436**:159-166

[119] Vecchi R, Marcazzan G, Valli G. A study on nighttime-daytime PM10 concentration and elemental composition in relation to atmospheric dispersion in the urban area of Milan (Italy). Atmospheric Environment. 2007;**41**(10):2136-2144

[120] Taneepanichskul N, Gelaye B, Grigsby-Toussaint DS, Lohsoonthorn V, Jimba M, Williams MA. Short-term effects of particulate matter exposure on daily mortality in Thailand: A case-crossover study. Air Quality, Atmosphere, and Health. 2018;**11**(6): 639-647

[121] Boman J, Wagner A, Gatari MJ. Trace elements in PM2.5 in Gothenburg, Sweden. Spectrochimica Acta Part B: Atomic Spectroscopy. 2010;**65**(6): 478-482. DOI: 10.1016/j.sab.2010.03.014

[122] Zhang J, Zhou X, Wang Z, Yang L, Wang J, Wang W. Trace elements in PM 2.5 in Shandong Province: Source identification and health risk assessment. Science of the Total Environment. 2018;**621**:558-577. DOI: 10.1016/j.scitotenv.2017.11.292

[123] Bellido-Martín A, Gómez-Ariza JL, Smichowsky P, Sánchez-Rodas D. Speciation of antimony in airborne particulate matter using ultrasound probe fast extraction and analysis by HPLC-HG-AFS. Analytica Chimica Acta. 2009;**649**(2):191-195

[124] Belis CA, Karagulian F, Larsen BR, Hopke PK. Critical review and meta-analysis of ambient particulate matter source apportionment using receptor models in Europe. Atmospheric Environment. 2013;**69**:94-108. DOI: 10.1016/j.atmosenv.2012.11.009

[125] Karagulian F, Belis CA, Dora CFC, Prüss-Ustün AM, Bonjour S, Adair-Rohani H, et al. Contributions to cities' ambient particulate matter (PM): A systematic review of local source contributions at global level. Atmospheric Environment. 2015;**120**: 475-483. DOI: 10.1016/j.atmosenv.2015.08.087

Chapter 2

Mobility of Trace Element Contaminants from Abandoned Gold Mine Dump to Stream Waters in an Agricultural Active Area

Godwin Okereafor, Elizabeth Makhatha, Lukhanyo Mekuto and Vuyo Mavumengwana

Abstract

In this study, the selected streams within the Blesbokspruit located in South Africa were characterised in this study. Because of prolonged mining activities coupled with ineffective management practices, several mine tailing (MT) dumps are widely distributed in this area. Metals and metalloid contamination from these tailing facilities have been reported to be major contributors to environmental hazards such as acid mine drainage (AMD). With increased agricultural activities in this area, an assessment of the general quality of water being utilised for irrigation purposes and feeding of farm animals becomes inevitable. A procedural method was implemented in a bid to identify relations between tailing and stream water contamination. Representative gold tailing sediments and water samples were collected, respectively. With the aid of X-ray fluorescence (XRF) and X-ray diffraction (XRD), the mineralogical characterisation of the tailing sediments was successfully carried out, while acid digestion using inductively coupled plasma-optical emission spectrometry (ICP-OES) was utilised in the determination of trace metal contents. Samples of different water sources were also characterised. There was a clear description of the link between tailings, water contamination and possible implications to animals and humans in the long run.

Keywords: contamination, livestock, acid mine drainage, water quality

1. Introduction

The mining of minerals such as gold, coal, titanium and other natural resources plays several roles in South Africa's emerging economy with many positive outcomes as well as negative consequences. Till date, the mining sector remains the largest employer of labour. Mining of various minerals is a common sight in the country but of late has become a public menace due to land degradation and contamination of major water bodies through the discharge of waste materials such as overburdens, waste rocks, tailings, slags, mine water and gaseous wastes.

The resource being mined, geology of an area and technology adopted are some of the factors that influence the nature, characteristic properties and level of mine waste generated at specific mines. Mining companies in South Africa, due to legislations and regulations pertaining to general waste management at mining sites, endeavour to manage waste generated during their operations in a bid to avoid the release of contaminants into the environment [1]. Unfortunately, some level of contaminants, which constitute crushed, sand-like materials, still finds their way into the ecosystem.

By definition, mine dumps, also known as residue stockpiles, refer to any debris, discard, tailing, slimes, screening, slurry, waste rock, foundry sand, beneficiation plant waste, ash or any other product derived from or incidental to a mining operation and which is stockpiled, stored or accumulated for potential reuse, or which is disposed of, by the holder of a mining right, mining permit, production right or an old order right [2, 3]. Generally, mine wastes are unwanted, currently uneconomic, solid and liquid materials found at or near mine sites that largely constitute one of the world's major waste streams due to their composition of high concentrations of elements and compounds that impact negatively on humans and animals, the surrounding environment and plants alike [4]. An alarming number of mine dumps often characterised by a complex mixture of metals and dust particles are located at several jurisdictions of the country. This implies that during windy conditions, exposure to dust could be high for close-by inhabitants. Other challenging issues associated with mine wastes include environmental impacts such as the loss of agricultural lands because of their conversion to waste storage facilities and the successive introduction of sediments and other impurities into adjoining surface and groundwater from water running over exposed chemically reactive wastes [5–8].

Most mine wastes have heavy metals (HMs) as a major component, and their continued circulation in soil and water puts the world in a nerve-wracking situation due to their toxic and deleterious effects [9]. HMs such as cadmium (Cd), thallium (Ti), zinc (Zn), lead (Pb), arsenic (As), mercury (Hg), chromium (Cr), copper (Cu), lead (Pb), iron (Fe) and nickel (Ni) are naturally occurring components of the earth's crust and are often referred to as metallic chemical elements with a relatively high density—at least five times the specific gravity of water [10]. At trace concentrations, HMs such as Cu, Cr + 3, Zn, Mn, Co and molybdenum (Mo) are of biological significance; however long-term exposures and high concentrations impact negatively on several biomolecules. It is noteworthy to mention that at very low concentrations, Hg, Cd, Cr + 6, As and Pb stand to be very toxic [11].

Sediments are vital components of aquatic habitat as they can retain contaminants that are not bioavailable to humans [12]. Pollutants are released into water bodies via processes such as sediment resuspension, desorption and redox reactions of sorptive substances. This further explains why sediments despite being reservoirs could at the same time act as likely source of pollutants in the aquatic system [12, 13]. Mine tailings (MT) having the possibilities of containing some HMs are seen in different areas of Blesbokspruit in Ekurhuleni Municipality, South Africa. Over the years, these mine tailings solidify, leaving mountainous structures made of very fine sand particles. Several agents of erosion such as wind and surface run-off after heavy rainfall transport sediments to surrounding water bodies (rivers). These rivers are major sources of water supply to agricultural lands and may constitute diverse environmental problems to humans, animals and plants [14].

A comprehensive understanding of the physicochemical features of soil and water pollutants provides a platform for careful soil management aimed at reducing the adverse effect of the pollutants on the ecosystem. Heavy metals from mine tailings in ionic solutions sometimes contaminate groundwater via leaching and are absorbed by plants; thus it is introduced to the food chain [15]. In another instance, animals drink from stream waters containing these HMs. The final consumption of these plants and animals is of great concern due to danger of HM toxicity.

South Africa's agricultural sector had suffered immense setbacks in the recent past, owing to issues of water shortages and the problem of acid mine drainage (AMD), amongst others. These issues over the years have impacted adversely on the quality and quantity of farm produce. Due to irregular municipal water supply, farmers suffer economically in the quest for portable water; thus, they end up relying on surrounding stream water for their animals and crops. Following the growing numbers of small-scale farming in the Blesbokspruit area of Ekurhuleni, this study seeks to investigate the physicochemical features of selected stream waters and sediments from an abandoned gold mine tailing site with the goal of assessing the distribution of HMs in stream water and sediments from surrounding mining sites and the ecological risks of such HMs.

2. Sampling and samples

2.1 Description of the sampling site

Due to outraging scarcity of water resources, South Africa is referred as a semi-arid nation. Study materials were obtained from a mine tailing dump facility located within the Blesbokspruit catchment (26°10′12″S 28°27′52″E), Heidelberg (34°01′59″S 18°52′28″E) and Suikerbosrand (26°29′46″S 28°21′00″E) rivers, respectively, all of which constitute part of the Vaal River Barrage secondary catchment, located in the eastern region of the Gauteng Province. The Blesbokspruit catchment has in it an important and international wetland covering an area of about 1858 km^2 with a Ramsar recognition dated as far back as 1986. This wetland has a gold mine tailing dump located very close to it and provides support and habitat to several species of birds and plants [16]. Extended part of the study area is used for agricultural activities such as cultivation of crops and rearing of animals. The sampling sites (**Figure 1a**) were selected due to what seems to be the ironic presence of an abandoned gold mine tailing dump that has located next to it an artificial wetland. The wetland was supposed to aid in remediation of the discharge from the mines. In addition, an informal settlement characterised with mostly rural–urban dwellers that engage in farming activities was also in proximity. The rivers being considered in this study are to the best of our knowledge the major water sources for farming activities such as irrigation and feeding of local cows. The subsistence farmers within this area created artificial channels in a bid to redirect water from the streams to special reservoirs that service mostly the animals (**Figure 1b**). During rainfall and often windstorms, sediments from the tailing dumps get eroded into the wetland and subsequently into the surrounding rivers as a result of overflow and surface run-offs (**Figure 1c**).

2.2 Collection of samples and pretreatment

Using sterile bottles, water samples were randomly obtained from four sampling sites located at various directions and distances within the Blesbokspruit catchment. The water samples were assigned names: wetland (WL), Blesbokspruit (BS), Heidelberg (HB) and Suikerbosrand (SB). Soil samples (mine tailing sediments) were collected using a Teflon-coated soil auger and put in labelled polyethylene bags with clear designation MT1–MT4, respectively. The collected water samples from identified streams were stored in an insulated icebox and transported to the laboratory for storage at 4°C prior to further analysis. Collected soil samples were oven

Figure 1.
(a) Map of South Africa showing the Blesbokspruit water catchment [17]; (b) study area showing tailing dump site, wetland and river; (c) transfer mechanism of trace metals/elements.

dried at 100°C and then stored in polyethylene bags for mineralogical and heavy metal analyses.

3. Analysis

3.1 Experimental analysis

Soil samples (tailing sediments) were oven dried at 100°C for 24 hours and mechanically pass through a sieve for particle distribution [18]. Approximately 2 g aliquots of the various soil samples were weighed into a Teflon crucible and then moistened with 100 mL of 1 M HCl acid in a microwave digestion system for the determination of the HCl-soluble fraction of heavy metals. The mixtures were covered and placed on a shaker for 12 hours at 130 rpm. The solutions were filtered using a Whatman filter paper and the filtrates stored in sterile bottles prior to being analysed for heavy metals using inductively coupled plasma-optical emission spectrometry (ICP-OES, GBC Quantima Sequential).

At very high pressure using a mould, 10 g of each of the representative soil samples were pelletised and then inserted in the sample compartment of an X-ray fluorescence (XRF; Rigaku ZSX PrismusII), to help in analysing the elemental compositions of the various minerals that make up the soil samples.

Morphological and mineralogical phase analyses of the tailing material were carried out using a scanning electron microscope (SEM, Tescan Vega 3 XMU) operated with an Oxford software and X-ray diffractometer (XRD; Rigaku Ultima IV), respectively.

In a soil-to-water suspension (1:2.5, w/w), the pH of representative soil samples was determined, while electrical conductivity (EC) was measured in a 1:5 soil-to-water suspension using a Crison multimeter model MM 41. Also carried out was an assessment of the total dissolved solids (TDS) in guidelines as stipulated by standard protocols of APHA [19].

3.2 Quality assurance and quality control

For precision analysis, all reagents used were of analytical standards, while apparatus and glassware inclusive were acid-washed with a 5% nitric acid. Multiple levels of calibration standard solutions are prepared from a Certipur ICP multi-element standard (Merck KGaA) by diluting the stock multi-elemental standard solution (1000 mg L^{-1}) in 0.5% (v/v) nitric acid. The calibration curves for all the studied elements were in the range of 0.01–1.0 mg L^{-1}. The conditions of the ICP were the same as described in a similar study as reported by Okereafor et al. [20].

4. Results and discussion

4.1 Particle size distribution of tailing sediments

Table 1 highlights the textural characteristics of the tailing sediments as gotten from mechanical sieve analysis. The main fractions of all tailing sediments were fine sand (0.150–0.075 mm) and clay (0.075–0.053 mm) with average composition of 66.03% for fine sand, 23.08% clay and 10.89% silt, respectively. With fine sand constituting a major part of the soil within the tailing dump site which are loose with little or no vegetation cover, wind and soil erosion particularly during continued rainfall are likely to occur with no restriction. The steep and inclined nature of

Sieve size (ASTM) % materials, retains (gms)	
No. 100	5.68
No. 140	45.51
No. 200	15.84
No. 270	10.25
PAN	22.72
Total	100
% Sand	67.03
% Silt	10.25
% Clay	22.72

Table 1.
Sieve analysis of the soil from the gold mine tailing.

the dump will support surface run-off of sediments which end up as discharge into the wetland.

4.2 Physicochemical analysis

Various indices such as electrical conductivity, pH, moisture, soil organic matter, texture, temperature, etc. contribute to the quality of soil and as such have great influence on its basic functions such as water retention, promotion of biodiversity, flood resistance, landslides, erosion and agricultural support [21]. However, the physicochemical properties of water to a large extent affect water usage. **Table 2** gives a summary of the observed physicochemical properties of water analysed in this study.

4.2.1 pH of stream water and soil samples

As a physical parameter, pH is a measure of the hydrogen ion concentration in the water/soil samples as ranked on a scale of 1.0–14.0. The lower the pH value of a material (soil/water), the more acidic it is, and the higher the pH value, the basic, or alkaline, it is. Generally, many chemical and biological processes are affected by pH as different organisms flourish better at different ranges of pH. A pH metre (Jenway model 3510) was used to determine the pH levels of the different water samples and sediments obtained from an abandoned golf mine dump. The observed pH values were within the limit of standard irrigation water (6.5–8.4), ranging from 6.31 to 6.35, thus validating the mobility and availability of heavy metals for plant uptake due to the presence of fewer H^+ ions [22]. A low pH of 4.21 was recorded for the tailing sediments which supports the heterogenous deposits of sulphidic residues from mining activities. The ability of plants to take up nutrients could be affected by the acidity of tailings, hence the sparse distribution of vegetation at the mine tailing site. The identified heavy metals (Zn^{2+}, Ni^{2+}, Pb^{2+}, AS^{2+}, Cu^{2+}) together with the observed pH of the tailings and wetland could result in acid mine drainage (AMD).

In the long run, the continuous erosion of the tailing sediments to surrounding water bodies could result in damage to metal pipes, tanks and fittings used for irrigation purposes by farmers. Consequently, the alkalinity (acid absorbing potential) of the various water samples based on the reported pH (6.31–6.35) is an indication that most of the available dissolved carbon dioxide have been converted into bicarbonate ion.

Sample names	pH	EC μS/cm	TDS mg/L
Wetland (WL)	6.31	948.00	606.72
Blesbokspruit (BS)	6.33	1040.00	665.60
Heidelberg (HB)	6.34	1235.00	790.40
Suikerbosrang (SB)	6.35	488.00	312.32
Mine tailing	4.21	132.00	84.48

Table 2.
Physicochemical analysis of sediments and water samples.

4.2.2 Electrical conductivity

The most influential water quality guideline on crop productivity is the water salinity hazard as measured by electrical conductivity (EC_w). EC measures salinity from all identified ions dissolved in a sample including negatively charged ions (e.g., Cl^-, NO^-_3) and positively charged ions (e.g., Ca^{++}, Na^+).

The values of electrical conductivity were determined by the concentration of ionic species contained in water. Using a standard of 84μS and a Crison multimeter (MM 41), water samples from Heidelberg (HB) recorded the highest electrical conductivity value of 1235 μS followed by (BS) 1040 μS, (WL) 948 μS and (SB) 488 μS.

A value of 132 μS was recorded for the tailing sediments. The higher electrical conductivity recorded could imply the presence of higher dissolved salt or ion concentration which suggests that the samples have higher conductivity.

The observed high EC_w water on crop productivity will result in plants' inability to compete with ions in the soil solution for water (physiological drought). The higher EC implies less water is available to plants, despite the soil appearing wet, thus a reduced yield potential. This is because plants can only transpire "pure" water as usable plant water in soil solution decreases as EC increases.

4.2.3 Total dissolved solids

Using the expression 0.64 × EC, a measure of the total dissolved amount of substance was obtained. The lowest value was observed in the MT (84.48 mg/L) with HB recording the highest (790.40 mg/L). WL and BS both had 606.72 mg/L and 665.60 mg/L, respectively, while 312.32 mg/L was recorded in (SB). Crop yield can be adversely affected by the higher concentration of salt in water, thereby leading to soil degradation and pollution of groundwater. This parameter however did not deviate from the international standards, but the high concentrations of dissolved solids could result in some technical effects. Dissolved solids can produce hard water, which leaves deposits and films on fixtures and on the insides of irrigation pipes.

4.3 Chemical composition of tailing sediments

The energy-dispersive X-ray (EDX) microanalysis is a technique of elemental analysis associated to electron microscopy based on the generation of characteristic X-rays that reveal the presence of elements present in specimens. EDX technique is useful in the study of environmental pollution as it carries a huge vantage in the detection of heavy metals because they are nonbiodegradable and they can accumulate in ecological systems, thus resulting in pollution.

The scanning electron microscope (SEM) and EDX analyses (**Figure 2**) indicate homogenous distribution of granules throughout mining tailing samples with EDX analysis, further confirming elements such as Si (33.58%), Fe (19.12%), O (54.25%), Al (5.33%), K (1.76%) and Mg (0.44%) which could be compared to elemental composition revealed from X-ray fluorescence (XRF).

A typical mineralogical composition of the tailing sediments is shown in **Figure 3** as determined by XRD. The XRD results confirm the presence of silicate minerals which are quartz (SiO_2), marcasite FeS_2, dialuminium silicate $Al(SiO_4)O$, pyrite (FeS_2) and gupeite (Fe_3Si). These could be linked with the elements identified from both XRF and ICP-OES.

Table 3 reports the partial compositional analysis of the tailing sediments collected from the abandoned gold mine dump. SiO_2 (81.82%) was shown to be the most abundant compound found in the tailing sediments. The oxides of Al, Fe and S were 6.93, 3.59 and 3.41%, respectively. Oxides of K were 1.98%, while those of Na, Mg, Ca, Mn, Zn, Pb and Cu were less than 1%.

ICP-OES analysis as illustrated in **Table 4** shows heavy metal concentrations in the sediments, wetland and surround streams. Filtrate from the tailing sediments showed very high concentration of various heavy metals with Cr recording the highest value of 43.13 mg/L, followed by Al 16.42 mg/L, As 10.17 mg/L, Pb 6.29 mg/L and Ni 1.34 mg/L, respectively. Considering the proximity and the fact that the artificial wetland and studied streams all get fed from the run-off water from the dump site during rainfall, it is not a coincidence that higher metal concentrations were observed. Many metal elements are essential nutrients for animals and crops but, in excess, may result in chronic or toxic effects.

Toxic substances are often in solution or as suspended solids in water which may affect the nutritional availability of toxic elements or substance in animals. Although short-term intake of toxic substance by animals has little or no noticeable effects, long-term exposure to those substances may result in serious damage. The extent of damage inflicted on animals by toxic elements may be determined by health status, age, and rate of consumption of toxic elements by the animals. However, the intake of toxic substances may not cause any measurable effect on growth, production, or reproduction yet may cause subcellular damage in farm animals which could be expressed as increased susceptibility to disease or to parasitic invasion.

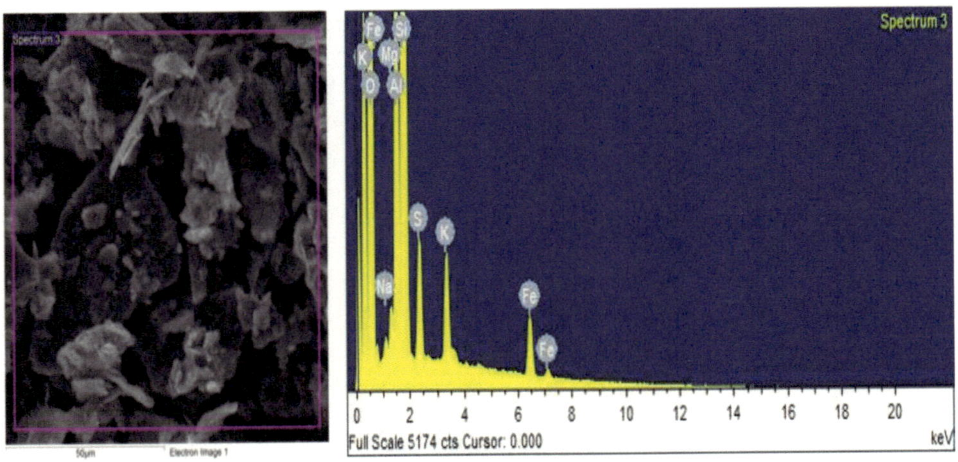

Figure 2.
SEM micrograph of tailing sediments.

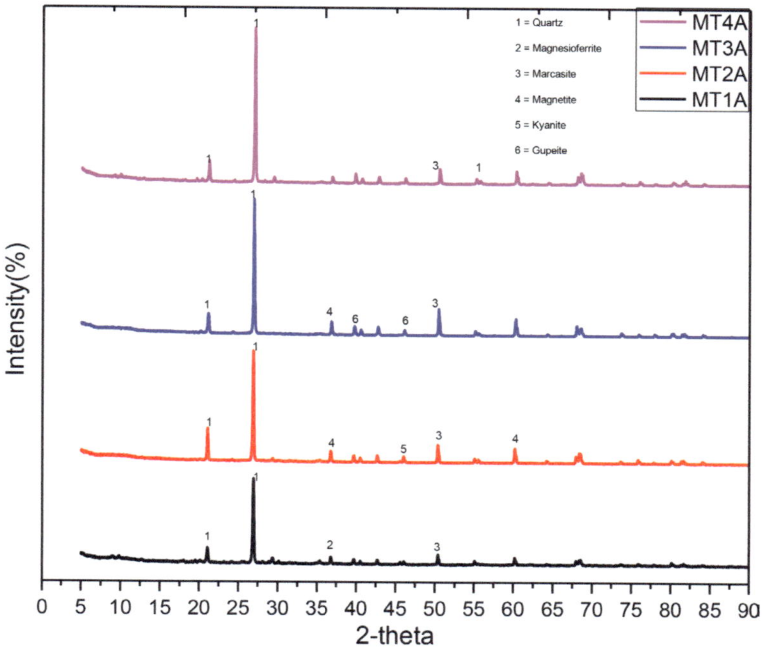

Figure 3.
XRD patterns of the filtered sediments.

Oxides	MT1 conc. (%)	MT2 conc. (%)	MT3 conc. (%)	MT4 conc. (%)	$MT_{Ave.}$ conc. (%)
SiO_2	69.42	86.87	86.35	84.63	81.82
Al_2O_3	9.09	6.03	5.18	7.42	6.93
Fe_2O_3	5.87	2.37	2.60	3.53	3.59
SO_3	8.94	1.44	2.32	0.92	3.41
K_2O	2.71	1.61	1.64	1.96	1.98
MgO	1.49	0.38	0.30	0.33	0.63
TiO_2	0.58	0.49	0.51	0.49	0.52
CaO	0.90	0.23	0.55	0.13	0.45
Na_2O	0.27	0.14	0.14	0.16	0.18
Cr_2O_3	0.15	0.17	0.15	0.14	0.52
PbO	0.04	0.03	0.04	0.04	0.04
NiO	0.11	0.01	0.02	0.01	0.04

Table 3.
Results of XRF analysis detailing composition of tailing sediments.

With agricultural activities taking place around the vicinity of the tailing dump, farmers employ water from the streams in irrigation and feeding of animals despite the high heavy metal concentrations that apparently exceed the maximum permissible level of the US Environmental Protection Agency water composition [23, 24]. Al, As, Zn, Cd, Ni, Cu, Pb and Cr were all above the required standard. There is a strong likelihood of the run-off water from the tailing dumps and fine particles being blown during severe windstorm introducing heavy metals such as Al, As,

Metal	MT (mg/L)	WL (mg/L)	SB (mg/L)	HB (mg/L)	BS (mg/L)	DWAF (1990) livestock	WHO (2011) domestic
Al	16.42	2.98	1.76	1.77	0.33	5.00	0.90
As	10.17	7.78	4.27	2.65	9.70	0.50	0.01
Cd	0.46	0.35	0.35	0.58	1.10	0.50	0.01
Pb	6.29	2.83	2.70	2.76	2.53	0.05	0.01
Ni	1.34	0.03	0.15	0.09	0.70	1.00	0.07
Zn	0.28	0.68	0.50	0.19	0.34	25.00	3.00
Cr	43.13	46.75	36.48	59.59	32.13	<0.05	<0.05
Cu	0.03	0.38	0.43	0.29	0.29	0.50	2.00

Table 4.
Results of ICP-OES analysis on stream water samples and filtrate from tailing sediments showing heavy metal concentration.

Cd, Pb, Ni, Zn, Cr and Cu into nearby rivers. Direct consumption of such waters by humans and animals such as cattle gets into the gastrointestinal system, leaving some adverse effects by increasing the gastrointestinal pH, resulting in surficial coating on the stomach [25]. In addition, some of the trace metals get leached into the water table through percolation and are absorbed by plants which the animals feed from.

In humans, several forms of cancer have been linked to arsenic, and chronic exposure to arsenic through drinking water has been associated to health effects such as nervous disorders, high blood pressure, diabetes and hyperkeratosis [26]. However, there are little or no reports on the effect of arsenic in drinking water on the health and/or effect of farm animals. Arsenic availability in soil can disturb normal functioning of plant metabolism, consequently leading to stunted growth and low crop productivity [27]. Previous studies indicated arsenic to be responsible for reduction in gas exchange attributes (photosynthetic rate, transpiration rate, stomatal conductance) and chlorophyll concentrations [28].

The total aluminium concentration in a human body is approximately 9 ppm (dry mass) with an approximately daily intake of 5 mg, of which only a small fraction is absorbed. The high aluminium content of the various water sources observed may have negative impacts on plants, humans and animals. Various ailments of the nervous system, such as Parkinson's disease, amyotrophic lateral sclerosis (Lou Gehrig's disease) and Alzheimer's disease as well as functional lung disorder, may be experienced in humans [29]. There are currently no reports on aluminium toxicity to ruminants. However, there are indications about the risks of inducing either a phosphorus deficiency or a condition known as grass tetany when ruminants consume large amounts of aluminium from soil, aluminium-rich forages or water high in aluminium content [30]. In general, more soluble forms of aluminium in plants may pose some risk such as the inhibition of root elongation [31].

Plant growth and development is often affected adversely by cadmium, a non-essential element due to its high toxicity and large solubility in water [32]. The uptake of minerals by plants has been reported to be altered by cadmium which impacts on the availability of minerals from the soil as well as a reduction in the population of soil microbes [33]. Stomatal opening, transpiration and photosynthesis have been reported to be affected by cadmium in nutrient solutions, but the metal is taken up into plants more readily from nutrient solutions than from soil [34]. The accumulation of cadmium in humans could lead to renal failure, decreased vitamin D synthesis and consequently osteoporosis. The high concentration of cadmium may adversely interfere with the metabolism of essential trace elements which in farm animals such as cattle could result in an unthrifty appearance; rough coat hair; dry scaly skin; dehydration; loss of hair from the legs, thighs, ventral chest, and brisket; mouth lesions; oedematous, shrunken scaly scrotum; sore and enlarged joints; impaired sight; extreme emaciation; and some atrophy of hindlimb muscles [35].

Despite plants being able to take up high levels of lead of up to 500 ppm from soils, as toxic pollutants, lead and some of its compounds can limit the synthesis of plant chlorophyll [36]. The growth of plants is often retarded by higher concentrations of lead. Crops cultivated in the study area stand a risk of suffering damage and reduced growth.

Besides gaining access into the food chain via plant uptake, humans, through water intake, consume more lead. With the alarming concentrations of lead shown in **Table 4**, inhabitants are likely to have excess lead intake which could, over time, result in paralysis, skin pigmentation and colic. Females may experience menstrual disorder, infertility and spontaneous abortion, while children may suffer lower IQs, behavioural changes and concentration disorder. Lead is the most common cause of cattle poisoning. Animals die or perform poorly after inadvertently ingesting lead either through feed or water. Gradual poisoning of the areas cannot be ruled out as evidenced from the tailing sediments. Lead when consumed by ruminants end up in the reticulum (fore stomach) which provides a reservoir from which it is absorbed into the bodies of cattle, sheep and goats. In older cattle and sheep, subacute lead poisoning is characterised by anorexia, rumen stasis, colic, dullness and transient constipation, frequently followed by diarrhoea, blindness, head pressing, bruxism, hyperesthesia and incoordination [37].

Chromium (Cr) occurs in the environment primarily in two valence states, trivalent chromium (Cr III) and hexavalent chromium (Cr VI), with the latter being more toxic.

The high concentration levels of Cr as contained in **Table 4** are worrisome as it is known from previous studies to be a toxic metal that can cause severe harm to plants depending on its oxidation state. Some of the toxic effects of Cr on plant growth and development include alterations in the process of germination and growth of roots, stem, and leaves, which may affect total dry matter production and yield [38]. Excessive Cr also impacts adversely on plant's physiological processes such as photosynthesis, water relations, mineral nutrition, oxidative imbalance and inhibition of enzymatic activities. Chromium can affect antioxidant metabolism in plant. The corrosive nature of some chromium (VI) compounds, when in excess, results in ulcerations, dermatitis and allergic skin reactions in humans. When inhaled on the other hand, it could lead to ulceration and perforation of the mucous membranes of the nasal septum, irritation of the pharynx and larynx, asthmatic bronchitis, bronchospasms and oedema [39]. In mammals, chromium (III) is an essential trace element involved in lipid and glucose metabolism [40]. Chromium (VI) as reported from previous studies adversely affected the developing embryo, causing retarded foetal development in cattle during gestation, resulting in reductions in the number of foetuses and foetal weight and a higher incidence of stillbirth and post-implantation loss [41].

5. Conclusion

This study revealed that the tailing sediments were largely comprised of fine sands that are loosely packed and prone to erosion, thus supporting the migration of trace metal contaminants. The ability of plants to survive in the area based on the recorded physicochemical data such as acidity and electrical conductivity of the tailing sediments and water from wetland is daisy. A continuous erosion and surface run-off of sediments from the tailing dump site increases the migration of Al, As, Pb and Cr, which were observed to be in elevated concentrations, into the wetland and streams, thus endangering sustainable agricultural activities within the surrounding farmlands as water sources to farm facilities are highly prone to contamination from heavy metals trapped in tailing sediments especially given their high concentrations. The dispersion of such tailing sediments not only affects the agricultural activities within the studied area but may also have a health-related effect on the human population that reside in proximity to this mine dump. There should be metal speciation studies carried out to ascertain the extent of metal toxicity exposures within these areas.

Acknowledgements

We thank the Mineral Processing and Technology Research Centre of the University of Johannesburg for providing the resources used in conducting this study.

Conflict of interest

The authors declare no conflict of interest.

Author details

Godwin Okereafor[1]*, Elizabeth Makhatha[1], Lukhanyo Mekuto[2] and Vuyo Mavumengwana[3]

1 Faculty of Engineering and the Built Environment, Department of Metallurgy, School of Mining, Metallurgy and Chemical Engineering, University of Johannesburg, Johannesburg, South Africa

2 Faculty of Engineering and the Built Environment, Department of Chemical Engineering, University of Johannesburg, Johannesburg, South Africa

3 Faculty of Science, Department of Biotechnology and Food Technology, University of Johannesburg, Johannesburg, South Africa

*Address all correspondence to: guokereafor@gmail.com

IntechOpen

© 2020 The Author(s). Licensee IntechOpen. This chapter is distributed under the terms of the Creative Commons Attribution License (http://creativecommons.org/licenses/by/3.0), which permits unrestricted use, distribution, and reproduction in any medium, provided the original work is properly cited. (cc) BY

References

[1] Matinde E, Simate GS, Ndlovu S. Mining and metallurgical wastes: A review of recycling and re-use practices. Journal of the Southern African Institute of Mining and Metallurgy. 2018;**18**(8):825-844. DOI: 10.17159/2411-9717/2018/v118n8a5

[2] MPRDA. Mineral and Petroleum Resources Development Act No. 28 of 2002. 2002

[3] Puell JO. Methodology for a dump design optimization in large-scale open pit mines. Cogent Engineering. 2017;**4**:1387955. DOI: 10.1080/23311916.2017.1387955

[4] Hudson-Edwards KA, Jamieson HE, Lottermoser BG. Mine wastes: Past, present, future. Elements. 2011;**7**:375-380

[5] Aznar-Sánchez JA, García-Gómez JJ, Velasco-Muñoz JF, Carretero-Gómez A. Mining waste and its sustainable management: Advances in worldwide research. Minerals. 2018;**8**(284):1-27. DOI: 10.3390/min8070284

[6] Jordan G, D'Alessandro M, editors. Mining, mining waste and related environmental issues: problems and solutions in Central and Eastern European Candidate Countries – Joint Research Centre of the European Commission, Ispra, EUR 20868 EN; 2004. 208 p. ISBN: 92-894-4935-7

[7] Sloss LL. Environmental and Other Effects of Coal Mining and Transport. IEA Clean Coal Centre; 2017. ISBN: 978-92-9029-604-1

[8] McKinnon E. The environmental effects of mining waste disposal at Lihir gold mine, Papua New Guinea. Journal of Rural and Remote Environmental Health. 2002;**1**(2):40-50

[9] Okereafor GU, Mulaba-Bafubiandi AF, Mavumengwana V. Possible heavy metals transportation and ground erosion during heavy rainfall in rural Mangweni (Tzaneen, Limpopo Province, South Africa). The Southern African Institute of Mining and Metallurgy. 2017:515-522

[10] Obodo G. The bioaccumulation of heavy metals in fish from Anambra River. Journal of Chemical Society of Nigeria. 2004;**29**:60-62

[11] Rahman Z, Singh V. Assessment of heavy metal contamination and Hg-resistant bacteria in surface water from different regions of Delhi, India. Saudi Journal of Biological Sciences. 2016;**25**(8):1687-1695. DOI: 10.1016/j.sjbs.2016.09.018

[12] Nemati K, Bakar NKA, Abas MRB, Sobhanzadeh E, Low KH. Comparison of unmodified and modified BCR sequential extraction schemes for the fractionation of heavy metals in shrimp aquaculture sludge from Selangor, Malaysia. Environmental Monitoring and Assessment. 2011;**176**:313-320

[13] Varol M. Assessment of heavy metal contamination in sediments of the Tigris River (Turkey) using pollution indices and multivariate statistical techniques. Journal of Hazardous Materials. 2011;**195**:355-364

[14] Alomary AA, Belhadj S. Determination of heavy metals (Cd, Cr, Cu, Fe, Ni, Pb, Zn) by ICP-OES and their speciation in Algerian Mediterranean Sea sediments after a five-stage sequential extraction procedure. Environmental Monitoring and Assessment. 2007;**135**:265-280

[15] Calace N, Di Muro A, Nardi E, Petronio B, Pietroletti M. Adsorption isotherms for describing heavy-metal retention in paper mill sludges.

Industrial and Engineering Chemistry Research. 2002;**41**:5491-5497

[16] Pillay K, Agjee NH, Pillay S. Modeling changes in land cover patterns in Mtunzini, South Africa using satellite imagery. Journal of Indian Society of Remote Sensing. 2014;**42**:51-60

[17] du Plessis A, Harmse T, Ahmed F. Quantifying and predicting the water quality associated with land cover change: A case study of the Blesbok Spruit catchment, South Africa. Water. 2014;**6**:2946-2968

[18] ASTM D422-63. Standard Test Method for Particle-Size Analysis of Soils. 2007

[19] Clescerl L, Greenberg A, Eaton A. Standard Methods for the Examination of Water and Wastewater. Washington, DC: American Water Works Association. Part 3000; 1999

[20] Okereafor U, Makhatha E, Mekuto L, Mavumengwana V. Dataset on assessment of pollution level of selected trace metals in farming area within the proximity of a gold mine dump, Ekuhurleni, South Africa. Data in Brief. 2019;**26**:104473. DOI: 10.1016/j.dib.2019.104473

[21] Tale KS, Ingole S. A review on role of physico-chemical properties in soil quality. Chemical Science Review and Letters. 2015;**4**(13):57-66

[22] Department of Water Affairs and Forestry (DWAF). South African Water Quality Guidelines. 2nd ed. Vol. 4. Agricultural Use: Irrigation; 1996

[23] Praveena SM, Yuswir NS, Aris AZ, Hashim Z. Contamination assessment and potential human health risks of heavy metals in Klang urban soils: A preliminary study. Environmental Earth Sciences. 2015;**73**:8155-8165

[24] Dubey M, Ujjania N. Assessment of water quality and sources of pollution in downstream of Ukai, Tapi River (Gujarat). Current World Environment: An International Research Journal of Environmental Sciences. 2015;**10**:350

[25] Ekosse GIE, Mulaba-Bafibiandi A. Granulometric evaluation of continental bentonites and kaolin for ceramic applications. Journal of Applied Sciences. 2008;**8**:1021-1027

[26] McDowell LR. Minerals in Animal and Human Nutrition. 2nd ed. Gainesville, FL, USA: Elsevier; 2003. ISBN: 9780444513670

[27] Nel A, Xia T, Mädler L, Li N. Toxic potential of materials at the nanolevel. Science. 2006;**311**:622-627

[28] Anjum SA, Tanveer M, Hussain S, Ashraf U, Khan I, Wang L. Alteration in growth, leaf gas exchange, and photosynthetic pigments of maize plants under combined cadmium and arsenic stress. Water, Air, and Soil Pollution. 2017;**228**:13

[29] Krewski D, Yokel RA, Nieboer E, Borchelt D, Cohen J, Harry J, et al. Human health risk assessment for aluminium, aluminium oxide, and aluminium hydroxide. Journal of Toxicology and Environmental Health, Part B. 2007;**10**:1-269

[30] Odette O. Grass tetany in a herd of beef cows. The Canadian Veterinary Journal. 2005;**46**(8):732-734

[31] Kochian LV, Pineros MA, Hoekenga OA. The physiology, genetics and molecular biology of plant aluminum resistance and toxicity. Plant and Soil. 2005;**274**:175-195

[32] Pinto AP, Mota AM, de Varennes A, Pinto FC. Influence of organic matter on the uptake of cadmium, zinc, copper and iron by sorghum plants.

Science of the Total Environment. 2004;**326**:239-247

[33] Benavides MP, Gallego SM, Tomaro ML. Cadmium toxicity in plants. Brazilian Journal of Plant Physiology. 2005;**17**(1):21-34. DOI: 10.1590/S1677-04202005000100003

[34] di Toppi LS, Gabbrielli R. Response to cadmium in higher plants. Environmental and Experimental Botany. 1999;**41**:105-130

[35] Olobatoke RY, Mathuthu M. Heavy metal concentration in soil in the tailing dam vicinity of an old gold mine in Johannesburg, South Africa. Canadian Journal of Soil Science. 2016;**96**:299-304

[36] Al-Akeel K. Lead uptake, accumulation and effects on plant growth of common reed *(Pharagmites Australis* (Cav.) Trin. ex Steudel) plants in hydroponic culture. International Journal of Advances in Agricultural and Environmental Engineering (IJAAEE). 2016;**3**(2):391-394. DOI: 10.15242/IJAAEE.C1216047

[37] Cowan V, Blakley B. Acute lead poisoning in western Canadian cattle—A 16-year retrospective study of diagnostic case records. The Canadian Veterinary Journal. 2016;**57**(4):421-426

[38] Bhalerao SA, Sharma AS. Chromium: As an environmental pollutant—A review. International Journal of Current Microbiology and Applied Sciences. 2015;**4**(4):732-746

[39] Agency for Toxic Substances and Disease Registry (ATSDR). Toxicological Profile for Chromium. Atlanta, US: US Department of Health and Human Services; 2000

[40] Department for Environment Food and Rural Affairs (DEFRA) and Environment Agency (EA). Contaminants in soil: Collation of toxicological data and intake values for humans. Chromium. R&D publications TOX 4. Bristol: Environment Agency; 2002

[41] Pechova A, Pavlata L. Chromium as an essential nutrient: A review. Veterinární Medicína. 2007;**52**(1):1-18

Chapter 3

The Provenance of Arsenic in Southeast Asia Discovered by Trace Elements in Groundwater from the Lowlands of Nepal

Barbara Mueller

Abstract

Arsenic concentrations in groundwater extracted from quaternary alluvial sediments pose a serious health issue for inhabitants living in several countries in Southeast Asia. A widely approved hypothesis states that reductive dissolution of Fe-bearing minerals releases As oxyanions to ground water and the original source of As has to be located in mafic rocks occurring across the entire Himalayan belt. Yet, recent trace element analyses of ground water from the lowlands (Terai) of Nepal show a clear decoupling of As and Fe. The positive correlation of K, Na, and trace elements like Li, B, and Mo with arsenic points out to clay minerals hosting the toxic element. This pattern of trace elements found in the ground water of the Terai also advocates against an original source of As in mafic rocks. The lithophile elements like Li, B, P, Br, Sr, and U reflect trace element composition typical for felsic rocks as an origin of As. All the mentioned elements are components of clay minerals found ubiquitously in some of the most characteristic felsic rocks of the Nepal Himalaya: metapelites and leucogranites—all these rocks exhibiting a high abundance of especially B, P, and As besides Cd and Pb.

Keywords: arsenic, groundwater, trace elements, felsic, Himalaya

1. Introduction

Arsenic concentrations found in the groundwater in quaternary alluvial sediments in the lowland Terai region of Nepal and other countries of South Asia (Bangladesh, India, Nepal, Myanmar, China, Vietnam, Cambodia, and China) often exceed the World Health Organization (WHO) drinking water guideline (10 μg/L). The oral intake of arsenic causes various detrimental health issues: skin lesions including pigmentation changes, mainly on the upper chest, arms, and legs, keratosis of the palms of the hands and soles of the feet, and as the most severe effect, cancer of the skin and internal organs [1–3]. The origin of the arsenic contamination is clearly geogenic, and its elevated concentrations in natural ground waters are considered to be due to natural weathering of the Himalayan belt [4–9]. These quaternary alluvial sediments are carried by the Ganga-Brahmaputra river system and build up the Himalayan foreland basin and the Bengal fan—one of the largest modern fluvial deltas of the world [10, 11]. Arsenic is not known as an element with a high abundance in the Earth's continental crust. Sulfide-bearing mineral deposits are the most common sources of As which also has a strong affinity for pyrite. Weathering of pyrite leads to formation of hydrous iron oxides containing As, but clay minerals are important hosts of As as well. Sediments usually containing 1–20 mg/kg (near crustal abundance) of arsenic can already be a cause of high dissolved arsenic (>50 μg/L) in groundwater if one or both of two potential "triggers"—an increase in pH above 8.5 or the begin of reductive iron dissolution—are initiated [12]. Pyrite represents the largest reservoir of As. Besides this iron sulfide, As is mainly concentrated in hydrous iron oxides and clay minerals. Arsenic can be easily solubilized in groundwaters depending on pH, redox conditions, temperature, and solution composition. The sediments of the Terai plain in Nepal are commonly reducing with a pH between 7.0 and 7.8. Beyond that, As in ground water is only weakly negatively or not correlated with Fe (decoupling of As) but positively correlated with lithophile elements like Na and K. Decoupling between aqueous As and Fe has also been described by [9, 13–15]. The mentioned elements are specifically derived from alumosilicates such as clay minerals during weathering. More immobile elements such as Fe and Al will therefore be concentrated in the remnants. Clay minerals that are extremely fine-grained are noted to be compatible of adsorbing arsenic to a high degree [16–18]. So far, there is basically a small number of source materials recognized as significant contributors to arsenic in the aquifer: for example, organic-rich or black shales, Holocene alluvial sediments with slow flushing rates, mineralized and mined areas (most often gold deposits), volcanogenic sources, and thermal springs. As there is no correlation found between As and Fe concentrations in groundwater in the Terai of Nepal, an obvious correlation between lithophile element concentration and As the initial source rocks is of felsic and not mafic composition. These trace elements in groundwater reflect the origin of the arsenic in the high Himalayas of Nepal.

2. Geology

All of the four major Himalayan tectonic units are exposed in the Terai (lowland of Nepal): (1) the Tethys Himalaya, delimited at the base by the South Tibetan Detachment system (STDS), (2) the Higher Himalayan Crystallines (HHC) delimited at the base by the Main Central Thrust I (MCT I), (3) the Lesser Himalaya (LH) divided into upper and lower Lesser Himalaya is delimited at the base by the Main Boundary Thrust (MBT), and (4) the Siwaliks, delimitated at its base by the Main Frontal Thrust (MFT) as much as the quaternary foreland basin. In the Terai once, east-west flowing rivers built up the Archean crystalline formations deep beneath the Alluvium of the Terai as well as the marine sedimentary deposits forming the high Himalayas and the Siwalik formation. The debris of these formations can be found within confined space [19].

These four units include a wide range of diverse rocks of metamorphic, sedimentary, and igneous in origin. Their differential erosion is accounting for some of the groundwater arsenic heterogeneity seen in the foreland and delta (e.g., [5, 7, 20, 21]). In the realm of provenance of the Terai sediments, the Tethys Himalaya is made up of 10 km of various metasedimentary rocks (limestones, calc-schists, shales, and quartzites) ranging from Cambrian to Jurassic. Leucogranites like the Manaslu leucogranite are also found emplaced within the Tethyan rocks (e.g., [22]).

The Terai region of Nepal itself is the direct prolongation of the Bengal Delta Plain (BDP), and it is the elongation of Indo-Gangetic trough. The Terai plain is composed of quaternary sediments including molasse units consisting of gravel, sand, silt, and clay and represents an active foreland basin. The rivers in the Terai generally flow from north to south. All major rivers have their origin in the high Himalayas, while minor rivers also emerge from the proximate Siwalik Hills, and hence sediments are deposited in the form of a fan along the flanks of the Terai basin. Organic material is widespread in the fine sediments which were once deposited in inter-fan lowlands, in wetlands, and in swamps [23–25].

The district of Nawalparasi is the most acute studied Terai province related to arsenic-contaminated groundwater in Nepal. The lithology of the Nawalparasi province sedimentary basin belongs to Holocene alluvium including the present-day alluvial deposits, channel sand, and gravel deposits as well as outwash deposits [26]. The major river, the Narayani/Gandaki, which has its source in the Higher Himalaya, flows along the eastern boundary of the Nawalparasi district and exerted a major influence on the underlying unconsolidated Holocene fluvial deposits that include the floodplain aquifer system. Atypical for the Terai, where finer sediments typically increase toward the south, in Nawalparasi, fines predominate in the north, and sand and gravels are found near the Nepal-India border [27]. In the areas with fine-grained sediments, elevated concentrations of As are typically recorded [15, 28, 29].

3. Material and methods

3.1 Sample collection

Around 20 years ago, co-workers from CAWST (Centre for Affordable Water Sanitation Technology) Calgary, Canada, in cooperation with ENPHO (Environment and Public Health Organization) Kathmandu, Nepal, began to install iron-assisted bio-sand filters built on the basis of arsenic removal from water using zero-valent iron (ZVI) media. The modified model now used in Nepal is known as Kanchan filter [30, 31]. Due to growing concerns about the malperformance of some of these filters, a groundwater sampling campaign was initiated by CAWST together with Eawag (Swiss Federal Institute for Environmental Science and Technology), Dübendorf, Switzerland. As reported in [32] and CAWST, the Kanchan filter efficiency under field conditions operating for a long period has scarcely been observed. Since a part of the Kanchan filters still had effluent arsenic concentrations exceeding the Nepal drinking water quality standard value (50 µg/l), groundwater, intermediately filtered, and effluent water for trace element analyses were sampled in October 2015 (post-monsoon). A second field campaign was arranged in pre-monsoon time (April 2017) in order to detect for differences in arsenic concentration in groundwater between the two seasons. Filters were also inspected at household levels. Measurements to improve the efficiency of the filters are under progress.

Thirty-five water samples from around Ramgram, the capital of the district Nawalparasi, were collected from hand pumps in October 2015 and again in April 2017. All pumps were thoroughly flushed before sample collection. Household for sample collections was selected referring to a register established by ENPHO including all groundwater samples exceeding the Nepal drinking water quality standard value (50 µg/L). Sampling sites consisted of groundwater from private tube wells located within the municipalities of Ramgram (former name, Parasi, the capital of the district Nawalparasi), Manari, Panchanagar, Sukauli, and Tilakpur (within proximity of Ramgram). Water samples were acidified with HNO_3 and sent to the laboratory in Switzerland for further examination.

3.2 Trace element analysis

All trace elements in the groundwater samples were determined by ICP-MS (Agilent Technology, 7500 Series, Agilent Technologies, Waldbronn, Germany) at Eawag, Dübendorf, Switzerland, after 1:2 dilution with 0.5 M HNO_3. Each measurement was conducted in triplicate. All ICP-MS determinations agreed to within 3–5% standard deviation [33].

4. Results and discussion

Figure 1a shows the correlation between As and Fe concentrations in the 35 groundwater samples taken in spring 2017. Results from analyses of samples collected autumn 2015 exhibit very similar trends and are therefore not included in this section. There is no visible correlation between As and Fe, and the correlation coefficient is only given for illustration. In contrast to the obvious decoupling between As and Fe, the concentration of As markedly and positively depends on the concentration of various lithophile elements (K shown as an example in **Figure 1b**). As is positively correlated with Na, Li, B, and Mo and negatively correlated with Ca, Mn, and As. Evidently, the reason for such a paradigm is mainly the derivation of As from silicates like clay minerals and feldspars and much less from Fe hydroxides and pyrite. Even though in [34] it is reported about a positive dependence of As on Fe in West Bengal and in [35] a positive correlation between As and Fe (r = 0.77) in the aquifer of the Nawalparasi district is also mentioned, a decoupling between aqueous As and Fe has also been observed in [13–15].

In [15] it is stated correctly that decoupling between Fe and As may result from sorption of Fe to other surfaces (i.e., clays) or precipitation of Fe(II) minerals, such as siderite. Anyways, the diagram in **Figure 1b** explicitly shows the correlation between As and the lithophile element K as an example. K as well as Na, Mg, and Sr (replacement of Na and K) can easily be dissolved from interlayers of clay minerals, Na, K, and Sr as well from alkali feldspars or the borosilicate tourmaline. Li, B, and Mo represent common trace elements found in micas; Li and B moreover are main components of tourmaline (general formula, (Ca,K,Na)

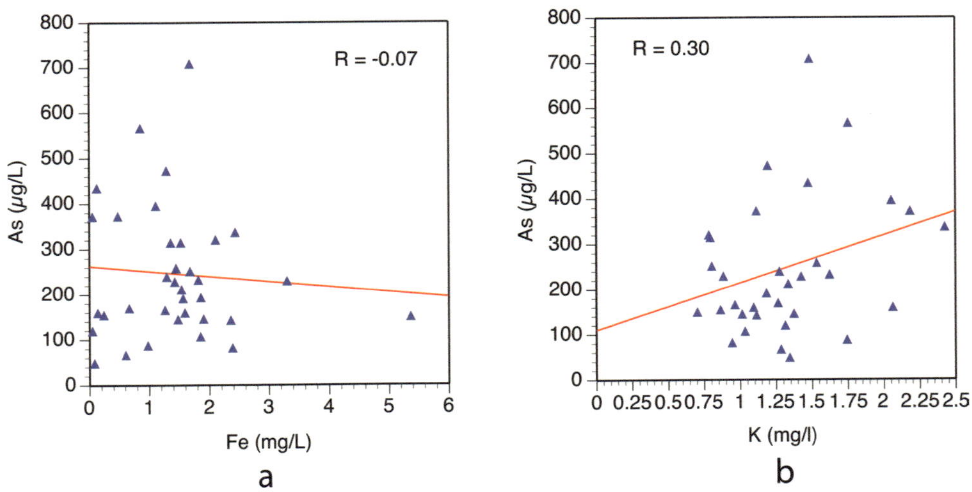

Figure 1.
(a) Apparent decoupling between Fe (mg/l) and As (µg/l). The correlation coefficient (r) is included in the diagram for illustration though it is not significant. (b) Correlation between K (mg/l) and As (µg/l).

$(Al,Fe,Li,Mg,Mn)_3(Al,Cr,Fe,V)_6(BO_3)_3(Si,Al,B)_6O_{18}(OH,F)_4)$. Besides this said, tourmaline is one of the very rare minerals fitting a significant amount of boron in its structure. As reported in [36], authors found tourmaline-containing aquifers enriched in As in West Bengal, India; in [37] it is clearly described how trace elements of groundwater in the Terai of Nepal refer to the origin of the mentioned soil minerals being decomposed from tertiary leucogranites (rich in B) in the High Himalayas. As Fe is predominantly contained in the octahedral layer of clay minerals such as biotite (general formula, $K(Mg,Fe^{2+},Mn^{2+})_3[(OH,F)_2|(Al,Fe^{3+},Ti^{3+})Si_3O_{10}])$ and fits hardly in the interlayer, so it is not possible to exchange Fe against K or Na. However, the likely source of As due to the positive correlation between Na, K, and As points out to silicate minerals as As can be readily adsorbed on the surface and edges of these minerals. Muscovite $KAl_2[(OH,F)_2|AlSi_3O_{10}]$ is another frequent clay mineral found in the sediments of Nawalparasi and is well known as a carrier of K and Li, B, and Mo as trace elements.

This obvious decoupling of As from Fe in the groundwater excludes Fe minerals like Fe hydroxides and pyrite being the source of arsenic in the alluvial sediments.

Above all, several trace elements (Li, B, P, V, Cr, Mn, Cu, Zn, Se, Br, Sr, Mo, Cd, P, and U) analyzed in the groundwater samples of Nawalparasi boreholes are among the most prominent besides arsenic which could be found in relevant concentrations. Particularly the presence of the lithophile Li, B, P, Mn, Br, Sr, and U in the groundwater is a striking feature. Siderophile elements like Cu or Zn could be found in minor concentrations or in a few samples only. The immediate attention was attracted by boron as this element is rarely found in significant amounts in common minerals like silicates. Tourmaline is one of the very rare minerals incorporating a significant portion of boron in its structure. As already mentioned in [38], boron in spring waters in the Peshawar basin and surroundings in the Himalayan foreland of Pakistan is closely associated with igneous complexes (most probably with the tourmaline-rich tertiary leucogranites). Even boron is widely known to be present in salt water; such an influence can be neglected in landlocked Nepal having no link to the ocean. Thermal springs can be located in various areas of the Nepal Himalaya, but their influence is considered to be marginal taking the widespread occurrence of tertiary leucogranites besides metapelites and black shales into account. Above all, in [39], elevated boron in metasedimentary rocks of the lesser Himalaya (up to 322 ppm) as well as in the Manaslu leucogranite (up to 950 ppm) where tourmaline represents the boron-containing mineral is reported. Tourmaline-containing aquifers enriched in As in West Bengal, India were detected in [36]. These authors conclude therefore that the heavy mineral assemblage of these aquifers (opaque minerals, garnet, tourmaline, kyanite, rutile, and zircon) prove a mixed metamorphic and igneous provenance for the eroded and deposited materials.

As the anion S was hardly detectable, the oxidation of pyrite in the sediments (suggested, e.g., by [40]) which would lead to an increased concentration of SO_4^{2-} in the groundwater can be clearly ruled out as an mechanism of arsenic release into groundwater. Despite these convincing facts, ophiolites were seen as the initial source of arsenic contained in arsenopyrite (e.g., [6, 16]). But ophiolites do not exist in the Nepalese Himalaya.

In conclusion, the diagram in **Figure 2** presents the most prominent trace elements in groundwater from Nawalparasi district in comparison with the few available data of the Macusani obsidian glass (peraluminous in composition, enriched in As-B-F-P). The data for comparison are taken from [41]. In [42], it was already declared that these volcanic rocks are compositionally and mineralogically equivalent to the Manaslu leucogranite. In article [41], it is reported for the first time that a significant accumulation of arsenic in a peraluminous glass from Macusani (SE Peru) which is representative of anatectic melts is derived from metasedimentary crustal

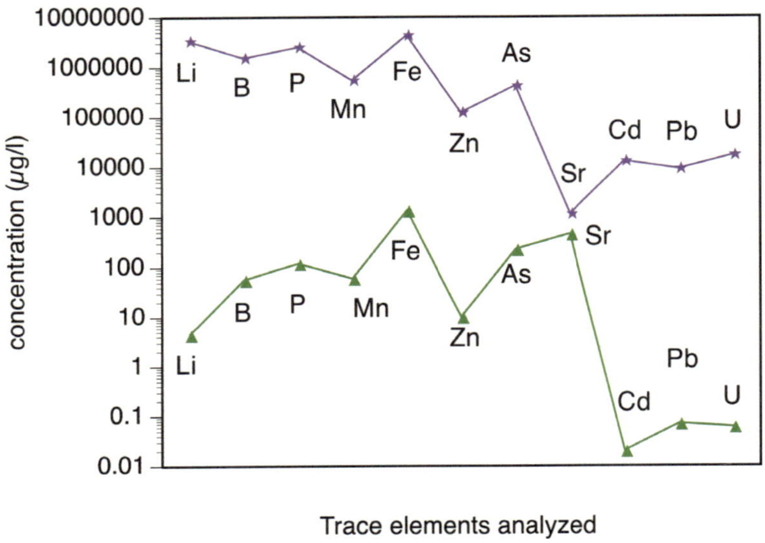

Figure 2.
The most noticeable trace elements in groundwater from Nawalparasi district (green triangles) compared with the few available data of the Macusani obsidian glass (peraluminous in composition, enriched in As-B-F-P). The data for comparison (purple stars) are taken from [41]. Note the logarithmic scale for comparison of concentrations.

protoliths. In this glass, the authors of [41] found accumulations by factors of 10–100 for As, compared with the mean continental crust values and other incompatible trace elements (e.g., Be, B, Rb, Sn, Sb, and Ta), and by factors of 100–200 for Li, Cd, and Cs. In a second article [43], the authors of reference [41] state that remarkable concentrations of Cd (up to ~300 ppm) could be found in quartz-hosted fluid and melt inclusions in hydrous peraluminous systems (pegmatites and leucogranites) for the first time ever. A distinct number of the groundwater samples used for this study show detectable concentrations of Cd. Significant amounts of Li, B, Zn, As, and Pb were also found in quartz-hosted fluid inclusions from the Huanuni tin deposit in Bolivia (hosted in peraluminous granites with ASI $\geq$ 1.1) [44]. The indicative trace elements of leucogranites (Li, B, P, Mn, Zn, As, Sr, Pb, and U) are similarly detected in the ground water in Nawalparasi. The high concentration of Sr in groundwater can be based by the frequent occurrence of calcium carbonates in the soil hosting the groundwater. In study [45], the authors mentioned lithologies like rhyolite and shale (greywackes) representing a greater risk of elevated As in groundwaters.

As frequently described in the literature, low-grade metapelites are often considered as protoliths of peraluminous granites (see, e.g., [22, 46, 47]) where concentrations of As, Sb, Be, B, Ba, and Rb by a factor of 5–10 higher than their average crustal abundances (2–5 ppm) [48–50] were not unusual. The leucogranites (two-mica or muscovite-tourmaline mica) found in the Himalayas of Nepal are undoubtedly peraluminous in composition (see, e.g., [22, 51, 52]) and coherently a comparison with the findings from [41] is warranted. Most of the leucogranites analyzed in [51] are peraluminous (ASI > 1.1) to strongly peraluminous (ASI $\geq$ 1.1). According to Ref. [53], lead represents one of the rare elements usually behaving incompatible during crustal melting. Pb can be significantly enriched in low T S-type granite melts, especially if the proportion of partial melting remains low. Moreover, muscovite, known as being a major mineral of metapelitic sources, can exhibit relatively high Pb contents. The leucogranites in the Nepal Himalayas are widely described to be of crustal origin and are derived from vapor-absent muscovite-dehydration melting of pelitic and psammitic protoliths during the Late Miocene (see, e.g., [47, 54–59]).

As described above, arsenic is primarily dissolved from micas as a major constituent of sediment hosting the groundwater in Nawalparasi. Keeping in mind that the Manaslu area is the watershed of the Nawalparasi water system, assuming the leucogranites as the original source of the arsenic is evidently warranted.

5. Conclusions

So far it was considered that As is mainly released from iron (hydr)oxides, but a major host for sorption and release of As are alumosilicates such as clay minerals (including micas) into groundwater. The apparent decoupling of the concentrations of Fe and As and the positive correlation between concentrations of Na, K, and As in the groundwater are a significant evidence that As cannot be released from Fe minerals predominantly—so silicates represent the fundamental source of As in sediments. Clay minerals preferentially lose Na and K from their interlayers during chemical weathering and hence become enriched in immobile elements such as Fe and Al. Moreover, the trace elements detected in the groundwater samples of Nawalparasi are well known to be contained in peraluminous obsidian glasses from Peru enriched in As-B-F-P. These glasses equal the leucogranites in the Manaslu area North of Nawalparasi in their geochemistry and represent the original host rocks of arsenic.

Acknowledgements

I am grateful for the help and analytical skills of Dr. Stephan Hug, Eawag, Dübendorf, Switzerland, and I thank Thomas Ruettimann, Eawag, for analysis of the samples by ICP-MS. My gratitude for support and valuable discussions is expressed to Tommy Ngai and Candice Young-Rojanschi from CAWST, Calgary, Canada; Bipin Dangol and Hari Boudhatoki from ENPHO, Kathmandu, Nepal; Gyan Prakash Yadav, Parasi, and to Som Rai, my loyal expedition and trekking guide in Nepal and responsible for all logistics over many years.

Conflict of interest

No conflict of interest.

Author details

Barbara Mueller
Bamugeobiochem, Ettenhausen, Switzerland

*Address all correspondence to: barbara.mueller@erdw.ethz.ch

IntechOpen

© 2020 The Author(s). Licensee IntechOpen. This chapter is distributed under the terms of the Creative Commons Attribution License (http://creativecommons.org/licenses/by/3.0), which permits unrestricted use, distribution, and reproduction in any medium, provided the original work is properly cited.

References

[1] Adhikari HJ, Ghimire TR. Prevalence of arsenicosis in Ramgram municipality, Nawalparasi, Nepal. International Journal of Health Research. 2009;**2**:183-188

[2] Thakur JK, Thakur KR, Ramanathan A, Kumar M, Singh SK. Arsenic contamination of groundwater in Nepal—An overview. Water. 2011;**3**:1-20

[3] Shrestha RK, Regmi D, Kafle BP. Seasonal variation of arsenic concentration in groundwater of Nawalparasi district of Nepal. International Journal of Applied Sciences and Biotechnology. 2014;**2**:59-63

[4] Acharyya SK, Lahiri S, Raymahashay BC, Bhowmik A. Arsenic toxicity of groundwater in parts of the Bengal basin in India and Bangladesh: The role of Quaternary stratigraphy and Holocene sea-level fluctuation. Environmental Geology. 2000;**39**:1127-1137

[5] Gurung JK, Ishiga H, Khadka M. Geological and geochemical examination of arsenic contamination in groundwater in the Holocene Terai Basin, Nepal. Environmental Geology. 2005;**49**:98-113

[6] Guillot S, Charlet L. Bengal arsenic, an archive of Himalaya orogeny and paleohydrology. Journal of Environmental Science and Health, Part A. 2007;**42**:1785-1794

[7] Guillot S, Garçon M, Weinman B, Gajurel A, Tisserand D, France-Lanord C, et al. Origin of arsenic in Late Pleistocene to Holocene sediments in the Nawalparasi district (Terai, Nepal). Environment and Earth Science. 2015;**74**:2571-2593. DOI: 10.1007/s12665-015-4277-y

[8] Mueller B. Arsenic in groundwater in the southern lowlands of Nepal and its mitigation options: A review. Environmental Reviews. 2015;**25**:296-305

[9] Mueller B, Hug SJ. Climatic variations and de-coupling between arsenic and iron in arsenic contaminated ground water in the lowlands of Nepal. Chemosphere. 2018;**210**:347-358

[10] France-Lanord C, Derry L, Michard A. Evolution of the Himalaya since Miocene time: Isotopic and sedimentologic evidence from the Bengal fan. In: Treloar PJ, Searle MP, editors. Himalayan Tectonics. London: Geol. Soc. Spec. Pub; 1993;74. p. 445-465

[11] Garzanti E, Vezzoli G, Ando S, France-Lanord C, Singh SK, Foster G. Sand Petrology and focused erosion in collision orogens: The Brahmaputra case. Earth and Planetary Science Letters. 2004;**220**:157-174

[12] Amini M, Abbaspour KC, Berg M, Winkel L, Hug SJ, Hoehn E, et al. Statisctical modeling of global geogenic arsenic contamination in groundwater. Environmental Science & Technology. 2008;**42**:3669-3675

[13] Dowling CB, Poreda RJ, Basu AR, Peters SL, Aggrawal PK. Geochemical study of arsenic release mechanisms in the Bengal Basin groundwater. Water, Air, and Soil Pollution. 2002;**38**:1173

[14] van Geen A, Zheng Y, Cheng Z, Aziz Z, Horneman A, Dhar RK, et al. A transect of groundwater and sediment properties in Araihazar, Bangladesh: Further evidence of decoupling between As and Fe mobilization. Chemical Geology. 2006;**228**:85-96

[15] Diwakar J, Johnston SG, Burton ED, Shrestha SD. Arsenic mobilization in an alluvial aquifer of the Terai region, Nepal. Journal of Hydrology: Regional Studies. 2015;**4**(A):59-79

[16] Stanger G. A palaeo-hydrogeological model for arsenic contamination in southern and south-east Asia. Environmental Geochemistry and Health. 2005;**27**:59-367

[17] Chakraborty S, Wolthers M, Chatterjee D, Charlet L. Adsorption of arsenite and arsenate onto muscovite and biotite mica. Journal of Colloid and Interface Science. 2007;**309**:392-401

[18] Uddin MK. A review on the adsorption of heavy metals by clay minerals, with special focus on the past decade. Chemical Engineering Journal. 2017;**308**:438-462

[19] Yadav IC, Devi NL, Sing S. Reductive dissolution of iron-oxyhydroxides directs groundwater arsenic mobilization in the upstream of Ganges River basin, Nepal. Journal of Geochemical Exploration. 2015;**148**:150-160

[20] Shah BA. Role of Quaternary stratigraphy on arsenic-contaminated groundwater from parts of Middle Ganga Plain, UP-Bihar, India. Environmental Geology. 2008;**53**:1553-1561

[21] van Geen A, Radloff K, Aziz Z, Cheng Z, Huq MR, Ahmed KM, et al. Comparison of arsenic concentrations in simultaneously-collected groundwater and aquifer particles from Bangladesh, India, Vietnam and Nepal. Applied Geochemistry. 2008;**23**:3244-3251

[22] Guillot S, Le Fort P. Geochemical constraints on the bimodal origin of High Himalayan leucogranites. Lithos. 1995;**35**:221-234

[23] Sharma CK. Shallow (Phreatic) Aquifers of Nepal. 1st ed. Kathmandu, Nepal: Sangeeta Publishing; 1995

[24] Kansakar DR. Geologic and geomorphologic characteristics of arsenic contaminated groundwater area in Terai, Nepal. In: D. R. Department of Irrigation, editor. Arsenic Testing and Finalization of Groundwater Legislation Project, Summary Project Report. Lalitpur, Nepal: HMG/Nepal; 2004. pp. 85-96

[25] Yadav IC, Dhuldhaj UP, Mohan D, Singh S. Current status of groundwater arsenic and its impacts on health and mitigation measures in the Terai basin of Nepal: An overview. Environmental Reviews. 2011;**19**:55-67

[26] Yadav IC, Devi NL. Sing spatial and temporal variation in arsenic in the groundwater of upstream of Ganges River Basin, Nepal. Environment and Earth Science. 2015;**73**:1265-1279. DOI: 10.1007/s12665-014-3480-6

[27] Shrestha SD, Brikowski T, Smith L, Shei TC. Grain size constraints on arsenic concentration in shallow wells of Nawalparasi, Nepal. Journal of Nepal Geological Society. 2004;**30**:93-98

[28] Brikowski TH, Neku A, Shrestha SD, Smith LS. Hydrologic control of temporal variability in groundwater arsenic on the Ganges floodplain of Nepal. Journal of Hydrology. 2014;**518**(C):342-353

[29] Brikowski TH, Smith LS, Shei TC, Shrestha SD. Correlation of electrical resistivity and groundwater arsenic concentration, Nawalparasi, Nepal. Journal of Nepal Geological Society. 2004;**30**:99-106

[30] Ngai TKK, Murcott SE, Shrestha RR, Dangol B, Maharjan M. Development and dissemination of Kanchan™ arsenic filter in rural Nepal. Water Science and Technology. 2006;**6**:137-146

[31] Ngai TKK, Shrestha RR, Dangol B, Maharjan M, Murcott SE. Design for sustainable development—Household drinking water filter for arsenic and pathogen treatment in Nepal. Journal of Environmental Science and Health.

Part A, Toxic/Hazardous Substances & Environmental Engineering. 2007;**42**:1879-1888

[32] Singh A, Smith LS, Shrestha S, Maden N. Efficacy of arsenic filtration by Kanchan Arsenic Filter in Nepal. Journal of Water and Health. 2014;**12**:596-599

[33] Wenk C, Kaegi R, Hug SJ. Factors affecting arsenic and uranium removal with zero-valent iron: Laboratory tests with Kanchan-type iron nail filter columns with different groundwaters. Environment and Chemistry. 2014;**11**:547-557

[34] McArthur JM, Ravenscroft P, Safiullah S, Thirlwal MF. Arsenic in groundwater: Testing pollution mechanisms for sedimentary aquifers in Bangladesh. Water Resources Research. 2001;**37**:109-117

[35] Bhattacharya P, Tandukar N, Neku A, Valero AA, Mukherjee AB, Jacks G. Geogenic arsenic in groundwaters from Terai Alluvial Plain of Nepal. Journal de Physique IV France. 2003;**107**:173-176

[36] Stueben D, Berner Z, Chandrasekharam D, Karma J. Arsenic enrichment in groundwater of West Bengal, India: Geochemical evidence for mobilization of As under reducing conditions. Applied Geochemistry. 2003;**18**:1417-1434

[37] Mueller B. Preliminary trace element analysis of arsenic in Nepalese groundwater may pinpoint its origin. Environment and Earth Science. 2018;**77**:35-40

[38] Yousafzai A, Eckstein Y, Dahl PS. Hydrochemical signatures of deep groundwater circulation in a part of the Himalayan foreland basin. Environment and Earth Science. 2010;**59**:1079-1098

[39] Rai SM. Distribution of boron in the rocks of central Nepal Himalaya. Journal of Nepal Geological Society. 2003;**28**:57-62

[40] Acharyya SK, Chakraborty P, Lahiri S, Raymahashay BC, Guha S. Bhowmik. Arsenic poisoning in the Ganges delta. Nature. 1999;**401**:545-547

[41] Borisova AY, Pokrovski GS, Pichavant M, Freydier R, Candaudap F. Arsenic enrichment in hydrous peraluminous melts: Insights from femtosecond laser ablation-inductively coupled plasma-quadrupole mass spectrometry, and in situ X-ray absorption fine structure spectroscopy. American Mineralogist. 2010;**95**:1095-1104

[42] Scaillet B, Pichavant M, Roux J. Experimental crystallization of leucogranite magmas. Journal of Petrology. 1995;**36**:663-705

[43] Borisova AY, Thomas R, Salvi S, Candaudap F, Lanzanova A, Chmeleff J. Tin and associated metal and metalloid geochemistry by femtosecond LA-ICP-QMS microanalysis of pegmatite–leucogranite melt and fluid inclusions: New evidence for melt–melt–fluid immiscibility. Mineralogical Magazine. 2012;**76**:91-113

[44] Mueller B, Frischknecht R, Seward TM, Heinrich CA, Camargo Gallegos W. A fluid inclusion study of the Huanuni tin deposit Bolivia: Some insights with LA-ICP-MS analysis. Mineralium Deposita. 2001;**36**:680-688

[45] McGrory ER, Brown C, Bargary N, Hunter Williams N, Mannix A, Zhang C, et al. Arsenic contamination of drinking water in Ireland: A spatial analysis of occurrence and potential risk. Science of the Total Environment. 2017;**579**:1863-1875

[46] Godin L, Grujic D, Law RD, Searle MP. Channel flow, ductile extrusion and exhumation in continental collision zones: An introduction. In: Law RD,

Searle MP, Godin L, editors. Channel Flow, Ductile Extrusion and Exhumation in Continental Collision Zones. Vol. 268. London: Geol. Soc. Spec. Pub; 2006. pp. 1-23

[47] Searle MP, Avouac JP, Elliott J, Dyck JE. Ductile shearing to brittle thrusting along the Nepal Himalaya: Linking Miocene channel flow and critical wedge tectonics to 25th April 2015 Gorkha earthquake. Tectonophysics. 2016. DOI: 10.1016/j.tecto.2016.08.003

[48] Onishi H, Sandell EB. Geochemistry of arsenic. Geochimica et Cosmochimica Acta. 1955;7:1-33

[49] Turekian KK, Wedepohl KH. Distribution of the elements in some major units of the earth's crust. Geological Society of America Bulletin. 1961;**72**:175-192

[50] Rudnick RL, Gao S. Composition of the continental crust. In: Holland HD, Turekian KK, editors. Treatise on Geochemistry. Vol. 3. Amsterdam: Elsevier; 2003. pp. 1-64

[51] Guo Z, Wilson M. The Himalayan leucogranites: Constraints on the nature of their crustal source region and geodynamic setting. Gondwana Research. 2012;**22**:360-376

[52] Carosi R, Montomoli C, Rubatto D, Visona D. Leucogranite intruding the South Tibetan Detachment in western Nepal: Implications for exhumation models in the Himalayas. Terra Nova. 2013;**25**:478-489

[53] Finger F, Schiller D. Lead contents of S-type granites and their petrogenetic significance. Contributions to Mineralogy and Petrology. 2012;**64**:747-755

[54] Deniel C, Vidal P, Fernandez A, Le Fort P, Peucat JJ. Isotopic study of the Manaslu granite (Himalaya, Nepal): Inferences on the age and source of Himalayan leucogranites. Contributions to Mineralogy and Petrology. 1987;**96**:78-92

[55] Inger S, Harris N. Geochemical constraints on leucogranite magmatism in the Langtang Valley, Nepal Himalaya. Journal of Petrology. 1993;**34**:345-368

[56] Ayres M, Harris N. REE fractionation and Nd-isotope disequilibrium during crustal anatexis: Constraints from Himalayan leucogranites. Chemical Geology. 1997;**139**:249-269

[57] Streule MJ, Searle MP, Waters DJ, Horstwood MSA. Metamorphism, melting, and channel flow in the Greater Himalayan Sequence and Makalu leucogranite: Constraints from thermobarometry, metamorphic modeling, and U-Pb geochronology. Tectonics. 2010;**29**. DOI: doi/10.1029/2009TC002533

[58] Rubatto D, Chakraborty S, Dasgupta S. Timescales of crustal melting in the Higher Himalayan Crystallines (Sikkim, Eastern Himalaya) inferred from trace element-constrained monazite and zircon chronology. Contributions to Mineralogy and Petrology. 2013;**165**:349-372

[59] Bikramaditya Singh RK. Origin and emplacement of the higher Himalayan leucogranite in the Eastern Himalaya: Constraints from geochemistry and mineral chemistry. Journal of the Geological Society of India. 2013;**81**:791-803

Section 2

Determination, Environmental Pollution and Health Effects

Chapter 4

New Methods in the Synthesis of (Meth)Acrylamides and Application Chelating Resin for Determination of Trace Metals in Certified Reference Materials and Waters

Cengiz Soykan

Abstract

The waste of trace metals are led the pollutions of water, soil and air. That's why the accurate and sensitive identification of amount of trace metals in food samples and environment are gained importance in analytical chemistry because of their toxicity to human health. Besides, the direct determination of trace metals, presented at very low concentration especially in real samples, is difficult. In this content, before determination of trace metals by instrumental methods can be achieved successfully using separation/preconcentration procedures. The different synthesized chelating polymer resin adsorbents are successfully used for the SPE of trace metals. The trace metals, such as Cu, Cr, Co, Mn, Zn and Fe, are necessary for human health, when it has been taken at certain limits. But, the trace metals, such as Hg, Cd, Pb, Ni and As are toxic metals for the human body. About this, there are studies on the determination of trace elements and mineral in food samples and environment. In this study, new methods in the synthesis of (meth)acrylamides and use as a sorbent in the trace metals extraction was identified.

Keywords: activated (meth)acrylamides, trace metals, separation/preconcentration, adsorbent, food samples

1. Introduction

Heavy metals [Co(II), Cu(II), Ni(II), Pb(II), Cd(II), etc.] have an important place between the water pollutants due to their toxic features. Organic pollutants in the water, biodecomposition at the end of the time may be ruined, heavy metals, biodegradable products are not converted into harmless products. That's why, heavy metals must be removed from the surrounding waters [1, 2]. Due to the toxic features of heavy metals in the environment, researches on these are increasing. One of the most partner and effective methods used to remove heavy metals from wastewater is the adsorption method. This fact that adsorption is a common and effective method has always grown the selection of a good adsorbent. Therefore,

IntechOpen

metal oxides, activated carbon, hydrogel polymers, ion exchange resins, and polymeric fibers, microspheres and natural polymers have been used in the literature as adsorbents to remove heavy metals from the aqueous environment. But, the adsorption capacity of some of these adsorbents is very low or a long time is necessary for the adsorption equilibrium. Some adsorbents are also difficult to revive and cannot be reused. Polymer resins with different functional groups and large surface area are broadly used for the separation of metal ions and enrichment of trace elements [3, 4]. Recently, the importance of removing heavy metal ions from aqueous media has been increasing for the purpose of control environmental pollution. Heavy metals in waste water may be removed by adsorption on solid materials. Different adsorbents such as metal oxides, activated carbon, peanut and rice shell, cotton, wool, chitosan, starch, sporopollen and wood shavings [5–14] are used as adsorbents in the removal of heavy metals from the aqueous environment. But, the adsorption capacity of some of these adsorbents is very low or they are broadly used in complexing with them. Recently, copolymer resin with functional groups have been used to remove heavy metal ions from waste waters. In the removal of heavy metal ions from aqueous media and enrichment of trace elements, the use of polymeric adsorbents with high adsorption rates and large surface area is increasing [15–20]. In a very short time, it is cited that the polymers of amidoxime group will find broad usage area as adsorbent in the recovery of uranium from ground water and sea [21–27]. The characteristics of the adsorbent, such as sorption capacity, selectivity to analyte, resistance to physical/chemical conditions, pH range, ease of availability, reusability and cost, are thought. In order to, new sorbents are continuously being examined and these features are expected to be advanced except in the literature [28, 29].

In the removal of heavy metal ions from aqueous media, functional groups such as ester, pyridine, amide, carboxylic acid, amidoxime and hydroxyl at the surface of the adsorbent are substantially effective. Adsorption of heavy metals from aqueous environmental can generally be controlled by functional groups on the adsorbent surface. Polymeric adsorbents containing N-methyl hydroxamic acid groups and amidoxime groups have been selectively used against some trivalent lanthanide metal ions [Nd(III), La(III), Gd(III), Sm(III), Tb(III)] and especially uranium ions.

The concentration range cited as edil "work concentration" it has changed over time in parallel with the improvements in artifact analysis techniques. Before 1940s, 10^{-1}–10^{-2}%, rarely 10^{-3}% of work sheets, 10^{-3}–10^{-5}% in the 1950s and 10^{-6}–10^{-8}% in the 1965s work is cited as skin. Kaiser suggested the systematic approach and first naming. Kaiser gave the explanation of ppm and ppb. According to today's general usage, 10^{-2}–10^{-6}% concentration range, and concentrations below10^{-6} are known as ultra works.

Recently, the most important parts of analytical chemistry, trace element analysis, soil, water and air pollution, such as pollution effects on the environment, such as the increasing importance of environmental issues, the elements of the human body and metabolism effects, the study of the chemical forms of the heavy metals in natural waters and determination of it has gained importance as the functions in many different areas are. The phrase trace element analysis is used for the determination of very small quantities of elements in the medium formed of large quantities of components. These environments can be minerals, metals, water, compounds, aqueous solutions, biological and organic substances. Often, the environment in which the trace element is available has a reverse effect on the determination. In such environments, accuracy and sufficient precision can not be obtained. The matter starts at the sampling phase in the study analysis. Especially, there may be a non-homogeneous trace element distribution in solid samples. Limitations at the limit of determination of the instrumental technique to be used in the determination of a suitable sample will be problematic. The concentration of trace element should be above the limit of determination. Otherwise an observable signal is not obtained. In such cases, separation and enrichment processes are carried out in order to concentrate the analyte into the appropriate medium or to collect it in small volumes. Various problems can be encountered during trace element analysis. Some of these problems are as follows;

1. The trace element concentration is too small to be directly determined.

2. In the very small amount of the initial sample, the main component, component and work analysis of elements.

3. Separation of trace elements to be determined from a very large sample.

4. To avoid interference with media and increase the assay capacity, to separate and collect a small volume.

This book chapter includes the synthesis of new acrylamide derived polymer resins and the use of these resins as adsorbents in the removal of heavy metal ions from environmental samples, and the investigation of their adsorption properties. In the literature, various substances are used as adsorbents, but methacrylamide-derived polymer resins are not available for this purpose and how it is used. The sections related to the examples of the studies carried out by our group have been made interesting. With this book chapter, it is aimed to introduce new methods with new adsorbent material for the removal of heavy metal ions from environmental samples. The amount of polymer resins used as adsorbents, their reusability and their ability to quantitatively remove heavy metal ions were the most important positive aspects of the studies.

2. Separation and enrichment methods

Separation of a substance between two phases in contact with varying proportions of distribution basis. In all separation methods solid-liquid, liquid-liquid, liquid-gas and solid-gas, there are two phases in the form. Trace element analysis generally has three separate applications of separation methods. These;

a. Macro-micro-separation: when the main component is removed from the sample trace components remains in solution.

b. Micro-macro separation: trace components from solid or thawed sample the main component remains in solution.

c. Micro-micro separation: the trace components are separated from the other components.

The first application is not used in the work analysis. Because the main component can be dragged along with trace elements. The other two applications are used more in work analysis. The trace elements are separated from the disturbing media components by enrichment methods. It is taken into smaller volume and concentrated. Two important criteria are used in the evaluation of enrichment methods.

The first one is the recovery efficiency. However, it is not always possible to achieve a large recovery value. At low concentrations, 90 or 95% recovery efficiencies are sufficient. The second term is the enrichment coefficient.

The enrichment methods in the work analysis undertake important functions. With enrichment techniques;

1. The ratio of the amount of trace element to the amount of media is increased.

2. Instead of the original media components (matrix), the new and the appropriate technique for the determination technique is created.

3. In the new environment, the analyte shows homogeneous distribution, especially with the enrichment process used in combination with the dissolving process.

4. The standard preparation for calibration is made easier.

The following criteria should be taken into account when selecting trace element enrichment methods;

- The method of determination following the enrichment method
- Minimum concentration limits of trace elements
- Number of desired trace elements
- Enrichment factor
- Recovery value of the method
- Contamination
- Number of samples
- Sample size
- Time for enrichment
- The complexity of the technique
- Price and laboratory conditions.

2.1 Enrichment by extraction

Extraction is the process of passing a chemical compound from a liquid phase into another liquid which is not mixed there with. The extraction of trace element into the organic phase is based on the incorporation of the trace element into a neutral structure. Neutral structures used in such a reaction; chelate, ion pair and covalent structures. In order to accurately measure trace elements, it is very important to separate and enrich the matrix before the analysis. Specific and sensitive methods are required for this.

2.1.1 Chelation extraction of trace elements

The trace elements at various levels in natural waters, wastewater, surface water and ground water are converted into chelates at a suitable pH and extracted into a small volume organic phase according to the sample. These elements taken into the organic phase are concentrated back to the aqueous phase directly or again and are determined by various determination techniques. Extraction is used not only in liquid samples but in solid samples to recover from matrix. The first step is to dissolve the sample. In chelate extraction the sample in the first aqueous phase is the appropriate reagent selection depending on the elements it contains. The reagent should form chelates with the desired trace elements and not form chelates with undesirable components. In this regard, chelates are also preferred in the extraction of trace elements, in particular in the alkali and alkaline earth compounds, or in samples containing them. Because chelating ligands do not generally form with alkali and alkaline earth alkalis.

2.1.2 Ion pair extraction

Ion pair complexes can also be extracted into organic solvents because they are neutral structures. It is difficult to derive quantitative equivalents or formulas in the extraction of ion pair complexes. Because ion pairs occur in the presence of extremely strong electrolytes. At the same time, a number of different ion strands can be found. Most inorganic complexes are used in extraction from ion pairs. Ion pairs usually occur between the anionic complexes formed by inorganic ligands and the H^+ ion. $[H^+, ML^-_{n+1}]$. In the extraction process, selectivity is achieved by utilizing variants such as pH, side reactions in the aqueous phase, ligand, solvent type and temperature. Extraction efficiency is defined as the size of the amount of material passing from the aqueous phase to the organic phase. The dispersion rate is determined by D.

$$D = C_{organic}/C_{aqueous} \quad (1)$$

When the balance is established, $C_{organic}$, total concentration of analyte in organic phase; $C_{aqueous}$, denotes the total concentration of the analyte in the aqueous phase. To increase efficiency in extraction processes; synergistic effect, co-extraction and salt effect processes are utilized.

2.1.3 Enrichment with co-precipitation

The use of precipitation methods in the separation of elements is based on the different solubility of the compounds in aqueous solutions. Collectors are used in quantitative separation of trace elements in solution by the method of co-precipitation. A large-surface precipitate with inorganic or organic character is formed, so that the adsorbed surface is adsorbed. The mechanism of co-precipitation depends on the physical and chemical properties of the trace element and the carrier and on the experimental conditions. According to this; three mechanisms in the form of confinement, mixed crystal formation or adsorption.

The carrier should be added to the sample solution so that sufficient amount of precipitate is formed. In order to prevent the adsorption of ions that may interfere, the amount of carrier should be as low as possible. Precipitation methods are mostly used for the separation of trace elements into single press as well as separating the main component from trace components. By controlling the precipitation pH, selectivity is ensured.

The use of the precipitation process to separate the main component from the work component is not common. Because when the main component collapses, it can also drag and drop work components together. This leads to substance loss.

2.1.4 Electrolytic enrichment

The electrolysis method is used to separate trace amounts of heavy metals from various solutions. The composition of the electrolyte and sample, the type and shape of the electrode, the electrolysis cell and other experimental variables greatly affect the electrolytic deposition of an element. Potentially controlled electrolysis, as well as stripping methods, are widely used in the enrichment of trace elements.

2.1.5 Ion exchange

Ion exchange technique is a method of enrichment of trace elements. The ion exchange method can be practiced in three ways. The first is the rinsing technique that the solution and the ion exchanger interact in the same container. The second is the filtration technique in which the sample solution is held by ion-exchanging paper or membrane filter and the trace elements are kept. The third technique is the column technique which is the most commonly used technique in the application. In the rinsing technique, the solution containing the trace element ion is rinsed with resin. It is expected to establish a balance of time distribution. The resin is separated from the solution by filtration. The resin can be reacted with suitable solutions and trace elements can be determined by solution techniques or directly by solid analysis techniques. In the filtration technique, the sample solution is passed through an ion exchanger paper or membrane filter. Filter elements are determined by direct or degradation. In the column technique, large-volume solutions containing trace elements from the filled column with ion-exchange resin are passed to selectively hold them. These trace elements are attached with a smaller volume of eluent. This final volume can be further reduced by evaporation. Selectivity of the functional groups in the selection of ion exchangers, changing capacity, exchange rate, regeneration of the ion exchanger and use of the appropriate eluent are the considerations.

2.1.6 Enrichment with evaporation

It is a very convenient method for some elements that can be converted to volatile or easily volatile components. In fly enrichment, the volatility difference between the matrix and the trace element must be large. The separation by blowing can be done in two ways by blowing both the matrix and the trace element. However, in inorganic analysis, it is not common to enrich the trace elements by flying.

2.1.7 Adsorption-enrichment

The change in the concentration of a solid or a liquid on the boundary surface is called adsorption. This phenomenon arises from the accumulation of molecules or ions of the solute from the gas, liquid or any solution by adhering to the surface of a solid substance. The increase in concentration is called positive adsorption. The decrease in concentration is also called negative adsorption. Physical and chemical properties of adsorbent substances are the most important factors affecting the adsorption process. Solids, metals and plastics have more or less adsorption power. Some natural solids with high adsorption power; coals, clays, zeolites and various metal sprouts, and artificial solids are activated charcoal, silica gel and special polymers. The amount of adsorbed substance in the solids having high adsorption power varies depending on the surface size and porous structure.

2.1.8 Solid phase extraction

The solid phase extraction method is applied in three ways: rinsing technique, semi-permeable retaining disc filtration technique and column technique [30–35].

2.1.8.1 Column technique

Solid phase extraction method is widely used in the column technique. Columns of different sizes can be used. This technique consists of four process steps: These process steps is given in **Figure 1**.

As column filler natural and artificial polymers, silica gel, alumina, florusil, octadecyl (C18); octyl (C8), ethyl (c2), cyano, phenyl, cyclohexyl, amino, diol, quaternary amine, aromatic sulfonic acid, carboxylic acid can be used.

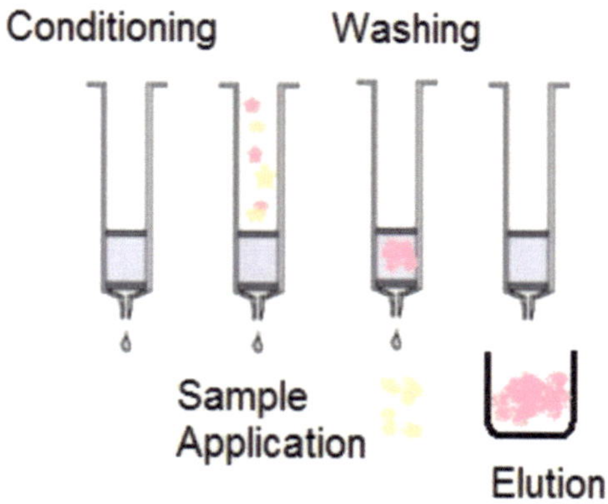

Figure 1.
Column technique processing steps.

3. Functional methacrylates from α-chloro-N-arylacetamides

3.1 Synthesis of α-chloro-N-arylacetamide

Synthesis of α-chloro-N-arylacetamide is as follows: potassium carbonate (K_2CO_3) (0.1 mol) and arylamine (1 mol) and are dissolved in 30 mL of anhydrous benzene. The reaction mixture is taken in a three-neck round bottom flask equipped with a magnetic stirrer, thermometer, and cooled to 0°C. After then chloroacetyl chloride (1.1 mol) are added dropwise to the solution and stirred at room temperature for 16 h. The organic layer is washed several times with diethyl ether and separation phase is filtered and dried over $MgSO_4$. α-Chloro-N-arylacetamide is crystallized from methanol. Yield: 80%. The synthesis reaction path is given in **Figure 2.**

The structure of the compoundα-chloro-N-arylacetamide is identified by the FT-IR techniques. FT-IR (cm^{-1}): 3340 (NH); 3100–2800 (C—H); 1680 (>C=O); 1580 (aromatic, C=C) [36, 37].

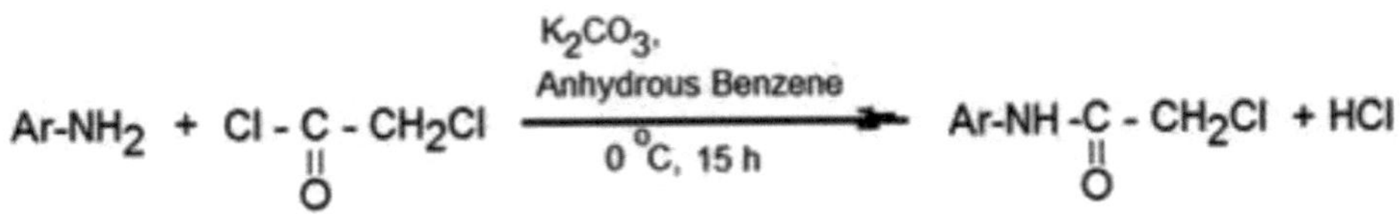

Figure 2.
Synthesis route of α-chloro-N-arylacetamide.

3.2 Synthesis of arylamido methyl methacrylate

Arylamido methyl methacrylate is synthesized as follows: 1.1 mol sodium methacrylate, 1 mol α-chloroacetamide, 0.1 mol NaI, and 0.1 mol TEBAC as catalyst are stirred in 100 mL acetonitrile at 80°C in a reflux condenser for 24 h in the presence of 100 ppm hydroquinone as the inhibitor. Then the solution is cooled to room temperature and neutralized with a 5% KOH solution. The organic layer is washed a few times with water, and the water layer is washed with diethyl ether several times. The diethyl ether layer and acetonitrile layer are aggregated and dried over anhydrous $MgSO_4$ overnight. Diethyl ether and acetonitrile are evaporated. The organic layers are collected and the residue was distilled at 130°C at 5 mmHg to give a colorless liquid. (Yield: 80%). The synthesis reaction path is given in **Figure 3**.

The structure of the monomer is confirmed by the FT-IR and ^{1}H- and ^{13}C-NMR spectroscopic techniques. FT-IR (cm^{-1}): 3325 (NH); 3100–2800 (C—H); 1680 (>C═O); 1630 (CH_2═C); 1580 (aromatic, C═C); 1230 (C—O—C). ^{1}H-NMR ($CDCl_3$, TMS): 9.1 (N-H); 8.0–6.7 (aromatics-H); 6.3–5.43 (CH_2═C); 1.8 (CH_3). ^{13}C-NMR ($CDCl_3$, TMS): 157.1–113.4 (aromatics-C); 134.4–124.2 (CH_2═C); 168.1 (>C═O); 18.1 (CH_3) [38, 39].

3.3 Synthesis of polymer resin

Arylamido methyl methacrylate-co-divinylbenzene polymer resin is prepared by copolymerizing of arylamido methyl methacrylate monomer (3.0 mmol) and divinylbenzene (3.0 mmol) as a crosslinker. The polymer resin is designed by the free radical solution polymerization technique in 1,4-dioxane and in a 50 mL sealed pyrex polymerization tube. The azobisisobutironitrile, AIBN is used at a 1:1 molar ratio as initiator. The reaction mixture is processed with nitrogen gas for 5 min in the polymerization tube and in an oil bath heated at 70 ± 1°C for 5 h. Later reaction, the formed polymer resin was kept at 25°C for cooling. The polymer resin was washed with ultra high pressure water, n-hexane and ultra high pressure water, respectively and then dried under vacuum at 40°C. The synthesis reaction path is given in **Figure 4**.

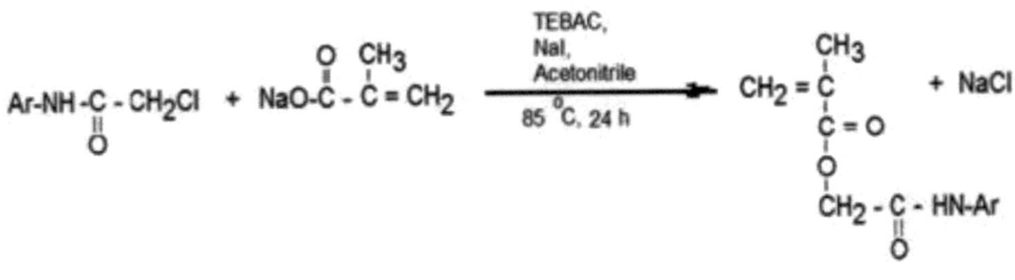

Figure 3.
Synthesis route of arylamido methyl methacrylate monomer.

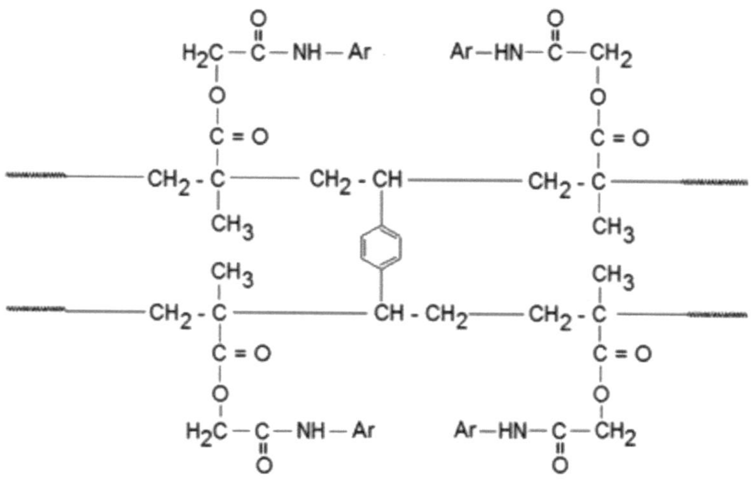

Figure 4.
Synthesis route of arylamido methyl methacrylate monomer resin.

4. Optimization of enrichment separation methods

In the optimization of the methods used in the studies, to the recovery value of the elements to be determined parameters such as pH, amount of carrier element, amount of complexing agent or precipitating reagent, sample volume and effect of matrix components are examined.

4.1 Effect of pH

The effect of pH on the precipitation of the studied trace elements such as Ni (II), Co (II), Cu (II), Mn (II), Cd (II), Zn (II), Pb (II) is investigated. The pH value plays an important role in the adsorption of sorbent-related ions and affects the state of sorption of heavy metals. Enrichment of metal ions pH effect is generally examined in the range of 1–8. Binding of analytes to the synthesized polymer resin in solid phase extraction (SPE) is known to be dependent on the pH of the samples solution. The pH value for quantitative SPE of the analytes is fixed by measurements of analytes in final solutions. All of the buffer solutions are prepared the pH 1–2 with 1 mol L^{-1} of KCl/HCl, pH 3–5 with 1 mol L^{-1} of CH_3COONa/CH_3COOH, pH 6–8 with 1 mol L^{-1} of CH_3COONH_4/CH_3COOH.

4.2 The effect of concentrations of eluents on the efficiency of analytes

Elution of metals from the column is carried out using acids. The influences of various eluents on the recoveries of analytes are usually examined using 1, 2 and 3 M from each of HCl, HNO_3 and H_2SO_4 and 10, 20, and 25 mL of these acid solutions.

4.3 Effect of flow rates of solutions

The contact time between the analytes and the complexing agent is the effect the recoveries of the analytes and the SPE time. This contact time is supplied to flow rate of the sample. The samples flow rates are usually examined in the range from 2 to 10 mL min^{-1}. The increase of sample flow rate, which reduces the interaction time between the analytes and the adsorbent. In order to achieve good precision, a sample flow rate is generally chosen at 5 mL min^{-1}. The analytes desorption on the adsorbent also adhere to on the flow rate of the elution solutions.

4.4 Effect of the amount of the complexing agent

The effects of the amounts of complexing agent on the adsorption of analytes keeping other parameters constant are investigated. For this purpose, the prepared model samples are generally passed through the column containing 1.0, 1.5, 2.0, 3.0 and 4.0 g of the complexing agent. The retained analytes on the complexing agent are usually eluted 10 mL of 1 mol L^{-1} HNO_3.

4.5 Effect of sample volume

The increase of sample volume will result in important increase in the preconcentration factors. The adsorption of the heavy metals are not affected by sample volume in the range 25–1000 mL of model solutions containing chromium, cadmium, copper, cobalt, iron, lead, manganese, nickel, and zinc ions.

4.6 Effect of matrix ions

The effect of matrix ions in the real samples on the simultaneous separation and preconcentration of heavy metals are generally tested. The model samples (25 mL) containing 2 μg of cadmium, and zinc, 10 μg of copper, cobalt, iron, manganese, nickel, 20 μg of chrome and lead ions. The known concentrations of matrix ions are prepared and work method is implemented to these prepared model samples. The tolerable levels of the matrix ions are given.

5. Conclusion

In this paper, reaction pathway for the synthesis of new methacrylamide resin and use as a sorbent in the trace metals extraction was described. This synthesized polymer resin can be used as column filler on the solid phase extraction for the effective separation and preconcentration of the trace elements in dam, waterfall, spring water, and river and dried eggplant, zucchini and tomato samples. The synthesized polymer resin is distinct in terms of precision and particularity for these analytes. The experimental parameters are advanced and the analytical parameters are identified [40–54].

Author details

Cengiz Soykan
Department of Materials Science and Nanotechnology, Faculty of Engineering, University of Uşak, Uşak, Turkey

*Address all correspondence to: cengizsoykan@usak.edu.tr

IntechOpen

© 2019 The Author(s). Licensee IntechOpen. This chapter is distributed under the terms of the Creative Commons Attribution License (http://creativecommons.org/licenses/by/3.0), which permits unrestricted use, distribution, and reproduction in any medium, provided the original work is properly cited. CC BY

References

[1] Randall JM, Reuter FW, Waiss AC Jr. Removal of cupric ion from solution by contact with peanut skins. Journal of Applied Polymer Science. 1975;**19**(6):1563-1571. DOI: 10.1002/app.1975.070190607

[2] Roberts EJ, Rowland SP. Removal of mercury from aqueous solutions by nitrogen-containing chemically modified cotton. Environmental Science and Technology. 1973;7(6):552-555. DOI: 10.1021/es60078a010

[3] Kojima T, Sowa T, Kodama S, Sato M, Shigetomi Y, Yamamoto Y. Adsorption behaviour and separation of copper(II) ions on cellulose triacetate polymer containing α-hydroxyoxime compounds. Analytica Chimica Acta. 1992;**264**(1):59-64. DOI: 10.1016/0003-2670(92)85296-I

[4] Chang X, Su Z, Luo X, Zhan G. Synthesis of poly (acrylamidrazone-hydrazide) chelating fiber and application of enrichment—Separation for traces of indium, tin, chromium, vanadium and titanium from solution samples. Talanta. 1993;**40**(4):527-532. DOI: 10.1016/0039-9140(93)80012-G

[5] Kavaklı PA, Uzun C, Güven O. Synthesis, characterization and amidoximation of a novel polymer: Poly(N,N′-dipropionitrile acrylamide). Reactive and Functional Polymers. 2004;**61**(2):245-254. DOI: 10.1016/j.reactfunctpolym.2004.06.005

[6] Kabay N, Egawa H. Kinetic behavior of lightly crosslinked chelating resins containing amidoxime groups for batchwise adsorption of UO2 $^{2+}$. Separation Science and Technology. 1993;**28**(11-12):1985-1993. DOI: 10.1080/01496399308016728

[7] Hancock RD, Martell AE. Ligand design for selective complexation of metal ions in aqueous solution. Chemical Reviews. 1989;**89**(8):1875-1914. DOI: 10.1021/cr00098a011

[8] Rivas BL, Pereira ED, Moreno-Villoslada I. Water-soluble polymer-metal ion interactions. Progress in Polymer Science. 2003;**28**(2):173-208

[9] Park IH, Suh JM. Preparation and uranyl ion adsorptivity of macroreticular chelating resins containing a pair of neighboring amidoxime groups in a monomeric styrene unit. Angewandte Makromolekulare Chemie. 1996;**239**(1):121-132. DOI: 10.1002/apmc.1996.052390111|

[10] Schenk HJ, Asthemier L, Witte EG, Schwochau K. Development of sorbers for the recovery of uranium from seawater. 1. Assessment of key parameters and screening studies of sorber materials. Separation Science and Technology. 1982;**17**(11):1293-1308. DOI: 10.1080/01496398208056103

[11] Suzuki T, Saito K, Sugo T, Ogura H, Oguma K. Fractional elution and determination of uranium and vanadium adsorbed on amidoxime fiber from seawater. Analytical Sciences. 2000;**16**(4):429-432. DOI: 10.2116/analsci.16.429

[12] Elliott JE, Macdonald M, Nie J, Bowman CN. Structure and swelling of poly(acrylic acid) hydrogels: Effect of pH, ionic strength, and dilution on the crosslinked polymer structure. Polymer. 2004;**45**(5):1503-1510. DOI: 10.1016/j.polymer.2003.12.040

[13] Bell CL, Peppas NA, Biopolymers II. Biomedical membranes from hydrogels and interpolymer complexes. Advances in Polymer Science. 1995;**122**:125-175

[14] Byrne ME, Park K, Peppas NA. Molecular imprinting within hydrogels. Advanced Drug Delivery Reviews. 2002;**54**(1):149-161. DOI: 10.1016/S0169-409X(01)00246-0

[15] Bajpai SK, Singh S. Analysis of swelling behavior of poly(methacrylamide-co-methacrylic acid) hydrogels and effect of synthesis conditions on water uptake. Reactive and Functional Polymers. 2006;**66**(4):431-440. DOI: 10.1016/j.reactfunctpolym.2005.09.003

[16] Kim B, Peppas NA. Poly(ethylene glycol)-containing hydrogels for oral protein delivery applications. Biomedical Microdevices. 2003;**5**(4):333-341

[17] Morita R, Honda R, Takahashi Y. Development of oral controlled release preparations, a PVA swelling controlled release system (SCRS). II. In vitro and in vivo evaluation. Journal of Controlled Release. 2000;**68**(1):115-120

[18] Mamada A, Tanaka T, Kungwatchakun D, Irie M. Photoinduced phase transition of gels. Macromolecules. 1990;**23**(5):1517-1519. DOI: 10.1021/ma00207a046

[19] Riley SL, Dutt S, De La Torre R, Chen AC, Sah RL, Ratcliffe A. Formulation of PEG-based hydrogels affects tissue-engineered cartilage construct characteristics. Journal of Materials Science. Materials in Medicine. 2001;**12**(10-12):983-990

[20] Sanchez-Chaves M, Felix A. Synthesis of amidoxime-containing modified dextran. Polymer. 1996;**37**(19):4403-4407. DOI: 10.1016/0032-3861(96)00269-8

[21] Lutfor MR, Silong S, Zin WM, Ab-Rahman MZ, Ahmad M, Haron J, et al. European Polymer Journal. 2000;**36**:2105-2113

[22] Seong-Ho C, Young CN, et al. Radiation Physics and Chemistry. 2000;**57**(2):187-193

[23] Pekel N, Güven O, et al. Colloids and Surfaces A: Physicochemical and Engineering Aspects. 2003;**212**(2-3):155-161

[24] Şahiner N, Pekel N, Akkas P, Güven O. Amidoximation and characterization of new complexing hydrogels prepared from N-vinyl-2-pyrrolidone/acrylonitrile systems. Journal of Macromolecular Science, Part A: Pure and Applied Chemistry. 2000;**37**(10):1159-1172. DOI: 10.1081/MA-100101146

[25] Liu X, Tong Z, Hu O. Swelling equilibria of hydrogels with sulfonate groups in water and in aqueous salt solutions. Macromolecules. 1995;**28**(11):3813-3817. DOI: 10.1021/ma00115a010

[26] Tong Z, Liu X. Swelling equilibria and volume phase transition in hydrogels with strongly dissociating electrolytes. Macromolecules. 1994;**27**(3):844-848. DOI: 10.1021/ma00081a033

[27] Durmaz S, Okay O. Acrylamide/2-acrylamido-2-methylpropane sulfonic acid sodium salt-based hydrogels: Synthesis and characterization. Polymer. 2000;**41**(10):3693-3704

[28] Coşkun R, Soykan C, Saçak M. Adsorption of copper(II), nickel(II) and cobalt(II) ions from aqueous solution by methacrylic acid/acrylamide monomer mixture grafted poly(ethylene terephthalate) fiber. Separation and Purification Technology. 2006;**49**(2):107-114. DOI: 10.1016/j.seppur.2005.09.002

[29] Şener S, Tokalıoğlu Ş, Özentürk İ, Soykan C. Preconcentration/separation of some trace metal ions from water samples by a new synthesized chelating

resin. Journal of AOAC International. 2014;**97**(2):598-604. DOI: 10.5740/jaoacint.12-479

[30] Islam A, Ahmad A, Laskar MA. Flame atomic absorption spectrometric determination of trace metal ions in environmental and biological samples after preconcentrationon a newly developed amberlite XAD-16 chelating resin containing p-aminobenzene sulfonic acid. Journal of AOAC International. 2015;**98**(1):165-175. DOI: 10.5740/jaoacint.11-429

[31] Jiang ZT, Yu JC, Liu HY. Simultaneous determination of cobalt, copper and zinc by energy dispersive X-ray fluorescence spectrometry after preconcentration on PAR-loaded ion-exchange resin. Analytical Sciences. 2005;**21**:851-854. DOI: 10.2116/analsci.21.851

[32] Milosavljevic SM. Infrared spectroscopoy. In: Strukturne Instrumentalne Metode. Vol. 2. Chemical Faculty, University of Belgrade; 1994

[33] Molaakbari E, Mostafavi A, Afzali D. Simultaneous separation and preconcentration of trace amounts of copper(II), cobalt(II) and silver(I) by modified Amberlyst®15 resin. International Journal of Environmental Analytical Chemistry. 2013;**93**(4):365-376. DOI: 10.1080/03067319.2012.663753

[34] Pourjavid MR, Arabieh M, Yousefi SR, Jamali MR, Rezaee M, Hosseini MH, et al. Study on column SPE with synthesized graphene oxide and FAAS for determination of trace amount of Co(II) and Ni(II) ions in real samples. Materials Science and Engineering. 2015;**C47**:114-122. DOI: 10.1016/j.msec.2014.11.028

[35] Omidi F, Behbahani M, Bojdi MK, Shahtaheri SJ. Solid phase extraction and trace monitoring of cadmium ions in environmental water and food samples based on modified magnetic nanoporous silica. Journal of Magnetism and Magnetic Materials. 2015;**395**:213-220. DOI: 10.1016/j.jmmm.2015.07.093

[36] Soykan C, Erol İ, Türkmen H, Akçamur Y. New poly(methacrylate)s containing benzylpiperazine and methylpiperidine moieties. Journal of Polymer Research. 2004;**11**(3):181-187. DOI: 10.1023/B:JPOL.0000043400.92054.92

[37] Ari H, Soykan C, Özpozan T. Preparation of organic/inorganic hybrid materials using aggregates of poly{2-methyl-N-[2-(phenylthio)phenyl] acrylamide-co-2-(trimethylsyloxy) ethyl methacrylate} as precursor and vibrational investigation of the polymerization. Journal of Macromolecular Science, Part A: Pure and Applied Chemistry. 2013;**50**(10):1022-1041. DOI: 10.1080/10601325.2013.821847

[38] Soykan C, Delibaş A, Coşkun R. Novel copolymers of N-4-bromophenyl-2-methacrylamide with glycidyl methacrylate: Synthesis, characterization, monomer reactivity ratios and thermal properties. Reactive and Functional Polymers. 2008;**68**(1):114-124. DOI: 10.1016/j.reactfunctpolym.2007.10.004

[39] Erol İ, Soykan C, Türkmen H, Tufan Y. Synthesis and characterization of novel methacrylates derived from morpholine and pyrrolidine: The determination of kinetic parameters with thermogravimetric analysis. Journal of Macromolecular Science, Part A: Pure and Applied Chemistry. 2003;**40**(11):1213-1225. DOI: 10.1081/MA-120024835

[40] Tokalıoğlu Ş, Arsav S, Delibaş A, Soykan C. Indirect speciation of Cr(III) and Cr(VI) in water samples by selective separation and preconcentration on a newly synthesized chelating resin. Analytica Chimica Acta.

2009;**645**:36-41. DOI: 10.1016/j.aca.2009.04.035

[41] Tokalıoğlu Ş, Yılmaz V, Kartal Ş, Delibaş A, Soykan C. Solid phase extraction of Pd(II) on a newly synthesized chelating resin prior to determination by flame atomic absorption spectrometry. Microchimica Acta. 2009;**165**:347-352. DOI: 10.1007/s00604-009-0141-7

[42] Tokalıoğlu Ş, Yılmaz V, Kartal Ş, Delibaş A, Soykan C. Synthesis of a novel chelating resin and its use for selective separation and preconcentration of some trace metals in water samples. Journal of Hazardous Materials. 2009;**169**:593-598. DOI: 10.1016/j.jhazmat.2009.03.146

[43] Saçmacı Ş, Saçmacı M, Soykan C, Kartal Ş. Synthesis and characterization of new chelating resin: Adsorption study of copper(II) and chromium(III) ions. Journal of Macromolecular Science Part A: Pure and Applied Chemistry. 2010;**47**(6):552-557. DOI: 10.1080/10601321003742055

[44] Saçmacı Ş, Kartal Ş, Saçmacı M, Soykan C. Novel solid phase extraction procedure for some trace elements in various samples prior to their determinations by FAAS. Bulletin of the Korean Chemical Society. 2011;**32**(2):444-450. DOI: 10.5012/bkcs.2011.32.2.444

[45] Saçmacı Ş, Kartal Ş, Yilmaz Y, Saçmacı M, Soykan C. A new chelating resin: Synthesis, characterization and application for speciation of chromium (III)/(VI) species. Chemical Engineering Journal. 2012;**181-182**:746-753. DOI: 10.1016/j.cej.2011.11.057

[46] Turan Ş, Tokalıoğlu Ş, Şahan A, Soykan C. Synthesis, characterization and application of a chelating resin for solid phase extraction of some trace metal ions from water, sediment and tea samples. Reactive and Functional Polymers. 2012;**72**(10):722-728. DOI: 10.1016/j.reactfunctpolym.2012.07.002

[47] Yılmaz S, Tokalıoğlu Ş, Şahan S, Ülgen A, Şahan A, Soykan C. On-line preconcentration/determination of zinc from water, biological and food samples using synthesized chelating resin and flame atomic absorption spectrometry. Journal of Trace Elements in Medicine and Biology. 2013;**27**:85-90. DOI: 10.1016/j.jtemb.2012.07.007

[48] Çetin T, Tokalıoğlu Ş, Ülgen A, Şahan S, Özentürk İ, Soykan C. Synthesis/characterization of a new chelating resin and on-line solid phase extraction for the determination of Ag(I) and Pd(II) from water, cream, anode slime and converter samples by flow injection flame atomic absorption spectrometry. Talanta. 2013;**105**:340-346. DOI: 10.1016/j.talanta.2012.10.054

[49] Çimen G, Tokalıoğlu S, Özentürk İ, Soykan C. Speciation and preconcentration of chromium from water and food samples by synthesized chelating resin. Journal of the Brazilian Chemical Society. 2013;**24**(5):856-864. DOI: 10.5935/0103-5053.20130113.

[50] Daşbaşı T, Saçmacı Ş, Çankaya N, Soykan C. A new synthesis, characterization and application chelating resin for determination of trace metals in honey samples by FAAS. Food Chemistry. 2016;**203**:283-291. DOI: 10.1016/j.foodchem.2016.02.078

[51] Daşbaşı T, Saçmacı Ş, Çankaya N, Soykan C. Synthesis, characterization and application of a new chelating resin for solid phase extraction, preconcentration and determination of trace metals in some dairy samples by flame atomic absorption spectrometry. Food Chemistry. 2016;**211**:68-73. DOI: 10.1016/j.foodchem.2016.05.037

[52] Yavuz E, Tokalıoğlu Ş, Erkılıç H, Soykan C. Novel chelating resin for

solid phase extraction of metals in certified reference materials and waters. Analytical Letters. 2017;**50**(2):364-378. DOI: 10.1080/00032719.2016.1181643

[53] Daşbaşı T, Muğlu H, Soykan C, Ülgen A. SPE and determination by FAAS of heavy metals using a new synthesized polymer resin in various water and dried vegetables samples. Journal of Macromolecular Science Part A: Pure and Applied Chemistry. 2018;**55**(3):288-295. DOI: 10.1080/10601325.2018.1424556

[54] Daşbaşı T, Soykan C, Çankaya N, Ülgen A. Determination of some trace metals with a new synthesized polymer resin by FAAS in various tea and herbal plants samples. Journal of Macromolecular Science Part A: Pure and Applied Chemistry. 2018;**55**(6):466-473. DOI: 10.1080/10601325.2018.1470464

Chapter 5

Evaluation of Trace Elemental Levels as Pollution Indicators in an Abandoned Gold Mine Dump in Ekurhuleni Area, South Africa

Godwin Okereafor, Mamookho Makhatha, Lukhanyo Mekuto and Vuyo Mavumengwana

Abstract

In the Blesbokspruit area of Ekuhurleni, South Africa, previous gold mining activities resulted in many tailings dump sites. 20 representative soil samples were used in describing the distribution of metals. The soils were very strongly acidic ranging from 3.86 to 4.34 with a low cation exchange capacity (CEC). Based on X-ray fluorescence (XRF) analysis, elemental composition of the soils revealed average values of major elements such as Na2O (0.18%), MgO (0.63%), Al2O3 (6.51%), SiO2 (81.83%), P2O5 (0.04%), SO3 (3.40%), K2O (1.98%), CaO (0.45%), TiO2 (0.51%), Cr2O3 (0.17%), MnO (0.04%), Fe2O3 (3.59%), NiO (0.04%), As2O3 (0.02%), with Rb2O and SrO falling below 0.01%. Trace metals (TM) contamination levels in the soils were evaluated using various pollution indices which revealed that over 60% of the soils were between the high degree and the ultra-high degree of contamination classes. The concentration of various trace metals varies from 860.3–862.6 mg/kg for Cr; 324.9–328.4 mg/kg for Al; 200.9–203.4 mg/kg for As; 130.1–136.2 mg/kg for Fe; 121.9–125.8 mg/kg for Pb; 27.3–30.2 mg/kg for Co; 23.8–26.8 mg/kg for Ni; 7.2–9.2 mg/kg for Ti; 7.1–9.2 mg/kg for Cd; 4.0–5.6 mg/kg for Zn and 0.1–0.6 mg/kg for Cu.

Keywords: mine tailings, trace metal, pollution, contamination factor, geoaccumulation index

1. Introduction

South Africa like other developing countries is faced with the challenges of environmental degradation via the continuous release into the environment of trace element-containing chemicals through urbanization, agricultural and mining activities, as well as industrialization. Trace metals (TM) are naturally occurring elements that have a high atomic weight and a density at least 5 times greater than that of water, and some of the commonly found ones particularly at contaminated sites include Arsenic (As), Cadmium (Cd), Chromium (Cr), Copper (Cu), Lead (Pb),

Mercury (Hg), Nickel (Ni) and Zinc (Zn) [1, 2]. Attempts towards the assessment, mechanism and the characteristics of trace metal pollution in surrounding areas of mines has been and continue being a theme of various scientific gatherings.

Globally, the extraction and distribution of minerals from ore deposits has been one of the actions that contribute to environmental degradation due to industrialization. The extraction and beneficiation processes often result in the release of tailings that end up in natural percolations within the earth crust, thus paving a way for various kinds of risk elements entering the ecosystem. Such practices result in serious environmental complications due to the elevated concentrations and accumulation of trace metals which poses risk for human health [3–7].

The mining and processing of gold is associated with certain elements such as Copper (Cu), Antimony (Sb), Nickel (Ni), Selenium (Se), Mercury (Hg), Thallium (Tl), Titanium (Ti), Zinc (Zn), Silver (Ag), Cobalt (Co), Lead (Pb) and Uranium (U). Most of these metals are somewhat released into the environment via trophic links ranging from agricultural soils to plants, animals and humans [8–10].

Pollutants from various anthropogenic activities ranging from mine effluents such as wastewaters, tailings, runoff from agricultural pesticides and atmospheric deposition often contaminate the surrounding soils and water bodies thus posing threat to the ecosystem and humans. This occurs via direct ingestion or contact with contaminated soil, the food chain (soil–plant-human or soil–plant–animal-human), drinking of contaminated ground water, reduction in food quality (safety and marketability) via phytotoxicity, reduction in land usability for agricultural production causing food insecurity, and land tenure problems [11, 12]. In humans, several health challenges such as abortion, cancer, kidney damage and sometimes death, are some of the consequences of prolonged exposure to extreme concentrations of trace metals [13].

The importance of soil cannot be over emphasized as it is characterized as a complex and dynamic system that is made up of sediments that are different in relation to their physical, chemical, mineralogical and biological constituents. Soil is an essential resource for natural living conditions of plants, animals and humans. The role of soil as a collector filter of both organic and inorganic residues helps in protecting groundwater and in the sequestration of toxic materials [14]. The accumulation of excess metals and metalloids in soils over an extended period exposes humans and other animals to toxicity [15]. Assessing the spatial distribution of trace metals is soil is crucial to obtaining basic information about areas of concerns and to prioritize site mitigation strategies [16]. However, the quantification of element concentrations in soil as a single parameter is not enough in evaluating the extent of contamination due to differentiation between natural background levels and anthropogenic enrichment [3]. Indexes including geoaccumulation index (I_{geo}) and contamination factor (CF) which are known to provide a better picture of the status of elemental contamination compared to the background concentration were used as pointers in identifying and quantifying the level of elemental pollution as well as the intensity of anthropogenic contaminants accumulated in the soil.

There are enormous impacts of mine tailings disposal sites with over 500,000 abandoned hard rock mines located in the United States, while Mexico alone is affected by 27.1 million hectares of mining activity [17–19]. Gold mine waste was reported in 2001 by South Africa's Department of Water Affairs and Forestry as the largest single source of waste constituting over 47% of mineral wastes generated in South Africa [20]. Previous studies indicate that there are close to 300 unlined and not vegetated tailings dumps covering over 400 km^2 surface area within the Witwatersrand Basin of the Republic of South Africa. With tailings dumps being a major source of contaminants, the Witwatersrand Basin's massive tailing dumps are a possible, environmental pollution threat [21]. Studies into the deposits in the mine regions of the Gauteng province of South Africa [17], revealed the deposits to be of great health concern; containing enormous amounts of toxic metals, such as U, As, Ra, Ni, Zn, etc.

Hence, this present study was aimed at determining the contamination level of identified trace metals in an abandoned mine tailing dump over time. In addition, findings from this study will assist the various stakeholders in resource management and policy implementation.

2. Materials and methods

2.1 Description of the study area

South Africa lies on the southernmost part of the African continent, and is known to have renowned varied topography, great natural beauty, and cultural diversity. It is a medium-sized country, with a total land area of 1,219,090 square kilometers. Ekurhuleni falls within the East Rand region and is characterized by rainfall known to be typical to the Highveld summer rainfall, which occurs from October to April. The average annual rainfall varies from 715 to 735 mm an indication that the study area has a distinct moisture deficit. Frost does occur frequently from mid-April to September, which makes temperatures below freezing common during winter times. This area is home to mild summers with temperatures seldom above 30°C. During spring and winter, northerly and north-westerly winds occur and during summer north-easterly to north–north-easterly winds occur [22]. There are many pans across the Ekurhuleni area. These pans cover a total area of 3559 hectares within the Ekurhuleni Metropolitan Municipality area and are mostly seasonal. There are also a few lakes created by mines, which are used for recreational parks. Germiston Lake, Benoni Lake and Boksburg Lake are the three main lakes used for recreational purposes within the Ekurhuleni Metropolitan Municipality area, but which fall outside the East Rand Basin area. The tailings dump has some informal settlements within its proximity with subsistence farming among the dwellers as shown in **Figure 1**. The specific description indicating coordinates of the

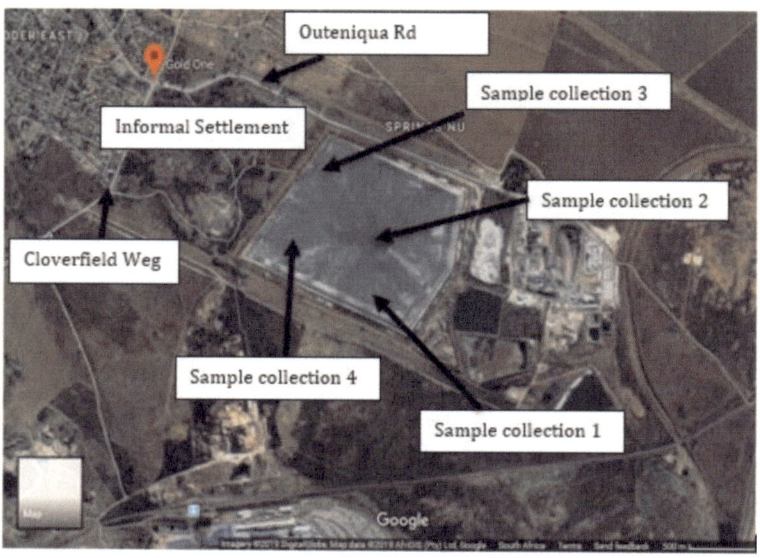

Figure 1.
Location of the sampling site.

Station no.	Latitude (S)	Longitude (E)
1	$26^0\ 10'$	$28^0\ 27'$
2	$26^0\ 15'$	$28^0\ 35'$
3	$26^0\ 04'$	$28^0\ 40'$
4	$26^0\ 17'$	$28^0\ 44'$
5	$26^0\ 21'$	$28^0\ 50'$
6	$26^0\ 30'$	$29^0\ 10'$
7	$26^0\ 00'$	$29^0\ 15'$
8	$26^0\ 27'$	$29^0\ 20'$
9	$26^0\ 09'$	$29^0\ 35'$
10	$26^0\ 38'$	$29^0\ 42'$
11	$26^0\ 43'$	$29^0\ 47'$
12	$26^0\ 34'$	$29^0\ 50'$
13	$26^0\ 13'$	$29^0\ 53'$
14	$26^0\ 19'$	$30^0\ 10'$
15	$26^0\ 48'$	$30^0\ 15'$
16	$26^0\ 36'$	$30^0\ 25'$
17	$26^0\ 40'$	$30^0\ 29'$
18	$26^0\ 14'$	$30^0\ 35'$
19	$26^0\ 23'$	$30^0\ 40'$
20	$26^0\ 54'$	$30^0\ 48'$

Table 1.
Location of the Blesbokspruit gold mine tailings sediment samples.

sampling site located along Outeniqua Road & Cloverfield Weg in Springs, Ekurhuleni are illustrated in **Table 1**.

2.2 Sampling (material) description

In a bid to assess the level of trace metal contamination in the mine tailings, about 2 kilograms of 20 representative tailing samples were obtained from the dump. Preceding the removal of top tailing samples (2 cm) using an auger, samples were taken at a depth of 10 cm for every 50 m horizontal interval for a wider coverage. The collected soil samples (tailings) were kept cool in an icebox (<4°C) and transported to the laboratory for further analyses in sterile plastic bags.

3. Analysis

3.1 Experimental analysis

20 representative tailing samples of about 5 g each were oven dried at 100°C for 24 hours and passed through a 2 mm sieve. Aliquots of approximately 2 g of the various tailing samples were weighed into a Teflon crucible and moistened with 100 mL of 1 M HCl acid for the determination of the HCl-soluble fraction of heavy metals. The mixtures were covered and placed on a shaker for 12 hours at 130 rpm.

The solutions were filtered through a Whatman filter paper, and the filtrates were stored in sterile bottles prior to analysis of minerals using inductively coupled plasma-optical emission spectrometry (ICP-OES).

10 g each of the representative tailing samples were pelletized using a mold at very high pressure and then placed in the sample compartment of the X-ray fluorescence spectrometer (XRF; Rigaku ZSX PrismusII). This was done to analyze the major and trace element oxides of the tailing samples.

Physicochemical properties such as pH and EC (electrical conductivity) of the soil samples (tailings) were measured in a soil-to-water suspension (1,2.5, w/w) and a 1:5 tailings-to-water suspension using a Crison multimeter (model MM 41) respectively [23]. Loss on Ignition (LOI) analysis was used to determine the organic matter content (% OM) of the various tailing's samples [24]. The grain size distribution of tailing samples was determined using the hydrometer method [25].

3.2 Quality assurance and quality control

Apparatus and glassware used were acid-washed with 5% nitric acid for precision analysis while reagents were of analytical standard. The trace metals were determined using ICP-OES (Model - GBC Quantima Sequential) operated under specific conditions of 1300 W RF power, 15 L min^{-1} plasma flow, 2.0 L min^{-1} auxiliary flow, 0.8 L min^{-1} nebulizer flow, 1.5 mL min^{-1} sample uptake rate. Multiple levels of calibration standard solutions prepared from a Certipur ICP multi-element standard (Merck KGaA) was used in the calibration of the ICP-OES. Metal determination was done using Axial view, while 2-point background correction and 3 replicates were employed in the measurement of analytical signal. The emission intensities were determined for the most sensitive lines free of spectral interference. By diluting the stock multi-elemental standard solution (1000 mg L^{-1}) in 0.5% (v/v) nitric acid, the calibration standards were prepared. The calibration curves for all the studied elements were in the range of 0.01 to 1.0 mg L^{-1}.

3.3 Data analyses

The history and degree of trace metal pollution in an environment can be ascertained from the surrounding sediments by comparing the pollutant metal concentration with an unpolluted reference material. The average shale concentration as an International standard reference for unpolluted sediment was utilized [26]. This study applied pollution indices such as (i) metal contamination factor, (ii) contamination degree, (iii) index of geoaccumulation, and (iv) pollution load index to assess heavy metal contamination.

3.3.1 Assessment according to contamination factor

By calculating the ratio of the concentration of a specific trace metal in the study area and the concentration of the background concentration of the corresponding metal, the contamination factor was determined. **Table 2** shows the various terminologies in describing contamination factor class and level [27]. CF is an effective tool for monitoring pollution over a period and for the respective metals was calculated using the equation as prescribed by [28].

$$\mathrm{CF} = \frac{(Mean\ metal\ concentration\ at\ contaminated\ site\ (Cm))}{(Level\ of\ pre-industrial\ concentration\ of\ individual\ metal\ (Cbackground))} \tag{1}$$

CF	Description
CF < 1	Low contamination factor
1 ≤ CF < 3	Moderate contamination factor
3 ≤ CF < 6	Considerate contamination factor
CF ≥ 6	Very high contamination factor

Table 2.
Terminologies used to describe contamination factor [27].

3.3.2 Assessment according to contamination degree

Contamination degree (CD) refers to the sum of all the contamination factor (CF) values of a specific sampling site. It is a diagnostic tool aimed at providing a measure of the degree of overall contamination in surface layers in a sampling site or core. In this study, CD was assessed using Eq. (2).

$$CD = \sum_{i=0}^{n} cf \tag{2}$$

A list of terminologies as prescribed by [29] used in describing the contamination degree of the site under investigation is summarized in **Table 3**.

3.3.3 Assessment according to geoaccumulation index

To quantify the level of heavy metal contamination associated with the study site, the geoaccumulation index (I-geo) was adopted. The I_{geo} is an important method used for the interpretation of the quality of sediments in the sampling site. It is used to assess impacts due to anthropogenic activities and was determined using Eq. (3) as prescribed by [30].

$$I_{geo} = \log_2 \frac{Cn}{1.5Bn} \tag{3}$$

where C_n is the measure of the metal concentration in the examined metal n in the sediment, Bn is the background concentration of the element (average shale concentration) or reference value of the metal n, and 1.5 is the correction factor due to the lithogenic effect that could result in variations in the background values for a given metal in the environment. There are seven grades (0–6) ranging from unpolluted to highly polluted in the geoaccumulation index scale as described by [30] (**Table 4**).

CD	Description
CD < 6	Low contamination degree
6 ≤ CD < 12	Moderate contamination degree
12 ≤ CD < 24	Considerate contamination degree
CD ≥ 24	Very high contamination degree

Table 3.
Terminologies used to describe contamination degree for soil [29].

I_{geo} Value	Class	Contamination Level
$I_{geo} \leq 0$	0	Uncontaminated
$0 < I_{geo} < 1$	1	Uncontaminated/moderately contaminated
$1 < I_{geo} < 2$	2	Moderately contaminated
$2 < I_{geo} < 3$	3	Moderately/strongly contaminated
$3 < I_{geo} < 4$	4	Strongly contaminated
$4 < I_{geo} < 5$	5	Strongly/extremely contaminated
$5 < I_{geo}$	6	Extremely contaminated

Table 4.
Classification for the geoaccumulation index (Igeo) [30].

3.3.4 Assessment according to pollution load index

Pollution load index, which is a useful tool in heavy metal pollution evaluation, refers to the number of times by which each heavy metal concentrations in the sediments (tailings) exceeded the background concentration in the soil, and it provides a summary of the overall level of heavy metal toxicity in a sample. The world average concentrations of metals using shale was used as background for identified heavy metals in this study [26]. The PLI can provide an estimate of the various metal contamination status and precautionary steps to be taking [31]. Using Eq. (4) as developed by [26], the PLI of the study site was calculated by obtaining the n-root from the n-CFs that was obtained for all the metals.

$$PLI = (CF1 \times CF2 \times CF3 \times \ldots \times CFn)^{1/n} \quad (4)$$

where CF is the contamination factor, CFn is the CF value of metal n, and n is the number of metals.

Interpretation of PLI values are categorized into two levels; polluted (PLI > 1) and unpolluted (PLI < 1) whereas PLI = 1 indicate trace metal loads close to the background level [32].

3.3.5 Assessment according to the United States environmental protection agency

The potential contamination of the tailing's sediments was evaluated using the proposed sediment quality guidelines by USEPA [19] **Table 5**. Illustrated the various criteria.

Metal	Not polluted	Moderately polluted	Heavily polluted	Present study
Cd	—	—	>6	7.1
Cr	<25	25–75	>75	860.3
Cu	<25	25–50	>50	0.1
Pb	<40	40–60	>60	121.9
Zn	<90	90–200	>200	3.9

Table 5.
USEPA guidelines for sediments (mg/kg dry weights) in comparison with gold mine tailings sediments.

4. Results and discussion

4.1 Soil physical properties

Textural properties obtained from sieve analysis of the gold mine tailings sediments using classification as prescribed by [25] are presented in **Table 6**. These results reveal that fine sand (0.150–0.075 mm) and clay (0.075–0.053 mm) were the principal fractions of all sediment samples, with an average composition of 66.03% for fine sand, 23.08% clay and 10.89% silt respectively. With the larger portion of the sediments being fine sand, there is a likelihood for nutrients accumulation is high due to the higher surface-to-volume ratios [33].

Geochemical properties of the sediments such as the pH, EC and carbonate content (see **Table 7**) helps in ascertaining vital information to comprehend the soils potential to withhold heavy metals [34]. The results obtained for the sediment pH measurements, showed that the study area is very strongly acidic ranging from 3.86 to 4.34. The low pH values in the study area were related with heterogeneous deposits of sulfidic residues from the mine surroundings, which resulted in low pH values that is attributed to microbial sulfide oxidation and the resultant formation

Sample no.	Sieve size (ASTM) % Materials; Retains (gms)								
	No. 100	No. 140	No. 200	No. 270	PAN	TOTAL	% Sand	% Silt	% Clay
1	5.68	45.51	15.84	10.25	22.72	100	67.03	10.25	22.72
2	5.75	46.82	13.79	10.58	23.41	100	66.01	10.58	23.41
3	5.40	46.52	13.61	10.62	23.85	100	65.53	10.62	23.85
4	5.37	45.84	14.71	11.25	22.83	100	65.92	11.25	22.83
5	5.42	45.93	13.93	11.81	22.91	100	65.28	11.81	22.91
6	5.39	47.88	13.01	11.20	22.52	100	66.28	11.20	22.52
7	5.42	48.23	11.87	10.78	23.70	100	65.52	10.78	23.70
8	5.88	46.38	13.42	10.44	23.88	100	65.68	10.44	23.88
9	5.94	46.82	13.00	10.32	23.92	100	65.76	10.32	23.92
10	5.66	44.46	16.15	10.58	23.15	100	66.27	10.58	23.15
11	5.86	47.20	14.22	9.88	22.84	100	67.28	9.88	22.84
12	5.42	45.30	15.83	11.32	22.13	100	66.55	11.32	22.13
13	5.38	45.92	13.68	11.84	23.18	100	64.98	11.84	23.18
14	5.62	46.34	13.74	10.68	23.62	100	65.70	10.68	23.62
15	5.48	46.82	13.81	10.31	23.58	100	66.11	10.31	23.58
16	5.23	46.92	14.81	10.22	22.82	100	66.96	10.22	22.82
17	5.98	48.22	11.62	11.69	22.49	100	65.82	11.69	22.49
18	5.36	48.80	11.78	11.38	22.68	100	65.94	11.38	22.68
19	5.92	48.24	11.34	11.75	22.75	100	65.50	11.75	22.75
20	5.68	47.36	13.37	10.94	22.65	100	66.41	10.94	22.65

Table 6.
Sieve analysis of the gold mine tailings sediment samples.

of sulfuric acid [35]. Nutrient uptake by plants may be inhibited by the level of acidity as most plant nutrients are optimally available to plants within 6.5 to 7.5 pH range which also support plant root growth [36]. The low CEC values which correlates with the high proportion of sand fragment is an indication that the sediments may likely not have reliable soil sorption capacity [37]. LOI of studied soils were in the range of (5.0–5.4%)-dry weight, which could be, attributed to growing plants within the tailing's sediments.

4.2 Metal content

The summary of the determined heavy metal concentrations within the sediments of the study area by using ICP-OES are presented in **Table 8**. The concentration of various heavy metal varies from 860.3–862.6 mg/kg for Cr; 324.9–328.4 mg/kg for Al; 200.9–203.4 mg/kg for As; 130.1–136.2 mg/kg for Fe; 121.9–125.8 mg/kg for Pb; 27.3–30.2 mg/kg for Co; 23.8–26.8 mg/kg for Ni; 7.2–9.2 mg/kg for Ti; 7.1–9.2 mg/kg for Cd; 4.0–5.6 mg/kg for Zn and 0.1–0.6 mg/kg for Cu. Chromium (Cr) was identified as the most abundant heavy metal in the sediment samples. Mean concentration of the metals were Cr: 861.5 mg/kg; Al: 326.8 mg/kg; As: 202.2 mg/kg; Fe: 134.3 mg/kg; Pb: 123.7 mg/kg; Co: 28.8 mg/kg; Ni: 25.4 mg/kg; Ti: 8.5 mg/kg; Cd: 8.3 mg/kg; Zn: 4.5 mg/kg and Cu: 0.2 mg/kg dry weights. The average order of metal concentration is Cr > Al > As > Fe > Pb > Co >

Station no.	pH	C.E (mS/cm)	CEC (meq/100 g)	LOI (%)
1	3.86	1.30	8.5	5.1
2	4.34	1.50	8.8	5.4
3	4.28	1.80	9.0	5.0
4	4.30	1.90	8.3	5.1
5	3.92	1.40	9.1	5.3
6	4.34	1.60	8.8	5.1
7	3.89	1.40	8.5	5.4
8	3.87	1.40	9.1	5.1
9	3.86	1.40	9.0	5.2
10	4.27	1.80	8.8	5.2
11	4.28	1.80	9.4	5.4
12	4.28	1.80	8.5	5.1
13	3.88	1.40	9.3	5.2
14	3.86	1.40	8.7	5.2
15	4.30	1.60	8.3	5.4
16	3.87	1.40	9.1	5.1
17	3.86	1.40	9.0	5.1
18	4.31	1.50	8.5	5.2
19	4.27	1.90	8.8	5.1
20	4.28	1.80	9.3	5.2

Table 7.
Geochemical properties of gold mine tailings sediments.

Ni > Ti > Cd > Zn > Cu. The mineral composition of the sediments and mining activities that took place within this region may be attributed to the high element concentrations in the soil samples.

In comparison to the interim sediment quality guidelines (ISQG) proposed by the Canadian Council of Ministers of the Environment [38], the elemental pollution status of the tailings (soil) were assessed **Table 8**. The heavy metals from the studied tailings sediments except for Zn and Cu all exceeded the ISQG. This implies that the sediments are toxic and could result in the introduction of sediment contaminants into the aquatic food web through predation by organisms at higher trophic levels.

In trace amounts, Arsenic is one of the priority toxic metals due to its several deteriorating effects to both plants and animals. The level of identified arsenic in the sediment is worrisome. As a non-essential element, Arsenic is not required for the growth of living organisms, though recent discovery reports a bacterium that

Station no.	Cr	Al	As	Fe	Pb	Co	Ni	Ti	Cd	Zn	Cu
1	862.6	327.4	201.7	134.1	125.6	28.4	26.1	9.0	9.2	4.7	0.6
2	860.4	327.9	203.4	136.2	122.9	30.2	25.3	8.3	8.8	4.0	0.1
3	861.3	328.0	202.9	133.7	123.1	29.5	26.4	9.2	8.1	4.1	0.2
4	862.4	328.4	202.4	130.1	124.7	28.8	24.7	8.1	7.9	3.9	0.2
5	862.1	326.5	202.1	132.5	121.9	29.6	23.8	8.7	7.2	5.6	0.3
6	861.5	325.7	201.7	134.9	122.1	29.3	25.1	7.9	8.3	4.3	0.1
7	860.6	324.9	203.0	135.3	123.5	28.7	25.7	8.5	7.5	5.2	0.1
8	861.1	328.1	201.9	135.1	123.2	29.2	26.3	9.0	7.9	4.9	0.3
9	860.7	327.9	202.6	135.9	124.1	27.5	26.8	9.2	9.0	4.2	0.1
10	860.3	326.3	202.1	132.7	124.8	29.1	25.2	8.1	8.5	5.1	0.1
11	860.6	325.4	201.7	136.0	122.3	27.3	25.7	8.3	7.1	5.3	0.1
12	861.0	326.7	200.9	131.8	122.5	27.9	23.9	8.7	8.7	5.0	0.2
13	862.1	326.1	201.2	135.9	124.9	28.7	24.3	7.6	8.3	5.5	0.6
14	860.5	327.9	201.4	134.1	123.1	28.3	26.0	9.1	9.1	4.5	0.1
15	862.5	328.2	203.0	133.7	122.7	28.0	25.8	9.2	8.3	5.0	0.2
16	862.3	326.3	202.6	134.9	125.8	29.1	24.6	8.3	8.5	5.3	0.3
17	862.4	325.9	201.5	133.5	123.7	29.5	24.2	8.0	8.1	4.1	0.1
18	861.9	327.4	202.3	134.2	125.1	29.7	26.5	7.2	8.0	4.8	0.4
19	861.6	326.1	201.9	134.9	124.3	28.1	25.7	8.4	8.5	4.4	0.5
20	862.0	325.6	203.1	135.7	125.3	28.4	26.3	8.7	8.8	4.6	0.1
Mean	861.5	326.8	202.2	134.3	123.7	28.8	25.4	8.5	8.3	4.5	0.2
Max	862.6	328.4	203.4	136.2	125.8	30.2	26.8	9.2	9.2	5.6	0.6
Min	860.3	324.9	200.9	130.1	121.9	27.3	23.8	7.2	7.1	3.9	0.1
B_n	90	88,000	13	47,200	20	19	50	4600	0.3	95	45
ISQG	52.3	NA	7.24	NA	30.2	NA	NA	NA	0.7	124.0	18.7

Table 8.
Heavy metals concentration (mg/kg dry weight) in gold mine tailings sediments.

replaces phosphorus with As for a number of cellular functions [39]. Plants often accumulate As by root uptake from soil or by absorption of airborne As deposited on their leaves [40]. Arsenate, a dominant specie of Arsenic in soils, based on its similarity to phosphate usually compete for the same uptake carriers in the root plasmalemma of most plants. In so doing interrupts with several metabolic processes that end up inhibiting plant growth and development through arsenic-induced phytotoxicity [41, 42]. Some of the toxicity symptoms may include inhibition of seed germination, decrease in plant height, depressed tillering, reduction in root growth and some necrosis, decrease in shoot growth, lower fruit and grain yield, reductions in chlorophyll and protein contents, and in photosynthetic capacity and even death [41–46]. Due to migration and expansion of residential areas into former mining territories, the danger of human exposure to soil As has risen in the last two decades which have affected adversely the health of many [47]. Continued exposure to As results in several clinical manifestations such as melanosis (hyperpigmentation), keratosis, and leukomelanosis (hypopigmentation) of which cutaneous lesions are the highest reported [48, 49]. As is also a well-known carcinogen, causing skin, lung, bladder, liver, and kidney cancers [50, 51].

The average concentration of Copper (Cu) being 0.2 mg/kg was within the maximum acceptable concentration of 6.6 mg/kg for agricultural soil and safe limit of the Republic of South Africa [52].

As an important micronutrient, Cu is required for the growth of both plants and animals. In humans, it aids in the production of blood hemoglobin while plants utilize it in seed production, disease resistance, and regulation of water. In high levels, Cu could cause anemia, liver and kidney damage, as well as stomach and intestinal irritation [53]. Cu typically occurs in drinking water from Cu pipes, as well as from additives intended to control algal growth. The interaction of Cu with the environment is complex, however different studies revealed that most Cu introduced into the environment rapidly becomes, stable and results in a form which does not pose a danger to the environment [54, 55].

Zinc is an important metal due to its enzymatic and regulatory functions in biological systems. Being a readily mobile element, Zinc (Zn) when in high doses exhibit toxic and carcinogenic effects that could result in serious hematological and neurologic complications, liver and kidney disorders, hypertension, gastrointestinal misery, loose bowels, pancreatic harm and a host of other ailments in both humans and animals [56]. On the earth crust, Zinc is found in an average concentration of 80 mg/kg in association with ores of other metals such as Pb, Cu and Cd [57].

Chromium (Cr) has an average concentration of 100 mg/kg in the earth crust and the only known ore of commercial value is chromite ($FeO.Cr_2O_4$). Contamination by Cr could result in toxicity in plants depending on its state of valency since Cr (VI) due to its being highly mobile is toxic, while Cr (III) as less mobile is less toxic. The subsequent uptake, translocation, and accumulation of Cr by plants is dependent on its speciation. Cr in its trivalent (III) and hexavalent (VI) forms are known to be of biological importance. Generally, Cr poses the greatest threat to humans, animals and plants. Decreased seed germination, reduction of growth, decreased yield, inhibition of enzymatic activities, impairment of photosynthesis, nutrient and oxidative imbalances, and mutagenesis are some of the symptoms of Cr toxicity in plants [58]. In a previous study by López-Luna et al. [59], the toxicity of Cr (VI), Cr (III), and Cr tannery sludge were compared with respect to Cr mobility in soil and toxicity in wheat, oat, and sorghum plants and findings were that Cr(VI) was more mobile in soil and caused higher toxicity on those plant seedlings, while tannery sludge was the least toxic [60]. In humans, prolonged exposure results in kidney and liver disorders [61].

Lead (Pb) is the largely known immobile nonessential element among the heavy metals with most of its compounds being noxious in nature. Pb on the earth crust has an average concentration of 0.1 mg/kg. There is a gradual phase out of Pb from the materials regularly used by humans due to it being a metal toxicant. Mostly via food chain, Pb penetrates human or animal metabolism. The observed Pb content within the samples was very high and have been reported globally to be very harmful to humans and other animals as a long-term exposure could result in the bioaccumulation and biomagnification that end up in serious neurological health challenges. In plants, concentrations above 5 mg/kg of Pb causes severe growth retardation, discoloration, and morphological deformities. Pb accumulates in the body organs (i.e., brain), which may lead to poisoning (plumbism) or even death. The presence of lead often affects the gastrointestinal tracts, kidneys, and central nervous system. Infants exposed to lead are likely to suffer impaired development, lower IQ, shortened attention span, hyperactivity, and mental deterioration [62]. Adults usually suffer decreased reaction time, loss of memory, nausea, insomnia, anorexia, and weakness of the joints when exposed to Pb [63]. Lead performs no known essential function in the human body, it can merely do harm after uptake from food, air, or water.

Industrial waste materials, lime, fertilizer and sewage sludge constitute the major sources of nickel into soils [64]. Till date, nickel (Ni) remains a heavy metal of environmental concern as a result of decreased soil pH, due to reduced use of soil liming in agricultural soils and mobilization arising from increased acid rain in industrialized areas [65]. With decreasing pH, Ni exhibits increased solubility and mobility, thus, soil pH is the major factor controlling its solubility, mobility and sorption, while clay content, iron- manganese mineral and soil organic matter are of secondary importance [66]. Nickel (Ni) concentrations were observed to be high which could result in toxic effects to both plants and animals due to its ability to replace other metal ions in enzymes, proteins or bind to cellular compounds [65]. Nickel (Ni) is reported to interact with at least 13 essential elements namely calcium, chromium, cobalt, copper, iodine, iron, magnesium, manganese, molybdenum, phosphorus, potassium, sodium and zinc [67]. As a result, prolong exposure of humans to oxides and sulfides of nickel is linked with possible risk to lung and nasal tumors, skin allergies, nasal sinusitis, rhinitis and dermatitis [68]. Symptoms of nickel toxicity in plants besides inhibited growth include chlorosis, stunted root growth and brown interveinal necrosis [69].

Cadmium (Cd) is being discussed on a global platform as one of the most eco-toxic metals with a tendency of adversely affecting biological activities, plant metabolism, soil health and human health. The usage of Cadmium (Cd) is widely seen in Ni/Cd batteries, as rechargeable or secondary power sources exhibiting high output, long life, low maintenance, and high tolerance to physical and electrical stress. Observed levels of Cadmium was high and of great concern because it is very biopersistent and, once absorbed by an organism, remains resident for many years. In humans, Cadmium is known to affect several enzymes. Previous research revealed that renal damage that results in proteinuria is the consequence of Cd adversely affecting enzymes responsible for reabsorption of proteins in kidney tubules [70]. A prolong exposure to this metal even at very low concentration also reduces the activity of delta-aminolevulinic acid synthetase, arylsulfatase, alcohol dehydrogenase, and lipoamide dehydrogenase, which often cause anemia, cardio-vascular disorders and hypertension whereas it enhances the activity of delta-aminolevulinic acid dehydratase, pyruvate dehydrogenase, and pyruvate decarbox-ylase [71].

Station no.	Contamination factor of single metal							Degree of contamination	
	Cr	As	Pb	Co	Ni	Cd	Zn		
1	9.58	15.52	6.28	1.49	0.52	30.67	0.05	64.11	Very high
2	9.56	15.65	6.15	1.59	0.51	29.33	0.04	62.83	Very high
3	9.57	15.61	6.16	1.55	0.53	27.00	0.04	60.46	Very high
4	9.58	15.57	6.24	1.52	0.49	26.33	0.04	59.77	Very high
5	9.58	15.55	6.10	1.56	0.48	24.00	0.06	57.33	Very high
6	9.57	15.52	6.11	1.54	0.50	27.67	0.05	60.96	Very high
7	9.56	15.62	6.18	1.51	0.51	25.00	0.05	58.43	Very high
8	9.57	15.53	6.16	1.54	0.53	26.33	0.05	59.71	Very high
9	9.56	15.58	6.21	1.45	0.54	30.00	0.04	63.38	Very high
10	9.56	15.55	6.24	1.53	0.50	28.33	0.05	61.76	Very high
11	9.56	15.52	6.12	1.44	0.51	23.67	0.06	56.88	Very high
12	9.57	15.45	6.13	1.47	0.48	29.00	0.05	62.15	Very high
13	9.58	15.48	6.25	1.51	0.49	27.67	0.06	61.04	Very high
14	9.56	15.49	6.16	1.49	0.52	30.33	0.05	63.60	Very high
15	9.58	15.62	6.14	1.47	0.52	27.67	0.05	61.05	Very high
16	9.58	15.58	6.29	1.53	0.49	28.33	0.06	61.86	Very high
17	9.58	15.50	6.19	1.55	0.48	27.00	0.04	60.34	Very high
18	9.58	15.56	6.26	1.56	0.53	26.67	0.05	60.21	Very high
19	9.57	15.53	6.22	1.48	0.51	28.33	0.05	61.69	Very high
20	9.58	15.62	6.27	1.49	0.53	29.33	0.05	62.82	Very high
Average	9.57	15.55	6.19	1.51	0.51	27.63	0.05	61.01	Very high

Table 9.
Contamination factor (CF) and degree of contamination at various sampling station at the Blesbokspruit abandoned gold mine tailings site.

4.3 Pollution status

The assessment of the overall contamination of the studied area was based on the contamination factor **Table 9**. The average contamination factor for single metal from this study revealed the sediments as slightly contaminated with Ni and Zn, moderately contaminated with Co and highly contaminated with Cr, As, Pb and Cd. The highest average contamination factor value was that of Cd (27.63). Overall, the degree of contamination values of the sediments from the study site indicate very high contamination.

The average index of geoaccumulation values and contamination levels from the various sampling points within the study area as shown on **Table 10** reveals an uncontaminated status for Co (0.01), Ni (−1.09) and Zn (−3.39) respectively. However, Cr and Pb within the area showed a moderately contamination level with average I_{geo} values being 1.85 and 1.42 respectively. The site was however moderately to strongly contaminated with As (2.34) and Cd (2.91).

As indicated in **Table 10**, Pollution load index (PLI) ranged from 2.56–2.75, with mean value 2.67. PLI values of the different stations are above 1 which strongly indicate that the sediments are all polluted by heavy metals, an indication of deterioration of the study site.

Station no.	Cr	As	Pb	Co	Ni	Cd	Zn	PLI	Description of PLI
1	1.85	2.34	1.43	0.00	−1.05	3.02	−3.51	2.72	Polluted
2	1.85	2.34	1.41	0.06	−1.09	2.97	−3.51	2.63	Polluted
3	1.85	2.34	1.41	0.04	−1.05	2.89	−3.51	2.61	Polluted
4	1.85	2.34	1.43	0.01	−1.11	2.87	−3.51	2.56	Polluted
5	1.85	2.34	1.40	0.04	−1.14	2.77	−3.22	2.67	Polluted
6	1.85	2.34	1.40	0.03	−1.11	2.91	−3.51	2.67	Polluted
7	1.85	2.34	1.42	0.01	−1.08	2.81	−3.22	2.64	Polluted
8	1.85	2.34	1.41	0.02	−1.05	2.87	−3.51	2.68	Polluted
9	1.85	2.34	1.42	−0.04	−1.02	3.00	−3.51	2.63	Polluted
10	1.85	2.34	1.43	0.02	−1.08	2.94	−3.22	2.68	Polluted
11	1.85	2.34	1.41	−0.04	−1.08	2.76	−3.22	2.66	Polluted
12	1.85	2.33	1.41	−0.02	−1.14	2.96	−3.22	2.65	Polluted
13	1.85	2.34	1.43	0.01	−1.14	2.91	−3.22	2.73	Polluted
14	1.85	2.34	1.41	−0.01	−1.05	3.01	−3.51	2.71	Polluted
15	1.85	2.34	1.41	−0.02	−1.08	2.91	−3.22	2.67	Polluted
16	1.85	2.34	1.43	0.02	−1.11	2.94	−3.22	2.75	Polluted
17	1.85	2.34	1.42	0.04	−1.14	2.89	−3.51	2.57	Polluted
18	1.85	2.34	1.43	0.04	−1.05	2.88	−3.51	2.69	Polluted
19	1.85	2.34	1.42	−0.01	−1.08	2.94	−3.51	2.68	Polluted
20	1.85	2.34	1.43	0.00	−1.05	2.97	−3.51	2.71	Polluted

Table 10.
Geoaccumulation index (Igeo) and pollution load index (PLI) at various sampling station at the Blesbokspruit abandoned gold mine tailings site.

It is evident from the present study that the abandoned gold mine tailings site is not polluted with Zn and Cu, but heavily polluted with Cd, Cr and Pb when evaluated by comparison with the sediment quality guideline proposed by USEPA.

5. Conclusion

The successful assessment of trace metal contamination of the abandoned gold mine tailings at Blesbokspruit-Ekurhuleni was done using indices such as geoaccumulation index, contamination factor, and degree of contamination and pollution load index. The sediment was mostly dominated by fine sand and silt/clay. Based on sediment quality guidelines proposed by the USEPA, the contamination of the sediment by Zn and Cu was negligible while Cd, Cr and Pb were detected at high concentrations. The evaluated pollution load index indicated that the sediments in the tailings dump are polluted while the geoaccumulation index revealed that Cr, Pb, and As contaminated the site, thus indicating very high degrees of contamination of the sediments at the mine dump. The high metal contaminants could be attributed to anthropogenic activities from previous extensive gold mining activities that took place within the area. Considering agricultural activities and human dwellers within the surrounding areas of the mine tailings, there are high tendencies of deleterious impacts. As a further precaution, this study strongly supports the call for analysis of the stream and drinking water quality, including the staple crops that are cultivated within the vicinity of the dump site, to ascertain the levels of heavy metals within such crops. Stringent mitigation plans or conversion of the tailings into value-added products should be considered.

Acknowledgements

We thank the Mineral Processing and Technology Research Centre of the University of Johannesburg for providing the resources used in conducting this study.

Conflict of interest

The authors declare no conflict of interest.

Author details

Godwin Okereafor[1*], Mamookho Makhatha[1], Lukhanyo Mekuto[2] and Vuyo Mavumengwana[3]

1 Department of Metallurgy, School of Mining, Metallurgy and Chemical Engineering, Faculty of Engineering and the Built Environment, University of Johannesburg, Johannesburg, South Africa

2 Department of Chemical Engineering, Faculty of Engineering and the Built Environment, University of Johannesburg, Johannesburg, South Africa

3 Department of Biotechnology and Food Technology, Faculty of Science, University of Johannesburg, Johannesburg, South Africa

*Address all correspondence to: guokereafor@gmail.com

IntechOpen

© 2019 The Author(s). Licensee IntechOpen. This chapter is distributed under the terms of the Creative Commons Attribution License (http://creativecommons.org/licenses/by/3.0), which permits unrestricted use, distribution, and reproduction in any medium, provided the original work is properly cited.

References

[1] Akoto O, Ephraim JH, Darko G. Heavy metal pollution in surface soils in the vicinity of abundant railway servicing workshop in Kumasi, Ghana. International Journal of Environmental Resources. 2008;**2**:359-364

[2] Wuana RA, Okieimen FE. Heavy metals in contaminated soils: A review of sources, chemistry, risks and best available strategies for remediation. International Scholarly Research Notices: Ecology. 2011;**2011**:20. DOI: 10.5402/2011/402647. Article ID: 402647

[3] Barbieri M. The importance of enrichment factor (EF) and geoaccumulation index (Igeo) to evaluate the soil contamination. Journal of Geology and Geophysics. 2016;**5**:237. DOI: 10.4172/2381-8719.1000237

[4] Barkouch Y, Pineau A. Evaluation of the impact of mine activity on surrounding soils of Draa Lasfar mine in Marrakech-Morocco. African Journal of Environmental Science and Technology. 2016;**10**(1):44-49

[5] Muhammad S, Tahir Shah M, Khan S. Heavy metal concentrations in soil and wild plants growing around Pb-Zn sulfide terrain in the Kohistan region, northern Pakistan. Microchemical Journal. 2011;**99**:67-75

[6] Figueroa F, Castro-Larragoitia J, Aragón A, García-Meza J. Grass cover density and metal speciation in profiles of a tailings-pile from a mining zones in Zacatecas, north-Central Mexico. Environment and Earth Science. 2010; **60**(2):395-407

[7] Sharma RK, Agrawal M, Marshall F. Heavy metal contamination of soil and vegetables in suburban areas of Varanasi, India. Ecotoxicology and Environmental Safety. 2007;**66**(2): 258-266

[8] Landers JE. Aquatic food webs and heavy metal contamination in the upper Blackfoot River, Montana. Graduate Student Theses, Dissertations, & Professional Papers. 10719; 2016. Available from: https://scholarworks.umt.edu/etd/10719

[9] Mann RM, Vijver MG, Peijnenburg WJGM. Metals and metalloids in terrestrial systems: Bioaccumulation, biomagnification and subsequent adverse effects. In: Ecological Impacts of Toxic Chemicals. Bentham Science Publishers Ltd; 2011. pp. 43-62

[10] Loureiro S, Ferreira A, Soares A, Nogueira A. Evaluation of the toxicity of two soils from Jales mine (Portugal) using aquatic bioassays. Chemosphere. 2005;**61**:168-177

[11] Song Y, Ji J, Mao C, Yang Z, Yuan X, Ayoko GA, et al. Heavy metal contamination in suspended soils of Changjiang River—Environmental implications. Geoderma. 2010;**159**: 286-295

[12] McLaughlin MJ, Zarcinas BA, Stevens DP, Cook N. Soil testing for heavy metals. Communications in Soil Science and Plant Analysis. 2000;**31** (11–14):1661-1700

[13] Okereafor GU, Mulaba-Bafubiandi AF, Sebola TE, Uche-Okereafor NC, Mavumengwana V. The effects of an acidic environment on selected geophagic clayey samples and its impact on the bioavailability of certain elements. Transaction of the Royal Society of South Africa. 2018;**73**(2): 180-185

[14] BIO Intelligence Service. Soil and water in a changing environment, Final Report prepared for European Commission (DG ENV), with support from HydroLogic; 2014

[15] Shafie NA, Aris AZ, Zakaria MF, Haris H, Yinglini W, Isa NM. Application of geoaccumulation index and enrichment factors on the assessment of heavy metal pollution in sediments. Journal of Environmental Science and Health, Part A. 2013;**48**(2): 182-190

[16] Xiao-san L, Shen Y, Yong-guan Z, Xiang-dong L. Trace metal contamination in urban soils of China. Science of the Total Environment. 2011; **421-422**:17-30. DOI: 10.1016/j.scitotenv.2011.04.020

[17] Ebenebe PC, Shale K, Sedibe M, Tikilili P, Achilo MC. South African mine effluents: Heavy metal pollution and impact on the ecosystem. International Journal of Chemical Sciences. 2017;**15**(4):198

[18] Schwegler F. Air quality management: A mining perspective: Air pollution XIV. In: JWS L, Brebbia CA, editors. WIT Transactions on Ecology and the Environment. In, Southampton, UK: WIT Press; 2006

[19] US Environmental Protection Agency. Abandoned mine lands team; 2004

[20] Oelofse SHH, Hobbs PJ, Rascher J. The Pollution and Destruction Threat of Gold Mining Waste on the Witwatersrand—A West Rand Case Study. Symposium on Environmental Issues and Waste Management in Energy and Mineral Production. Bangkok; 2007

[21] AngloGold Ashanti. Case studies. Woodlands Project—good progress being made with phytoremediation project. Environment—AngloGold Ashanti Report to Society; 2004

[22] EMM (Ekurhuleni Metropolitan Municipality). Environmental management framework for Ekurhuleni. Ekurhuleni; 2007

[23] Aris AZ, Ismail FA, Ng HY, Praveena SM. An experimental and modelling study of selected heavy metals removal from aqueous solution using *Scyllaserrataas* biosorbent. Pertanika Journal of Science & Technology. 2014;**22**(2):553-566

[24] Robertson S. Direct Estimation of Organic Matter by Loss on Ignition: Methods. SFU Soil Science Lab; 2011

[25] ASTM D422–63. Standard Test Method for Particle-Size Analysis of Soils; 2007

[26] Turekian KK, Wedepohl DH. Distribution of the elements in some major units of the Earth's crust. Geological Society of America Bulletin. 1961;**72**(2):175-192

[27] Hakanson L. An ecological risk index for aquatic pollution control a sedimentological approach. Water Research. 1980;**14**(8):975-1001

[28] Tomlinson DL, Wilson JG, Harris CR, Jeffrey DW. Problems in the assessment of heavy metal levels in estuaries and the formation of a pollution index. Helgoländer Wissenschaftliche Meeresuntersuchungen. 1980;**33**: 566-575. DOI: 10.1007/BF02414780

[29] Ahdy HHH, Khaled A. Heavy metals contamination in sediments of the Western part of Egyptian Mediterranean Sea. Australian Journal of Basic and Applied Sciences. 2009; **3**(4):3330-3336

[30] Muller G. Index of geo-accumulation in sediments of the Rhine River. GeoJournal. 1969;**2**:108-118

[31] Angula E. The Tomlinson pollution index applied to heavy metal, mussel-watch data: A useful index to assess coastal pollution. Science of the Total Environment. 1996;**187**:19-56

[32] Cabrera F, Clemente L, Barrientos DE. Heavy metal pollution of soils affected by the guadiamar toxic flood. Science of the Total Environment. 1999;**242**(1–3):117-129

[33] Kaye JP, Groffman PM, Grimm NB, Baker LA, Pouyat RV. A distinct urban biogeochemistry. Trends in Ecology & Evolution. 2006;**21**:192-199

[34] Maas S, Scheifler R, Benslama M, Crini N, Lucot E, Brahmia Z, et al. Spatial distribution of heavy metal concentrations in urban, suburban and agricultural soils in a Mediterranean city of Algeria. Environmental Pollution. 2010;**158**:2294-2301

[35] Bernhard D. Evolution of acid mine drainage formation in Sulphidic mine tailings. Minerals. 2014;**4**:621-641. DOI: 10.3390/min4030621

[36] Matsumoto S, Shimada H, Sasaoka T, Miyajima I, Kusuma GJ, Gautama RS. Effects of acid soils on plant growth and successful Revegetation in the case of mine. In: Soil pH for Nutrient Availability and Crop Performance. 2017. pp. 9-27. DOI: 10.5772/intechopen.70928

[37] Lambooy AM. Relationship between cation exchange capacity, clay content and water retention of Highveld soils. South African Journal of Plant and Soil. 1984;**1**(2):33-38. DOI: 10.1080/02571862.1984.10634106

[38] Canadian Council of Ministers of the Environment. Canadian sediment quality guidelines for the protection of aquatic life, polychlonnated dioxins and furans (PCDD/Fs). In: Canadian environmental quality guidelines. Winnipeg: Canadian Council of Ministers of the Environment; 2001, 1999

[39] Wolfe-Simon F, Blum JS, Kulp TR, Gordon GW, Hoeft SE, Pett-Ridge J. A bacterium that can grow by using arsenic instead of phosphorus. Science. 2010;**332**(6034):1163-1166

[40] United States Environmental Protection Agency (U.S. EPA). An Exposure and Risk Assessment for Arsenic. Washington, DC: Office of Water Regulations and Standards, EPA-440/4–85-005; 1982

[41] Marin AR, Masscheleyn PH, Patrick WH Jr. The influence of chemical form and concentration of arsenic on rice growth and tissue arsenic concentration. Plant and Soil. 1992;**139**:175-183

[42] Marin AR, Pezeshki SR, Masscheleyn PH, Choi HS. Effect of dimethylarsinic acid (DMAA) on growth, tissue arsenic and photosynthesis of rice plants. Journal of Plant Nutrition. 1993;**16**:865-880

[43] Abedin MJ, Cottep-Howells J, Meharg AA. Arsenic uptake and accumulation in rice (Oryza sativa L.) irrigated with contaminated water. Plant and Soil. 2002;**240**:311-319

[44] Carbonell-Barrachina AA, Burlo-Carbonell F, Mataix-Beneyto J. Arsenic uptake, distribution and accumulation in tomato plants: Effect of arsenic on plant growth and yield. Journal of Plant Nutrition. 1995;**18**:1237-1250

[45] Jahan I, Hoque S, Ullah SM, Kibria MG. Effects of arsenic on some growth parameters of rice plant. Dhaka University Journal of Biological Sciences. 2003;**12**:71-77

[46] Kang LJ, Li XD, Liu JH, Zhang XY. The effect of arsenic on the growth of rice and residues in a loam paddy soil. Journal of Jilin Agricultural University. 1996;**18**:58-61

[47] Mandal BK, Chowdhury TR, Samanta G, Mukherjee D, Chanda CR, Saha KC. Impact of safe water for drinking on five families for 2 years in

West Bengal, India. Science of the Total Environment. 1998;**218**:185-201

[48] Das NK, Sengupta SR. Arsenicosis: Diagnosis and treatment. Indian Journal of Dermatology, Venereology and Leprology. 2008;**74**(6):571-581

[49] WHO. Guidelines for Drinking-Water Quality, Addendum to Volume 1: Recommendations. Geneva; 1998

[50] IARC (International Agency for Research on Cancer). A Review of Human Carcinogens: Arsenic, Metals, Fibres, and Dusts. Lyon: World Health Organization Press; 2012

[51] Program NT. Report on Carcinogens. 12th ed. Research Triangle Park: NTP; 2011

[52] Herselman JE. The concentration of selected trace metals in South African soils. Dissertation presented for the degree of Doctor of philosophy, Department of Soil Science, University of Stellenbosch; 2007

[53] Wuana RA, Okieimen FE. Heavy metals in contaminated soils: A review of sources, chemistry, risks and best available strategies for remediation. ISRN Ecology. 2011;**2011**:402647. DOI: 10.5402/2011/402647

[54] Martínez CE, Motto HL. Solubility of lead, zinc and copper added to mineral soils. Environmental Pollution. 2000;**107**(1):153-158

[55] Eriksson J, Andersson A, Andersson R. The state of Swedish farmlands. In: Swedish Environmental Protection Agency. Stockholm, Sweden: Tech. Rep; 1997. p. 4778

[56] Roa MS, Gopalkrishnan R, Venkatesh BR. Medical geology – An emerging field in environmental science. In: National Symposium on Role of Earth Sciences, Integrated and Related Societal Issues, Lucknow, India, 2–4 November, GSI Spl. Pub. No. 65. 2001. pp. 213-222

[57] Ghazban F, Parizanganeh A, Zamani A, Taghilou B. Assessment of heavy metal pollution in water and sediments from the Ghalechay River, Baychebagh copper mine area, Iran. Soil and Sediment Contamination. 2014; **24**(2):172-190

[58] Helena O. Chromium as an environmental pollutant: insights on induced plant toxicity. Journal of Botany. 2012;**2012**:375843. DOI: 10.1155/2012/375843

[59] López-Luna J, González-Chávez MC, Esparza-García FJ, Rodríguez-Vázquez R. Toxicity assessment of soil amended with tannery sludge, trivalent chromium and hexavalent chromium, using wheat, oat and sorghum plants. Journal of Hazardous Materials. 2009;**163**(2–3): 829-834

[60] Kanu S, Sudipta P. Metabolic adaptability in hexavalent chromium-treated renal tissue: An in vivo study. Clinical Kidney Journal. 2018;**11**(2): 222-229. DOI: 10.1093/ckj/sfx069

[61] Scragg A. Environmental Biotechnology. 2nd ed. Oxford, UK: Oxford University Press; 2006

[62] Baldwin DR, Marshall WJ. Heavy metal poisoning and its laboratory investigation. Annals of Clinical Biochemistry. 1999;**36**(3):267-300

[63] NSC. Lead Poisoning, National Safety Council; 2009

[64] Mcllveen WD, Negusanti JJ. Nickel in terrestrial environment. Science of the Total Environment. 1994;**148**: 109-138

[65] Cempel M, Nikel G. Nickel: A review of its sources and environmental toxicology. Polish Journal of

Environmental Studies. 2006;**15**(3): 375-382

[66] Tye AM, Young S, Crout NMJ, Zhang H, Preston S, Zhao FJ, et al. Speciation and solubility of Cu, Ni and Pb in contaminated soils. European Journal of Soil Science. 2004;**55**:579-590

[67] Nielsen FH. Interaction of nickel with essential minerals. In: Nriagu JO, editor. Nickel in the Environment. New York: John Wiley & Sons; 1980a. pp. 611-634

[68] Rahman SH, Khanam D, Adyel TM, Islam MS, Ahsan MA, Akbor MA. Assessment of heavy metal contamination of agricultural soil around Dhaka export processing zone (DEPZ), Bangladesh: Implication of seasonal variation and indices. Applied Sciences. 2012;**2**:584-601. DOI: 10.3390/app2030584

[69] Uren N. Forms, reactions and availability of Ni in soils. Advances in Agronomy. 1992;**48**:141-195

[70] Fasinu PS, Orisakwe OE. Heavy metal pollution in sub-Saharan Africa and possible implications in cancer epidemiology (a review). Asian Pacific Journal of Cancer Prevention. 2013; **14**(6):3393-3402

[71] Manahan SE. Toxicological Chemistry and Biochemistry. 3rd ed. Boca Raton, Florida: CRC Press, Limited Liability Company (LLC); 2003

Chapter 6

Trace Elements in Volcanic Environments and Human Health Effects

Diana Paula Silva Linhares, Patrícia Ventura Garcia and Armindo dos Santos Rodrigues

Abstract

Trace elements play an essential role in the normal metabolism and physiological functions of living beings. The distribution and concentration of trace elements in the environment results from both anthropogenic and natural origins; this chapter will focus on volcanism as one of the major natural sources of trace elements. In volcanic areas, the emissions and deposits of volcanogenic elements are key factors for geochemical mobility of trace elements and their distribution in the environment and, their effects on animals and human health. Volcanic areas have been associated with increased incidence of several diseases, such as fluorosis or even some types of cancer, leveraging the studies on the potential of this natural phenomenon as a promoter of diseases. As the Azores Archipelago is a volcanic area, with several manifestations of active volcanism, this region presents itself as an ideal study scenario for a multidisciplinary approach on environmental health problems, such as the exposure to toxic and/or deficient levels of trace elements. This chapter will present an integrated approach, describing the occurrence, the monitoring of trace elements and their characterization, the biological role in human body, and the human biomonitoring and health risk assessment, using case studies as examples.

Keywords: human health, biomonitoring, risk assessment, volcanism, trace elements

1. Introduction

Quality of life is based on an intricate relationship of various factors that include having sufficient nutrition, adequate accommodation and environment, social and psychological fulfillment, and health. Not neglecting the importance of these factors, environment stands out as it plays a crucial role in people's physical, mental, and social well-being.

Given the link between environment and health, as environmental chemicals affect not only the surroundings but also the quality of food and water, there has been a growing concern in the scientific community in the last decades, and consequently an increase in studies characterizing the environmental availability of elements, particularly in the soil. These recent studies have added substantial knowledge regarding elemental availability in soils, particularly for the biogeochemistry of trace elements [1]. The assessment of the concentrations of trace elements in soil is very important not only for environmental purposes, such as quantifying the contamination level, but also to help solve problems associated with elemental toxicity or deficiency in humans and plants.

Trace elements (TEs) are dietary minerals present in living tissues in small amounts; some of them are known to be nutritionally essential, playing a vital role in the normal metabolism and physiological functions of animals and humans [2]. The TEs' essentiality for the human body has been a matter of discussion throughout time and the term "trace elements" has never been clearly defined, being used both in geochemistry and biological sciences for chemical elements that occur in the Earth's crust in amounts less than 0.1% (1000 mg/kg) [1]. Despite their "low" content in the human body, TEs are components of a complex physiological system involved in the regulation of vital functions at all stages of development of the living organism [3].

Throughout time, limited attempts have been made for classifying trace elements. In 1973, WHO [4] classified 19 trace elements into three groups: (i) essential elements: zinc, copper, selenium, chromium, cobalt, iodine, manganese, and molybdenum; (ii) probably essential elements; and (iii) potentially toxic elements. Shortly after, Frieden [5] considered 29 types of elements present in the human body and classified them into five groups:

i. Group I: basic components of macromolecules such as carbohydrates, proteins, and lipids. Examples include carbon, hydrogen, oxygen, and nitrogen;

ii. Group II: nutritionally important minerals also referred to as principal or macroelements. The daily requirement of these macroelements for an adult person is above 100 mg/day. Examples include sodium, potassium, chloride, calcium, phosphorus, magnesium, and sulfur;

iii. Group III: essential trace elements. The trace elements are also called minor elements. An element is considered a trace element when its requirement per day is below 100 mg/day. The deficiency of these elements is rare but may prove fatal. Examples include copper, iron, zinc, chromium, cobalt, iodine, molybdenum, and selenium;

iv. Group IV: additional trace elements. Their role is yet unclear and they may be essential. Examples include cadmium, nickel, silica, tin, vanadium, and aluminum. This group may be equivalent to probably essential trace elements in the WHO classification;

v. Group V: these metals are not essential and their functions are not known. They may produce toxicity in excess amounts. Examples include gold, mercury, and lead. This group is equivalent to potentially toxic elements defined in the WHO classification.

More recently, Frieden [6] proposed a biological classification of trace elements based on their amount in tissues: (i) essential trace elements: boron, cobalt, copper, iodine, iron, manganese, molybdenum, and zinc; (ii) probably essential trace elements: chromium, fluorine, nickel, selenium, and vanadium; and (iii) physically promoter trace elements: bromine, lithium, silicon, tin, and titanium.

As TEs play a significant role in the regulation of many important adaptive mechanisms, including the functioning of all vital systems of the organism, the balance of each element in an optimum range of concentrations is fundamental. The chronic deficiency of essential TEs can, therefore, result in metabolic disturbances and distinct clinical and morphological changes; on the other hand, we must not disregard that all TEs can be toxic if consumed at high levels for long periods, disturbing the normal function of vital systems.

2. Trace elements in soil

The main advances in trace element research have been made in soil sciences since soils are considered the most important environmental compartment functioning as a sink for TEs [7–9]. Trace elements are usually distributed over different soil compartments and their retention will depend on several soil characteristics, as well as the parent rock material.

The main soil characteristics include pH, cation exchange capacity (CEC), particle size distribution, electrical conductivity, and organic matter content [10, 11]. These soil properties can promote the accumulation of TEs in soils or their depletion. The most adequate pH for the maximum TE availability is within 6.0–8.0; however, some TEs such as manganese, iron, boron, copper, and zinc are more available to plants when the soil is acidic (pH between 4.5 and 6.5), which contributes to manganese and boron toxicities in plants growing on acidifying soils [12].

The CEC is also a very important soil property as it can influence the soil structure stability, nutrient availability, soil pH, and the soil reaction to fertilizers and other ameliorants [13]. For example, negatively charged sites increase the CEC, holding H^+, Ca^{2+}, Mg^{2+}, Na^+, and NH^{4+}, while the positively charged sites increase the retention of OH^-, SO_4^-, NO_3^-, and PO_4^- [14]. All these soil properties, either combined or isolated, can promote the accumulation or the leakage of TEs in soils.

The parent rock material also assumes high importance in TE availability; when parent materials have high trace element concentrations, the resulting soils also have high or even higher TE concentrations, particularly when the former also result from anthropogenic activities, such as agriculture [15].

Considering that specific soil characteristics can affect the TE availability, the use of universal background concentrations is inadequate, as it may not reflect the "normal" values for specific regions. In this way, each country should determine the background levels for each region with different geological substrates and establish normative values for environmental legislation based on these studies, avoiding misinterpretation of abnormally low or high TE contents [16].

2.1 Measurement of trace element levels in soils

The soil background concentrations/levels will depend on the mineralogical composition of the parent rock material and on the weathering processes that have led to its formation, the granulometry fractions, and the organic matter content [17–19]. These background measurements, which represent natural concentrations in unpolluted pristine soils, are very difficult to assess because they require a soil free of contamination. Given this difficulty, the measurement usually applied is the geochemical baseline concentration that represents an expected range of element concentrations around medium normal sample mean [20]. Although the TE baseline concentration levels in the soil may differ between countries and/or geographical regions, their assessment has been recognized as the only means to establish reliable worldwide elemental concentrations in natural materials [21, 22].

The measurement of TE in soils requires well-planned sampling strategies to achieve accurate data. There are several defined protocols for soil sampling and many digestion techniques to optimize the TE quantification [23]. The conventional methods are based on a regular soil sampling design, with soil sample collection at a depth of 0–20 cm and subsequent chemical analysis of the sampled soils in the laboratory, followed by geostatistical interpolation of the data to obtain the spatial distributions of soil heavy metal content. For the assessment of TE in agricultural areas, the protocol of geochemical mapping of agricultural soils and grazing land of Europe (GEMAS) is the most used as the aim of the project is to provide harmonized geochemical data of arable land and of land under permanent grass cover at the continental European scale. The application of this protocol for meadows requires that all samples will be taken as rather large (2–2.5 kg) composite samples from one extensive field; the minimum size of field should be about 25 × 50 m [24]. The sampling stage is critical and it must take into account what we want to measure and the geological attributes of the site. Also, in order to avoid cross-contamination in the sampling of TEs, metal tools should not be used in the field or in the lab. The sample preparation and storage in the lab often require that the soil samples are air-dried and sieved to less than 2 mm [25]. Afterward, the total TE contents or the extractable fraction can be determined.

Given that the application of these methods has some disadvantages, since they are time-consuming and costly and cannot provide accurate estimates of soil heavy metal content over large areas, new approaches such as remote sensing are starting to be widely used as they can rapidly lead to spatially explicit estimates of soil heavy metal content and monitor their dynamics at a regional scale with low cost [26]. By capturing electromagnetic radiation reflected from the target, remote sensing can be used in the detection of heavy metals in soil and vegetation [27]. However, soil's properties cannot be easily assessed using hyperspectral sensing so the monitoring heavy metal contamination in soils has not been assessed comprehensively and it needs further studies [28].

The assessment of the total concentration of trace elements is required to: (i) determine the background (natural) TE levels in the soil; (ii) assess the total metal content; and (iii) evaluate if there has been TE accumulation over time [29, 30]. To assess the total concentration of TE, soils need to be digested to break down the primary silicate structures of the more resistant quartz and feldspar soil minerals and release the TE into solution. The most common types of digestion are carried with concentrated nitric acid and hydrogen peroxide or with a mixture of *aqua regia* concentrated nitric and hydrochloric acids.

Finally, the elemental concentrations of the digest solutions can be determined by spectroscopic methods, such as atomic absorption spectroscopy (AAS), inductively coupled plasma optical emission spectroscopy (ICP-OES), or by plasma mass spectrometry (ICP-MS) [31]. ICP-OES and ICP-MS have more advantages when compared with the AAS, as they allow one to obtain numerous data from running the sample just once and have very low detection limits [32]. While ICP-OES is based on the measurement of excited atoms and ions at the wavelength characteristic for the specific element being measured, ICP-MS measures an atom's mass by mass spectrometry (MS). These distinct approaches result in different lower detection limits; the lower detection limit in ICP-OES is in parts per billion (ppb) while in ICP-MS can be extended to parts per trillion (ppt) [33]. On the study of trace elements in environmental samples, ICP-OES is more commonly used since it may be applied for samples with high total dissolved solids or suspended solids and is, therefore, more robust for analyzing groundwater, wastewater, soil, and solid waste. It is therefore, usually used to measure contaminants for environmental safety assessment and elements with a higher regulatory limit [31]; if the trace elements in study have very low regulatory limits, ICP-MS is adequate for quantification.

2.2 Sources of trace elements

Trace elements can enter the soil by natural or anthropogenic sources [34], and their behavior and fate in soils differ according to their source and species.

The most important anthropogenic sources of trace elements for soils include mining and metallurgical activities, commercial fertilizers, biosolids, irrigation water, coal combustion residues, and auto emissions [34, 35]. Mining and metallurgical activities are recognized as the most important producers of waste and environmental pollution; the metalliferous mines, processing plants, and smelters generate huge amounts of mine tailings that can be transported mainly in the form of wastewater and airborne dust particles [36–38]. Using fertilizers in intensive agriculture can increase the TEs in the soil given that fertilizers contain trace amounts of several elements [15]. Sewage sludges and effluents also contain variable amounts of trace elements of various nature and various anthropogenic origins, although they can be an interesting way to enrich the soil with organic matter [39]. The direct use of sewage sludge can lead to heavy metal phytotoxicity problems due to the lack of stability of organic matter that could be obtained through an appropriate composting process. The TEs also enter the soil trough the water used for irrigation or even by the atmospheric deposition from industrial and urban combustion emissions. The TEs can travel long distances in either gaseous form or in particle phase before deposition; therefore, the contaminated zones can extend up to a great distance from the contamination site. This is easily observed with the automobile exhaust emissions that contaminate not only the roadside soils but, depending on the location, traffic intensity and predominant wind conditions (direction and frequency); the contamination can be observed in soils hundred meters apart from the road [40].

There are also several natural sources (and processes) that contribute to TEs' deposition in soils. The main natural source is the soil parent material as the weathering of rocks and mineral deposits produces metals in dust, sediments, groundwaters, and surface waters [17]. Forest fires with the release of metal and mineral particulate matter in ash, gas, and aerosols or even sea spray with metals in the water droplets can also contribute to the deposition of TEs in soils, not only in the vicinity of their sources but also in areas far away. Another natural factor of major importance is geological activities; they are rather volcanic activities, earthquakes, landslides, debris flows, etc. all of which introduce major and trace elements into the environment [41]. The concentration of metallic TEs is the most affected by volcanic activity [42], resulting in high baseline concentrations of some metals in volcanic soils [15, 43].

The excess of metals in soils may affect the surrounding ecology and human health as these elements are non-biodegradable and therefore environmentally persistent [44]. While some of these trace elements are essential for plant growth and development, the monitoring of TE baseline values in soils is fundamental since at elevated levels they can become toxic. The toxicity of TEs will depend on the dose, exposure pathway, and duration of exposure [45, 46]. Regarding the bioavailability of TE in soils, the context is much wider as it includes chemical availability to a variety of biota [47]. Even with no universal definition, it is assumed that for the study of the environmental risk assessment of TEs in soils, the bioavailability assessment should include their soluble and solid-phase-associated labile fractions. Although the dissolution-desorption, transport, and uptake are very complex processes, the Committee on Bioavailability of Contaminants in Soils and Sediments [48] considered that bioavailability is the fraction of the total concentration of contaminants present in soil (solid and solution phases) which is potentially available for plant uptake or absorption by soil-dwelling organisms.

Considering that the excess of TEs in soil due to volcanic activity cannot be controlled, and that TE ingestion, inhalation, or dermal contact can cause damage to several human systems, the monitoring and baseline determination of TEs in volcanic environments assume particular importance for the inhabitants of these areas.

2.3 Trace elements in volcanic soils

There are several hazards that result from a volcanic eruption; the most immediate and threatening hazards are pyroclastic falls, pyroclastic density currents, lava (flows and domes), lahars and flooding, debris avalanches, and volcanic gases. Nevertheless, volcanic emissions also occur in the post-eruptive phase and in quiescent volcanoes, continuing to affect the ecosystem and consequently human health.

Volcanic regions step-up important scenarios for the study of TEs in soils because: (i) they are densely inhabited in some areas of the Earth and (ii) due to the physicochemical properties, volcanic soils retain TEs, acting as a reservoir and affecting agriculture [15].

2.4 Human health effects

The concern regarding the health effects of environmental exposure to TEs from natural sources has driven the development of tools and methods for assessing the impact of emissions in water, soil, and air. Biological monitoring or biomonitoring is the most commonly applied method to measure human exposure to xenobiotics [49]. There are several studies worldwide that establish the association between concentrations of TE in volcanic soils and its effects on human health.

One of the most well-detailed health problem associated with volcanic activity is fluorosis, which results in high fluoride (F) concentrations in groundwater. The problem was first recognized in Japan and was called "Aso volcano disease" [50], but during the course of the year, high fluoride concentrations (greater than the WHO guideline value of 1.5 mg/l) were also found in Africa, where the crater lakes of western Uganda often have high F concentrations (e.g.4.5 mg/l F in Lake Kikorongo) [51]. High concentrations of F were also found in eastern Turkey, near the Tendurek Volcano, where the natural waters contained fluoride levels between 2.5 and 12.5 ppm [52], and in oceanic islands, such as Tenerife Island (Canary-Spain) [53] and São Miguel Island (Azores-Portugal) [54–56].

Another health problem that has been proven to be strongly associated with the exposure to volcanic environments is thyroid cancer, in particular papillary thyroid cancer (PTC). In 2009, Pellegriti et al. [57] in a register-based epidemiologic study showed that the residents of Catania, a province in the vicinity of Mt. Etna, presented a higher incidence of papillary thyroid cancer than elsewhere in Sicily. More recently, these results were reinforced in the study conducted by Malandrino et al. [58]. In this study, the authors evaluated the environmental pollution and bio-contamination in a volcanic area of Sicily and compared the data with the thyroid cancer epidemiology data obtained from the Sicilian Regional Registry for Thyroid Cancer. Their results indicated that the residents in Mt. Etna volcanic area had significantly higher levels of several TEs in their urine when compared to the control area [Cd (×2.1), Hg (×2.6), Mn (×3.0), Pa (×9.0), Th (×2.0), V (×8.0)] and that thyroid cancer incidence was 18.5 and 9.6 per 105 inhabitants in the volcanic and the control areas, respectively; the observed thyroid cancer incidence was exclusively from the papillary histotype.

Besides these, that pose as the most studied TE and linkages to human health effects, there are several other health issues that have an increased risk due to environmental exposure. In Sicily, Italy, various soil types developed from different parent materials were analyzed to compare heavy metal distribution under different geopedological conditions, evidencing that the former depended on the parent rocks [59]. In Turkey, an association was established between the volcanic soil and the high prevalence of upper gastrointestinal cancer rates in the Van region as the fruit and vegetable samples produced in those soils contained potentially carcinogenic levels of heavy metals [60]. More recently, Rodriguez-Espinosa and co-authors [61] analyzed the elemental composition of 25 soils and ash samples after the eruptions of the Popocatépetl Volcano in México, observing that the concentrations of TEs such as Zn, Pb, Ni, Hg, Cr, Cd, Cu, and As were significantly higher than those observed in older samples from eruptions in 1997, suggesting that the naturally highly volatile and mobile metals leach into nearby freshwater sources. In the Azorean volcanic islands, there are also some studies focusing in the concentrations of TEs in soils [15, 62] and its effects on local organisms [39, 63, 64] and human health [50, 53], the latter being particularly focused on fluoride [65–67].

In 2016, Linhares and co-authors [65] verified that there are areas in São Miguel Island-Azores that even with modern water treatment systems present fluoride concentrations slightly above the WHO recommendations [68]. Considering that the main sources of human exposure to fluoride are diet, especially through the ingestion of water and, that in volcanic regions fluoride is continuously released into the environment, these authors developed a biomonitoring study to investigate the feasibility of urine and nail clippings as biomarkers of exposure. Nail clippings revealed to be a more reliable biomarker of chronic exposure to fluoride than urine for populations of different age classes (children vs. adults), with a positive correlation between the fluoride daily intake and fluoride content in nail clippings in children ($r_s = 0.475$; $p < 0.001$), and in adults ($r_s = 0.495$, $p < 0.001$). More recently, Linhares et al. [66] assessed the risk of skeletal fluorosis from environmental exposure to fluoride in hydrothermal areas, using wild mice (*Mus musculus*) as bioindicator species. *Mus musculus* were collected in Furnas village (a village located inside the caldera of Furnas volcano), an area where volcanic activity is marked by active fumarolic fields, hot and cold CO_2-rich springs, and soil diffuse degassing phenomena [69, 70]. The results demonstrated that mice from Furnas village had higher concentrations of fluoride in bones when compared with mice from an area without volcanic activity (616.5 ± 129.3 mg F/g vs. 253.8 ± 10.5 mg F/g, respectively), reinforcing that chronic exposure to fluoride may lead to the development of not only dental fluorosis but also of skeletal fluorosis.

3. Azores as a volcanic scenario

The Azores archipelago is located in the North Atlantic Ocean, in the triple junction of the North American, African, and Eurasian plates [71–73]. The archipelago is formed by nine islands of volcanic origin that represent the emerged part of the Azores Plateau, a thick and irregular area of the oceanic crust roughly limited by the 2000-m bathymetric curve [74]. As a result of the Azores archipelago's location on an active plate boundary, frequent seismic and volcanic activity occurs, including volcanic eruptions and secondary manifestations of volcanism, such as hydrothermal vents and soil degassing processes.

São Miguel, the largest island of the archipelago, is formed by five active volcanic systems, including three central active volcanoes (Sete Cidades, Fogo and Furnas), separated by two fissure systems (Picos and Congro), and two extinct volcanic systems (Povoação and Nordeste) [75, 76].

Since the volcanic activity on the island contributes to a distinct soil elemental profile, resulting in a higher baseline for elements, the study of the baseline levels of TE is fundamental. When bioavailable to plants, animals, and humans, TEs can cause several diseases due to their elevated concentrations in soils. In addition, the high concentrations of some TEs in soils can inhibit the bioavailability of other elements, promoting the deficiency in elements that can be essential for plants, animals, and human health.

Considering that in this island, volcanic activity is usually in a quiescent phase, it presents itself as an ideal study scenario for an approach on environmental health problems, such as the exposure to toxic levels of TE and/or their deficiency.

3.1 Distribution of TEs in São Miguel island soils

In São Miguel island, with five active volcanic systems, all the soils have TE inputs resulting from the volcanic activity along the island. Nevertheless, in an island where the main income is agriculture, with the production of dairy, meat, and horticulture, little is known about the TE profile in these young volcanic soils.

In 2006, Amaral et al. [62] undertook a study to determine some baseline levels of trace elements in soils with different ages from active (Furnas; S. Miguel Island) and inactive volcanic sites (Santa Maria Island) of the Azores archipelago. These authors observed that, except for SiO_2, Na_2O, K_2O, and Zn, the concentrations of major and trace elements were higher in Santa Maria soils. The authors point that these differences may be related to the higher capability of Santa Maria soils to retain the elements, given that these soils are richer in fine grain size particles, to retain those elements. In the study by Amaral et al. [63], using the earthworm *Lumbricus terrestris* as a model, the authors found that even though the volcanic site showed lower levels for most of the analyzed metals, the earthworms presented higher concentrations of the same TEs than those from the site without volcanic activity. These earthworms, with higher levels of trace metals, responded to this environment with higher bioavailability of TEs with a reduction of the thickness of the chloragogenous tissue and intestinal epithelium [63]. The higher bioavailability of TEs in these soils can be explained by the lower pH and clay content, as the authors suggested. Later on, a higher risk for uptake of potentially toxic metals in the active volcanic area was observed by Amaral et al. [77] when studying the scalp hair of men living in Furnas and in Santa Maria Island. The authors found that the scalp hair of men from Furnas had higher concentrations of Cd (96.9 ppb), Cu (16.2 ppm), Pb (3417.6 ppb), Rb (216.3 ppb), and Zn (242.8 ppm) when compared with men living in Santa Maria Island.

More recently, Parelho et al. [15] collected and analyzed soil samples from the farms of the main producers of vegetables in São Miguel island; these farms were located in the Picos Fissural Volcanic System, in the western half of the island. Results revealed that the TE background values fitted in the average values for European volcanic soils. However, this work showed that in addition to agricultural input there are elements of volcanogenic origin and that these specific soils tend to accumulate some trace metals due to their physicochemical properties.

Although these studies gather some information regarding the TE profile in the island volcanic soils, they were limited to areas of island without active volcanism and, therefore, TE contents may be even higher in the soils from where active manifestations of volcanism occur. Lately, there have been some studies that focused on the distribution of several TEs in all the volcanic complexes of São Miguel Island, evidencing a depletion of some TEs in the soils, such as iodine [55] and cobalt [78] and elevated concentrations of others, such as manganese [78].

3.2 Iodine

Iodine is a vital micronutrient required at all stages of life, with the fetal stage and early childhood being the most critical phases of requirement [79]. The connection between geological materials and TE deficiency is well documented for iodine since an inadequate intake of iodine results in disease conditions collectively known as Iodine Deficiency Disorders (IDD) [80]. The iodine overload is less common, but can cause thyrotoxicosis as hyperthyroidism, chronic thyroiditis, Hashimoto's thyroiditis, and even may increase the risk of thyroid gland cancer [81–83].

Since the 1980s, the existence of health problems associated with iodine deficiency has been acknowledged. In 1986, Oliveira et al. [84] made a survey for endemic goiter on the island of São Miguel-Azores and observed that the median iodine intake ranged from 10 to 49 mg iodine/g creatinine, with a goiter prevalence usually greater than 20%. Later on, the studies by Limbert et al. [85, 86] established that urinary iodine deficiency was not only observed in children, but also in pregnant women. The most noteworthy fact of these studies was that the deficient iodine intake was not the same in all the islands of the Azores, with a positive highlight for the population of the island of Santa Maria, with mild deficiencies in iodine intake and, a negative one for the population of São Miguel, with severe deficiencies in iodine intake.

Since it is recognized that the ocean is the main reservoir of iodine and that the Azorean islands have geographical and climatic features that are clearly oceanic, Linhares et al. [55] investigated the environmental availability of iodine and bioavailability to human populations, especially in children at school age. This study reinforced the observations obtained in the previous studies [84, 86], revealing a deficient intake of iodine in the resident population of São Miguel Island, but it went further in the establishment of the causes. In this study, Linhares and the co-authors observed that the environmental availability of iodine was different in the soils from both islands, being significantly higher in the soils of Santa Maria than in São Miguel (58.12 ppm ± 40.94 vs. 14.53 ppm ± 11.79, respectively). The volcanic activity of São Miguel island; the islands' geomorphology; and consequently climate characteristics, such as orography and rainfall, are the main causes for the lower content of iodine in its soils. It must also be taken into account that the iodine soil content results from the complex dynamic balance of three processes: incorporation from the atmosphere, fixation, and volatilization. Soil characteristics, such as soil organic and inorganic components and the clay fraction, can affect iodine fixation. Higher concentrations of the organic and inorganic components and a higher clay content, as observed in Santa Maria island, provide a strong fixation of iodine in the soil, reducing the volatilization; therefore, in more mature volcanic soils (like those from Santa Maria) higher deposits of iodine are expected.

The outcome of this last study reinforces the risk of iodine deficiency in São Miguel Island, emphasizing the necessity of introducing an iodine supplementation program in the population of this island, to overcome the low environmental availability of this halogen and its continued vigilance by periodic urinary iodine surveys.

3.3 Cobalt

Cobalt is usually found in the environment combined with other elements such as oxygen, sulfur, and arsenic. Small amounts of these chemical compounds can be found in rocks, soil, plants, and animals. The concentrations of cobalt in soil range from about 1 to 40 ppm and the amount of cobalt in the air is less than 2 ng/m^3. This specific TE has some notorious differences when compared with the remaining TEs; whereas the other elements are required in ionic form and are then converted into their metabolically active species, the body requires Co in a preformed compound, vitamin B_{12}. The ability to synthesize vitamin B_{12} is only found in some bacteria, algae, and in some ruminants. Grazing cattle can synthesize vitamin B_{12} in the rumen, but in humans the main source of vitamin B_{12} is animal-related foods.

There are reports of health problems related to B12 deficiency, such as pernicious anemia and nerve damage [87, 88], and even psychical disorders, like impaired memory, irritability, depression, dementia, and psychosis [89].

Like iodine, cobalt deficiency has been long ago identified in São Miguel Island, particularly in grazing ruminants [90, 91]. However, it has been a subject of more interest only a few years ago, when Pinto [92] verified that approximately 40% of the dairy cattle in his study had deficient Co intake. These findings are extremely important for volcanic regions like the Azores, where livestock industries, such as dairy cattle farms, rely mostly on pasture grazing, where cattle are raised outdoors with an almost 100% natural diet of grass available in pastures.

To better understand the distribution of Co and its baseline levels on the island of São Miguel, Linhares et al. [78] collected soil samples from grazing sites through the island and observed a distinct pattern in the distribution. The highest concentrations of Co in soils were observed in the volcanic regions of Nordeste and Picos. These differences are related to the volcanic bedrock characteristics of the island; the Co content in the parental volcanic rocks is higher (20–58 mg/kg Co) in the low-silica rocks of Nordeste and Picos when compared to the high-silica rocks of the other volcanic regions. The differences within the volcanic regions of the island lie in the pedogenesis of parental volcanic materials; distinct geochemical compositions related to different degrees of magmatic evolution at depth result in different types of magma: (i) low-silica magmas (basalts and trachybasalts) as in Nordeste and Picos volcanic regions, that have high concentrations of iron, magnesium, chromium, nickel, and cobalt and; (ii) high-silica magmas (trachytes and rhyolites), such as Sete Cidades, Fogo/ Congro, and Povoação, that have low concentrations of these elements [93–95]. Linked to this, the existence of other TEs in the soils can also restrict the Co availability, as it happens with manganese (Mn) when present in high concentrations in the soils of São Miguel Island.

The study by Linhares et al. [78] revealed that the soils' volcanic origin (related to the parent rocks) and soil-forming processes affect the Co availability and, therefore, it is expected that severe Co deficiency can occur in most animals, especially the ones grazing in areas such as Furnas/Congro and Povoação.

The human dietary cobalt deficiency is unusual in individuals that consume animal-related food, fish, nuts, leafy green vegetables, such as broccoli and spinach, and cereals, including oats, since these are good food sources of cobalt [96].

In São Miguel Island, the lack of Co in soil assumes particular importance as the basis of the population feeding relies on locally produced agricultural products, and on meat and dairy products from grazing ruminants that are mainly fed with the available pasture grass in the grazing sites. Therefore, it is expected that, as observed in the ruminants, there might be some defined populated areas where the residents will have a deficient intake of Co and consequently may be more prone to develop several health problems associated with the lack of Co availability.

3.4 Manganese

Manganese (Mn) is a bioelement that has a cofactor function in the enzymatic processes [97]. It takes part in the functioning of antioxidant, musculoskeletal, immune, and reproductive systems and in detoxification processes [98]; nevertheless, excessive quantities of Mn cause toxic effects, especially in the central nervous system resulting in neurological diseases [99]. Manganese is ubiquitous in the environment, and human exposure arises from both natural and anthropogenic activities.

The Mn concentrations in soils strongly depend on the parent rock composition; the Mn contents in rocks can go from 174 mg/kg in sandstones to 1300 mg/kg in basalts, with an overall mean of 733 mg/kg Mn in the upper continental crust [100]. The Mn oxides in soils have very high sorption ability and they can accumulate ions from the soil solution; these oxides have a strong affinity for Co ions, which can reduce Co availability to plants.

The human exposure to Mn occurs mainly by ingestion; this TE is naturally present in a wide variety of foods, including whole grains, clams, oysters, mussels, nuts, soybeans and other legumes, rice, leafy vegetables, coffee, tea, and many spices [101, 102]. Manganese is absorbed in the small intestine and, after absorption, some Mn remains free, but most of it is bound to transferrin, albumin, and plasma alpha-2-macroglobulin. Mn deficiency is very rarely observed in humans but cases of Mn toxicity have been reported. Mn toxicity can be related to the dietary Mn intake and to chronic environmental exposure in welding and mining sites, as the inhalation of Mn dust can be toxic [103]. Mn toxicity affects the central nervous system and can cause tremors, muscle spasms, tinnitus, and hearing loss [101, 102]. Mn toxicity can also cause "manganism," a neurodegenerative disease with symptoms that resemble Parkinson's disease [104] and "Machado Joseph Disease (MJD)," a progressive spinocerebellar ataxic disorder [105].

In the Azores, there are few studies focusing on Mn availability and its effects. The existing studies focus on the assessment of Mn concentration in hydrothermal vents, as tracer of hydrothermal activity intimately related to mid-ocean ridge processes [106] and the Mn bioaccumulation in marine species, such as *Cystoseira abies-marina* [107]. Regarding the assessment of Mn in soils, the most recent studies are orientated to vineyard soils in the islands of Terceira, Graciosa, and Pico. Lima et al. [108] revealed that the Mn concentration in the soils of cultivated vines was 692.5 mg/kg in Pico, 1023.8 mg/Kg in Terceira, and 2041.6 mg/kg in Graciosa, evidencing significant differences between these islands.

More recently, in a survey to access cobalt concentration in volcanic soils to predict the risk of cobalt deficiency [78], the concentration of Mn was also determined as it can affect Co bioavailability. These authors verified an uneven distribution of Mn in the defined volcanic regions of the island; the highest Mn concentrations were observed in Nordeste and Picos (1782.50 mg/kg ± 108.98 and 1461.11 mg/kg ± 63.93, respectively), while the lowest concentrations were observed in the soils of Povoação and Furnas/ Congro (874.88 mg/kg ± 78.52 and 746.25 mg/kg ± 209.07, respectively). As observed for Co, the Mn concentration in soil is strongly associated with their content in the parental volcanic rocks. Therefore, low-silica magmas (basalts) of Nordeste and Picos have higher contents of Mn when compared to Furnas/ Congro and Povoação parent volcanic rocks formed by high-silica magmas (trachytes).

The most noteworthy aspect of the Mn concentration in the volcanic soils of São Miguel island is that all the volcanic regions have concentrations above the estimated background mean of the European soils (524 mg/kg) [24] and that in most grazing sites the measured concentrations were higher than 900 mg/kg (upper limit threshold) [Linhares et al. (unpublished data)]. Considering that high concentrations of Mn can be associated with the Machado Joseph disease [105, 108], this scenario undertakes a singular meaning in the Azorean context. There are two major ancestral origins: (1) a worldwide-spread haplotype, TTACAC or the Joseph lineage, and (2) a more recent one, GTGGCA or the Machado lineage, seen mostly in Portuguese people [109, 110], being associated with families with MJD from the Portuguese Azorean islands of Flores and São Miguel (respectively, the birthplace of the Joseph and the Machado kindreds). Nowadays, the Azorean group remains the most important cluster of this disease, with 32 extended MJD families with origins in Flores, S. Miguel, Terceira, and Graciosa islands [111]. In Flores Island, the prevalence of the disease has been decreasing through the years but MJD still reaches its highest worldwide value (1:239), constituting a public health problem [112].

In 2004, Purdey [105] established a common abnormal hallmark of high manganese (Mn)/low magnesium (Mg) status and suggested that this aberrant mineral ratio inactivates the Mn/Mg catalyzed endonuclease-1 enzyme. The high Mn/low Mg rate observed in all volcanic regions of São Miguel Island reinforces the need for further studies in these elements as they are intimately related to MJD.

These studies evidence that the Azores archipelago presents itself as an ideal scenario for the study of TE availability and possible health effects. However, the total TE concentration in soil is a relatively weak measure of their bioavailability. Given that the bioavailability depends on specific soil characteristics, such as organic matter content and pH, similar concentrations of TEs in different soils may not have the same bioavailability.

The assessment of TE bioavailability is fundamental, as the bioavailable fraction of trace elements is the fraction most likely to harm plants and animals. Consequently, the impact of TEs on soil and the surrounding environment cannot be predicted by measuring the total concentration of elements *per se*, since only the soluble and mobile fraction has the potential to leach or to be taken up by plants and enter the food chain [113]. Future studies should consider the assessment of the bioavailable part of the TEs in volcanic environments, to define remediation strategies in order to prevent health problems associated with TE depletion or excessive intake.

4. Conclusions

The assessment of the concentrations of TE in the soil is very important, not only for environmental purposes but also to help solve problems associated with human health and plant toxicity. Trace elements' profiles in soil result essentially from the weathering of geologic parent materials since their concentrations are directly linked to the parent material based on their immobile nature.

Given the volcanic origin of the Azorean soils, the natural enrichment of some elements, such as manganese, and an uneven and reduced distribution of others, such as cobalt and iodine, can contribute to the development of health problems in plants, animals, and humans. Due to the natural enrichment of some chemical elements in volcanic soils, resulting from the volcanic activity, which cannot be controlled, a very tight control of the possible sources of anthropogenic contamination is crucial to prevent the occurrence of toxic levels that prejudice plants, animals, and human's health. Regardless of substantial progress in the study of soil TEs, the application of critical exposure concentrations and the associated health risks are yet scarce and not fully clarified. Although the data obtained for the Azorean soils pinpoints to possible toxicity of manganese and deficiency of cobalt in some areas, more thorough studies, such as the ones developed for iodine, are required. Only with further information, focusing primarily on the bioavailability and bioaccessibility of the trace elements, it will be possible to adequately predict the health risks of exposure to soil TEs, which is particularly relevant in the areas where the environmental risk is greater, such as the volcanic environments.

Author details

Diana Paula Silva Linhares[1*], Patrícia Ventura Garcia[2,3]
and Armindo dos Santos Rodrigues[1,2]

1 Institute of Volcanology and Risks Assessment (IVAR), University of the Azores, Ponta Delgada, Portugal

2 Faculty of Sciences and Technology, University of the Azores, Ponta Delgada, Portugal

3 Centre for Ecology, Evolution and Environmental Changes (cE3c), and Azorean Biodiversity Group, University of the Azores, Ponta Delgada, Portugal

*Address all correspondence to: diana.ps.linhares@uac.pt

IntechOpen

© 2020 The Author(s). Licensee IntechOpen. This chapter is distributed under the terms of the Creative Commons Attribution License (http://creativecommons.org/licenses/by/3.0), which permits unrestricted use, distribution, and reproduction in any medium, provided the original work is properly cited.

References

[1] Kabata-Pendias A, Mukherjee AB. Trace elements of group 12. In: Trace Elements from Soil to Human. Berlin: Springer; 2007. pp. 283-319

[2] Prashanth L, Kattapagari KK, Chitturi RT, Baddam VR, Prasad LK. A review on role of essential trace elements in health and disease. Journal of Dr. NTR University of Health Sciences. 2015;**4**:75-85

[3] Skalnaya GM, Skalny AV. Essential Trace Elements in Human Health: A Physician's View. Tomsk: Publishing house of Tomsk State University; 2018. 224 p

[4] World Health Organization (WHO). Trace-Elements in Human Nutrition. Report of a WHO Expert Committee. Geneva, Switzerland: World Health Organization; 1973. WHO Technical Report Series, No. 532

[5] Frieden E. The evolution of metals as essential elements (with special reference to iron and copper). In: Friedman M, editor. Protein-Metal Interactions. Vol. 48. New York, NY, USA: Springer; 1974. pp. 1-31. (advances in experimental medicine and biology)

[6] Frieden E. New perspectives on the essential trace elements. Journal of Chemical Education. 1985;**62**:917-923

[7] Senesil GS, Baldassarre G, Senesi N, Radina B. Trace element inputs into soils by anthropogenic activities and implications for human health. Chemosphere. 1999;**39**(2):343-377

[8] Wuana RA, Okieimen FE. Heavy metals in contaminated soils: A review of sources, chemistry, risks and best available strategies for remediation. ISRN Ecology. 2011: 20 p. Article ID: 402647

[9] Gupta N, Yadav KK, Kumar V, Kumar S, Chadd RP, Kumar A. Trace elements in soil-vegetables interface: Translocation, bioaccumulation, toxicity and amelioration—A review. Science of the Total Environment. 2019;**651**(2):2927-2942

[10] Adugna A, Abegaz A. Effects of soil depth on the dynamics of selected soil properties among the highlands resources of northeast Wollega, Ethiopia: Are these sign of degradation? Solid Earth Discussions. 2015;**7**:2011-2035

[11] Ramos TB, Horta A, Gonçalves MC, Pires F, Duffy D, Martins JC. The INFOSOLO database as a first step towards the development of a soil information system in Portugal. Catena. 2017;**158**:390-412

[12] Kidd PS, Proctor J. Why plants grow poorly on very acidic soils: Are ecologists missing the obvious? Journal of Experimental Botany. 2001;**52**:791-799

[13] Hazelton PA, Murphy BW. Interpreting Soil Test Results What Do all the Numbers Mean? Melbourne: CSIRO Publishing; 2007

[14] Mukhopadhyay S, Masto RE, Tripathi RC, Srivastava NK. Application of soil quality indicators for the Phytorestoration of mine spoil dumps. In: Vimal Chandra Pandey Kuldeep Bauddh, editor. Phytomanagement of Polluted Sites, Market Opportunities in Sustainable Phytoremediation. Elsevier. 2019. pp. 361-388

[15] Parelho C, Rodrigues AS, Cruz JV, Garcia P. Linking trace metals and agricultural land use in volcanic soils—A multivariate approach. Science of the Total Environment. 2014;**496**:241-247

[16] Baize D, Sterckeman T. Of the necessity of knowledge of the natural pedo-geochemical background content in the evaluation of the contamination

of soils by trace elements. Science of the Total Environment. 2001;**264**(1-2):127-139

[17] Sayyed MRG. Lithological control on the mobility of elements during chemical weathering. Comunicações Geológicas. 2014;**101**(1):63-69. ISSN: 0873-948X; e-ISSN: 1647-581X

[18] Santos-Francés F, Martinez-Graña A, Alonso Rojo P, García SA. Geochemical background and baseline values determination and spatial distribution of heavy metal pollution in soils of the Andes Mountain range (Cajamarca-Huancavelica, Peru). International Journal of Environmental Research and Public Health. 2017;**14**(8):859

[19] Ferreira EP, Coelho RM, Valladares GS, Dias LMS, Assis ACC, Silva RC, et al. Mineralogy and concentration of potentially toxic elements in soils of the São Francisco Sedimentary Basin. Revista Brasileira de Ciências do Solo. 2018;**42**:e0170088

[20] Bech J, Tume P, Longan L, Reverter F. Baseline concentrations of trace elements in surface soils of the Torrelles and Sant Climent municipal districts (Cataolini, Spain). Environmental Monitoring and Assessment. 2005;**108**:309-322

[21] Kabata-Pendias A, Pendias H. Trace Elements in Soils and Plants. 2nd ed. Boca Raton: CRC Press; 1992

[22] Tack FMG, Verloo MG, Vanmechelen L, Van Ranst E. Baseline concentration levels of trace elements as a function of clay and organic carbon contents in soils in Flanders (Belgium). Science of the Total Environment. 1997;**201**(2):113-123

[23] Albright E. Background concentrations of trace elements in soil and rocks of the Georgia Piedmont [MSc. thesis].University of Georgia; 2004

[24] EuroGeoSurveys Geochemical Mapping of Agricultural and Grazing Land Soil of Europe (GEMAS) - Field Manual. NGU Report 2008.038. 2008

[25] Lawrence GB, Fernandez IJ, Hazlett PW, Bailey SW, Ross DS, Villars TR, et al. Methods of soil resampling to monitor changes in the chemical concentrations of forest soils. Journal of visualized experiments. 2016;**117**:54815

[26] Liu Z, Lu Y, Peng Y, Zhao L, Wang G, Hu Y. Estimation of soil heavy metal content using hyperspectral data. Remote Sensing. 2019;**11**(12):1464

[27] Liu M, Liu X, Wu L, Duan L, Zhong B. Wavelet-based detection of crop zinc stress assessment using hyperspectral reflectance. Computers and Geosciences. 2011;**37**:1254-1263

[28] Wang F, Gao J, Zha Y. Hyperspectral sensing of heavy metals in soil and vegetation: Feasibility and challenges. Journal of Photogrammetry and Remote Sensing. 2018;**136**:73-84

[29] Salminen R, Plant JA, Reeder S, Salminen R. Geochemical atlas of Europe. Part 1. Background information, methodology and maps. Espoo: Geological Survey of Finland; 2005. pp. 67-79

[30] Reimann C, Flem B, Fabian K, Birke M, Ladenberger A, Négrel P, et al. Lead and lead isotopes in agricultural soils of Europe—The continental perspective. Applied Geochemistry. 2012;**27**(3):532-542

[31] Rutzke MA. Atomic absorption, inductively coupled plasma optical emission spectroscopy, and infrared spectroscopy. In: White WM, editor. Encyclopedia of Geochemistry. Encyclopedia of Earth Sciences Series. Cham: Springer; 2018

[32] National Research Council (NRC). Commitee on Scientific Assessment

of Bullet Lead Elemental Composition Comparison; Forensic Analysis: Weighing Bullet Lead Evidence. Washington, USA: National Academies Press; 2004

[33] Baralkiewicz D, Gramowska H, Hanć A, Krzyzaniak I. A comparison of ICP-OES and ICP-MS in the determination of elements in lake water. Atomic Spectroscopy -Norwalk Connecticut. 2007;**28**(5):164-170

[34] Marchant BP, Saby NPA, Arrouays D. A survey of topsoil arsenic and mercury concentrations across France. Chemosphere. 2017;**181**:635-644

[35] Adriano DC. Trace Elements in the Terrestrial Environment. New York: Springer-Vedag; 1986

[36] Lim HS, Lee JS, Chon HT, Sager M. Heavy metal contamination and health risk assessment in the vicinity of the abandoned Songcheon Au–Ag mine in Korea. Journal of Geochemical Exploration. 2008;**96**:223-230

[37] Li Z, Ma Z, Van der Kuijp TJ, Yuan Z, Huang L. A review of soil heavy metal pollution from mines in China: Pollution and health risk assessment. Science of the Total Environment. 2014;**468-469**:843-853

[38] Cai LM, Xu ZC, Qi JY, Feng ZZ, Xiang TS. Assessment of exposure to heavy metals and health risks among residents near Tonglushan mine in Hubei, China. Chemosphere. 2015;**2015**(127):127-135

[39] Parelho C, Rodrigues AS, Bernardo F, Barreto MC, Cunha L, Poeta P, et al. Biological endpoints in earthworms (*Amynthas gracilis*) as tools for the ecotoxicity assessment of soils from livestock production systems. Ecological Indicators. 2018;**95**:984-990

[40] Werkenthin M, Kluge B, Wessolek G. Metals in European roadside soils and soil solution—A review. Environmental Pollution. 2014;**189**:98-110

[41] Yu S, Chen Z, Zhao K, Ye Z, Zhang L, Dong J, et al. Spatial patterns of potentially hazardous metals in soils of Lin'an City, southeastern China. International Journal of Environmental Research and Public Health. 2019;**16**:246

[42] Adumitroaei MV, Iancu GO, Rățoi BG, Doru CS, Sandu CM. Spatial distribution and geochemistry of major and trace elements from Mohoș peatland, Harghita Mountains, Romania. Holocene. 2018;**28**:1936-1947

[43] Memoli V, Eymar E, García-Delgado C, Esposito F, Panico SC, Marco AD, et al. Soil element fractions affect phytotoxicity, microbial biomass and activity in volcanic areas. Science of the Total Environment. 2018;**636**:1099-1108

[44] Peña-Fernández A, González-Muñoz MJ, Lobo-Bedmar MC. Establishing the importance of human health risk assessment for metals and metalloids in urban environments. Environment International. 2014;**72**:176-185

[45] Tchounwou PB, Yedjou CG, Patlolla AK, Sutton DJ. Heavy metal toxicity and the environment. Experientia. Supplementum. 2012;**101**:133-164

[46] Tran THM, Nguyen KG. Metal and metalloid concentrations in soil, surface water, and vegetables and the potential ecological and human health risks in the northeastern area of Hanoi, Vietnam. Environmental Monitoring and Assessment. 2018;**190**:624

[47] Naidu R, Bolan NS, Megharaj M, et al., editors. Chemical bioavailability. In: Terrestrial Environments, Development in Soil Science. Vol. 32. London: Elsevier; 2008

[48] National Research Council (NRC). Bioavailability of Contaminants in Soils and Sediments: Processes, Tools and Applications. Washington DC: National Academic Press; 2002

[49] Dongarrà G, Varrica E, Tamburo D, D'Andrea D. Trace elements in scalp hair of children living in differing environmental contexts in Sicily (Italy). Environmental Toxicology and Pharmacology. 2012;**34**(2):160-169

[50] Kawahara S. Odontological observations of Mt. Aso-volcano disease. Fluoride. 1971;**4**:172-175

[51] Mugoma S. The alkaline, saline lakes of Uganda: A review. Hydrobiologia. 1990;**208**(1-2):75-80

[52] Oruc N. Occurrence and problems of high fluoride waters in Turkey: An overview. Environmental Geochemistry and Health. 2008;**30**(4):315-323

[53] Hardisson A, Rodriguez MI, Burgos A, Flores LD, Gutierrez R, Varela H. Fluoride levels in publicly supplied and bottled drinking water in the island of Tenerife, Spain. The Bulletin of Environmental Contamination and Toxicology. 2001;**67**(2):163-170

[54] Baxter P, Baubron J, Coutinho R. Health hazards and disaster potential of ground gas emissions at Furnas volcano, São Miguel, Azores. Journal of Volcanology and Geothermal Research. 1999;**92**:95-106

[55] Linhares D, Garcia P, Almada A, Ferreira T, Queiroz G, Cruz JV, et al. Iodine environmental bioavailability and human intake in oceanic islands: Azores as a case study. Science of the Total Environment. 2015;**538**:531-538

[56] Linhares D, Garcia P, Goulart S, Sebastião C, Mota Preto P, Rodrigues A. Fatores de Risco e Composição Química de Cálculos Urinários na População Açoriana (Ilha de São Miguel - Portugal): Um Estudo Preliminar. Acta Urológica Portuguesa. 2018;**35**(1-2):32-38

[57] Pellegriti G, De Vathaire F, Scollo C, Attard M, Giordamno C, Arena G, et al. Papillary thyroid cancer incidence in the volcanic area of Sicily. Journal of the National Cancer Institute. 2009;**101**:1575-1583

[58] Malandrino M, Russo A, Ronchi C, Minoia D, Cataldo C, Regalbuto C, et al. Increased thyroid cancer incidence in a basaltic volcanic area is associated with non-anthropogenic pollution and biocontamination. Endocrine. 2015;**53**(2):471-479

[59] Palumbo B, Angelone M, Bellanca A, Dazzi C, Hauser S, Neri R, et al. Influence of inheritance and pedogenesis on heavy metal distribution in soils of Sicily, Italy. Geoderma. 2000;**95**:247-266

[60] Türkdoğan MK, Kilicel F, Kara K, Tuncer I, Uygan I. Heavy metals in soil vegetables and fruits in the endemic upper gastrointestinal cancer region in Turkey. Environmental Toxicology and Pharmacology. 2002;**13**:175-179

[61] Rodriguez-Espinosa PF, Jonathan MP, Morales-García SS, et al. Metal enrichment of soils following the April 2012-2013 eruptive activity of the Popocatépetl volcano, Puebla, Mexico. Environmental Monitoring and Assessment. 2015;**187**:717

[62] Amaral A, Cruz JV, Cunha RT, Rodrigues A. Baseline levels of metals in volcanic soils of the Azores (Portugal). Soil and Sediment Contamination. 2006;**15**:123-130

[63] Amaral A, Soto M, Cunha R, Marigómez I, Rodrigues A. Bioavailability and cellular effects of metals on Lumbricus terrestris inhabiting

volcanic soils. Environmental Pollution. 2006;**142**(1):103-108

[64] Parelho C, Rodrigues A, Barreto MC, Ferreira N, Garcia PV. Assessing microbial activities in metal contaminated agricultural volcanic soils—An integrative approach. Ecotoxicology and Environmental Safety. 2016;**129**:242-249

[65] Linhares D, Garcia P, Amaral L, Ferreira T, Cury J, Vieira W, et al. Sensitivity of two biomarkers for biomonitoring exposure to fluoride in children and women: A study in a volcanic area. Chemosphere. 2016;**155**:614-620

[66] Linhares D, Camarinho R, Garcia PV, Rodrigues AR. *Mus musculus* bone fluoride concentration as a useful biomarker for risk assessment of skeletal fluorosis in volcanic areas. Chemosphere. 2018;**205**:540-544

[67] Linhares D, Garcia P, Rodrigues A. Fluoride in volcanic areas: A case-study in medical geology. Environmental Health. 2019. DOI: 10.5772/intechopen.86058

[68] World Health Organization (WHO). Guidelines for Drinking-Water Quality: Recommendations. 3rd ed. Geneva, Switzerland. 2004. ISBN 92 4 154638 7

[69] Cruz J, Coutinho RM, Carvalho MR, Oskarsson N, Sigurdur RG. Chemistry of waters from Furnas volcano, Sao Miguel, Azores: Fluxes of volcanic carbon dioxide and leached material. Journal of Volcanology and Geothermal Research. 1999;**92**:151-167

[70] Ferreira T, Gaspar JL, Viveiros F, Marcos M, Faria C, Sousa F. Monitoring of fumarole discharge and CO_2 soil degassing in the Azores: Contribution to volcanic surveillance and public health risk assessment. Annals of Geophysics. 2005;**48**:4-5

[71] Searle R. Tectonic pattern of the Azores spreading centre and triple junction. Earth and Planetary Science Letters. 1980;**51**:415-434

[72] Madeira J, Ribeiro A. Geodynamic models for the Azores triple junction: A contribution from tectonics. Tectonophysics. 1990;**184**:405-415

[73] Vogt PR, Jung WY. The Terceira rift as hyper-slow, hotspot dominated oblique spreading axis: A comparison with other slow-spreading plate boundaries. Earth and Planetary Science Letters. 2004;**218**:77-90

[74] Needham H, Francheteau J. Some characteristics of the rift valley in the Atlantic Ocean near 36° 48′ north. Earth and Planetary Science Letters. 1974;**22**:29-43

[75] Pacheco JM, Ferreira T, Queiroz G, Wallenstein N, Coutinho R, Cruz JV, et al. Notas sobre a geologia do arquipélago dos Açores. In: Dias R, Araújo A, Terrinha P, Kullberg JC, editors. Geologia de Portugal. Vol. 2. Lisboa: Escolar Editora; 2013. pp. 595-690

[76] Gaspar JL, Guest JE, Queiroz G, Pacheco, Pimentel A, Gomes A, et al. Eruptive frequency and volcanic hazards zonation in São Miguel Island, Azores. In: Gaspar JL, Guest JE, Duncan AM, Barriga FJAS, Chester DK, editors. Volcanic Geology of São Miguel Island (Azores Archipelago). Geological Society, London, Memoirs 44; 2015. pp. 155-166

[77] Amaral A, Arruda M, Cabral S, Rodrigues A. Essential and non-essential trace metals in scalp hair of men chronically exposed to volcanogenic metals in the Azores, Portugal. Environment International. 2008;**34**:1104-1108

[78] Linhares D, Pimentel A, Borges C, Cruz JV, Garcia P, Rodrigues A. Cobalt distribution in the soils of São Miguel

Island (Azores): From volcanoes to health effects. Science of the Total Environment. 2019;**684**:715-721

[79] Gilbert ME, Rovet J, Chen Z, Koibuchi N. Developmental thyroid hormone disruption: Prevalence, environmental contaminants and neurodevelopmental consequences. Neurotoxicology. 2012;**33**:842-852

[80] World Health Organization (WHO). Global Health Risks: Mortality and Burden of Disease Attributable to Selected Major Risks. Geneva, Switzerland: World Health Organization; 2009

[81] McConahey WM, Keating FR Jr, Beahrs OH, Woolner LB. On the increasing occurrence of Hashimoto's thyroiditis. The Journal of Clinical Endocrinology and Metabolism. 1962;**22**:542

[82] Maitra A. Thyroid gland. In: Chmidt W, Gruliow R, editors. Robbins and Cotran Pathologic Basis of Disease. 8th ed. Philadelphia: Saunders Elsevier; 2010. pp. 1107-1130

[83] Ross DS, Burch HB, Cooper DS, Greenlee MC, Laurberg P, Maia AL, et al. 2016 American Thyroid Association guidelines for diagnosis and management of hyperthyroidism and other causes of thyrotoxicosis. Thyroid. 2016;**26**(10):1343-1421

[84] Oliveira A, Gonçalves MJ, Sobrinho LG. Endemic goiter in the island of S. Miguel (the Azores). Ata Endocrinologia. 1986;**11**:200-203

[85] Limbert E, Prazeres S, Pedro MS, Madureira D, Miranda A, Ribeiro M, et al. Iodine intake in Portuguese pregnant women: Results of a countrywide study. European Journal of Endocrinology. 2010;**163**:631-635

[86] Limbert E, Prazeres S, Madureira D, Miranda A, Ribeiro M, Abreu FS, et al. Aporte do iodo nas Regiões Autónomas da Madeira e dos Açores. Revista Portuguesa de Endocrinologia Diabetes e Metabolismo. 2012;**7**(2):2-7

[87] Yadav MK, Manoli NM, Madhunapantula SV. Comparative assessment of vitamin-B12, folic acid andhomocysteine levels in relation to p53 expression in megaloblastic anemia. PLoS One. 2016;**11**(10):e0164559

[88] Tun AM, Thein KZ, Myint ZW, Oo TH. Pernicious anemia: Fundamental and practical aspects in diagnosis. Cardiovascular and Hematological Agents in Medicinal Chemistry. 2017;**15**(1):17-22

[89] Oh R, Brown DL. Vitamin B12 deficiency. American Family Physician. 2003;**67**(5):979-986

[90] Barbosa VFM. Será o Tomadiço uma Doença de Carência? Boletim da Comissão Reguladora dos Cereais do Arquipélago dos Açores. Separata n° 9. São Miguel, Açores. 1949

[91] Toste JM. Doença da Volta ou Tomadiço na Ilha de São Miguel, Açores. Relatório de Tese de Licenciatura em Medicina Veterinária; 1953. pp. 16-23

[92] Pinto C. Hematúria enzoótica bovina: contribuição para seu estudo etiopatogénico [PhD thesis]. Universidade Técnica de Lisboa - Faculdade deMedicina Veterinária; 2010. 229 pp

[93] Neall VE. Volcanic soils. In: Verheye W, editor. Land Use and Land Cover, Encyclopaedia of Life Support Systems (EOLSS). Oxford, UK: EOLSS Publishers with UNESCO; 2006

[94] Pimentel A, Zanon V, De Groot LV, Hipólito A, Di Chiara A, Self S. Stressinduced comenditic trachyte effusion triggered by trachybasalt intrusion: Multidisciplinary study of the AD 1761 eruption at Terceira Island

(Azores). Bulletin of Volcanology. 2016;**78**(3):22

[95] Porreca M, Pimentel A, Kueppers U, Izquierdo T, Pacheco J, Queiroz G. Event stratigraphy and emplacement mechanisms of the last major caldera eruption on Sete Cidades volcano (São Miguel, Azores): The 16 ka Santa Bárbara formation. Bulletin of Volcanology. 2018;**80**:76

[96] Barceloux DG, Barceloux D. Cobalt. Journal of Toxicology. Clinical Toxicology. 1999;**37**(2):201-216

[97] Skalny AV. Bioelementology as an interdisciplinary integrative approach in life sciences: Terminology, classification, perspectives. Journal of Trace Elements in Medicine and Biology. 2011;**25S**:S3-S10

[98] Williams M, Todd GD, Roney N, et al. Toxicological Profile for Manganese. Atlanta (GA): Agency for Toxic Substances and Disease Registry (US); health effects; 2012

[99] O'Neal SL, Zheng W. Manganese toxicity upon overexposure: A decade in review. Current Environmental Health Reports. 2015;**2**(3):315-328

[100] Wedepohl KH. Handbook of Geochemistry. Berlin: Springer; 1969-1978

[101] Nielsen FH. Manganese, molybdenum, boron, chromium, and other trace elements. In: Erdman JW Jr, Zeisel SH, editors. Present Knowledge in Nutrition. 10th ed. Oxford: Wiley-Blackwell; 2012. pp. 586-607

[102] Buchman AR. Manganese. In: Catharine Ross A, Caballero BH, Cousins RJ, Tucker KL, Ziegler TR, editors. Modern Nutrition in Health and Disease. 11th ed. Baltimore, MD: Lippincott Williams & Wilkins; 2014. pp. 238-244

[103] Baker MG, Simpson CD, Stover B, Sheppard L, Checkoway H, Racette BA, et al. Blood manganese as an exposure biomarker: State of the evidence. Journal of Occupational and Environmental Hygiene. 2014;**11**(4):210-217

[104] Boudissa SM, Lambert J, Muller C, Kennedy G, Gareau L, Zayed J. Manganese concentrations in the soil and air in the vicinity of a closed manganese alloy production plant. Science of the Total Environment. 2006;**361**:67-72

[105] Purdey M. The pathogenesis of Machado Joseph disease: A high manganese/low magnesium initiated CAG expansion mutation in susceptible genotypes? Journal of the American College of Nutrition. 2004;**23**(6):715S-729S

[106] Aballéa KM, Appriou J, Bougault P, Charlou H, Donval JL, Etoubleau JP, et al. Manganese distribution in the water column near the Azores triple junction along the mid-Atlantic ridge and in the Azores domain. Deep Sea research part I. Oceanographic Research Papers. 1998. pp. 1319-1338

[107] Wallenstein F, Torrão DF, Neto AI, Wilkinson M, Rodrigues AS. Effect of exposure time on the bioaccumulation of Cd, Mg, Mn and Zn in Cystoseira abies-marina samples subject to shallow water hydrothermal activity in São Miguel (Azores). Marine Ecology. 2009;**30**(s1):118-122

[108] Lima MTR, Kelly MT, Cabanis MT, Cassana G, Matos L, Pinheiro J, et al. Determination of iron, copper, manganese and zinc in the soils, grapes and wines of the Azores. Journal International des Sciences de la Vigne et du Vin. 2004;**38**(2):109-118

[109] Kiloh LG, Lethlean AK, Morgan G, Cawte JE, Harris M. An endemic neurological disorder in tribal Australian aborigines. Journal

of Neurology, Neurosurgery and Psychiatry. 1980;**43**(8):661-668

[110] Martins S, Calafell F, Gaspar C, et al. Asian origin for the worldwide spread mutational event in Machado-Joseph disease. American Neurological Association. 2007;**64**(10):1502-15081

[111] Bettencourt C, Santos C, Ka T, et al. Analysis of segregation patterns in Machado–Joseph disease pedigrees. Journal of Human Genetics. 2008;**53**:920-923

[112] Lima M, Mayer FM, Coutinho P, Abade A. Origins of a mutation: Population genetics of Machado–Joseph disease in the Azores (Portugal). Human Biology. 1998;**70**:1011-1023

[113] Robinson B, Bolan N, Mahimairaja S, Clothier B. Solubility, mobility and bioaccumulation of trace elements: Abiotic processes in the rhizosphere. In: MNV P, Sajwan K, Naidu R, editors. Trace Elements in the Environment: Biogeochemistry, Biotechnology and Bioremediation. New York: CRC Press; 2005. p. 744

Chapter 7

Dermatologic Toxicities and Biological Activities of Chromium

Jumina Jumina and Harizal Harizal

Abstract

Chromium is a versatile metal with various industrial applications and biological activities. However, as a transition metal, this element forms several species, i.e. oxidation states of −4 to +6, with different degrees of toxicities that affect ecosystems and organisms including human beings. The skin is the outermost organ that usually interacts directly with chromium species in nature. These contact and interaction induce the formation of several acute and chronic negative effects including contact dermatitis, skin cancer, allergy, etc. In this chapter, toxicity and biological activity of several chromium species, such as chromium zero-valent, trivalent, hexavalent, will be reviewed to obtain better comprehension in chromium toxicity. Sources and routes of exposure, toxicity and possible treatment, and biological activity on the skin are arranged and explained systematically.

Keywords: chromium, skin, toxicities, biological activities, treatments

1. Introduction

Chromium is a unique transition metal with relatively high abundance on earth crust (1.4×10^{-2}%). Chromium can form several species with different oxidation states from −4 to +6. Chromium with 0, +3, and +6 oxidation states are most commonly found and utilized in ambient conditions [1]. In industrial sectors, chromium-based materials have been used as coating material for corrosion resistant, pigments and dyes, wood preservatives, tanning agent, catalyst, and medical apparatus [1]. Chromium, especially for chromium(III), also showed certain biological activity especially in regulating carbohydrates and lipid metabolism [2, 3]. As an essential micronutrient, a low dietary of chromium will exhibit several adverse effects such as glucose intolerance, growth disorders, diminished longevity, etc. [3, 4].

Chromium toxicity has been a controversial problem due to its status as an essential micronutrient [5]. Various studies have shown that numerous acute and chronic adverse effects can be caused by any dermal or systemic exposure of chromium species in human organ systems [6]. The toxicity and biological activity of chromium seem to be correlated directly with the concentration of corresponding chromium species [7]. In this case, chromium species have its optimum concentration to produce beneficial effects. Meanwhile, accumulation of less toxic chromium species in relatively high concentration will still produce a negative effect in the accumulation site [8, 9]. Chromium picolinate, for instance, has been mainly used as food supplement. Chromium(III) in this compound tend to accumulate in male

Sprague-Dawley rats' cells over the period of investigation [10] and may be oxidized to more carcinogenic chromium(V) and chromium(VI) within the cells [11].

As the outermost organ that protects the human body from various pollutants, the skin is usually exposed to various sources of chromium, and it causes many dermatological acute and chronic negative effects such as contact dermatitis [12], systemic contact dermatitis [13], and possibly skin cancer [14]. In the same way, any topical or systemic administration of chromium compounds also can exhibit a beneficial effect for the skin such as antiacne [15], rapid wound healing [16], and anti-aging [17]. In this chapter, both toxicity and biological activity of chromium species in the skin are described starting from the source and route of exposure, toxicity and its possible treatment, and biological activity.

2. Source and exposure route of chromium in the skin

In modern life, chromium has been used in many forms and applications with Cr(0), Cr(III), and Cr(VI) as the main oxidation states. Various sources of chromium that affect or may affect the skin have been identified and tabulated in several review [12, 18, 19]. In general, exposure route of chromium that comes from these sources can be classified into two pathways including dermal and systemic pathways. In these cases, direct dermal exposure would cause contact dermatitis, irritation, and skin cancer, while systemic administration would elicit systemic contact dermatitis and skin tumor.

Dermal exposure (**Figure 1**) is initiated from direct contact of chromium sources on the skin. Chromium species are then accumulated on the skin surface or penetrated into the skin layers mediated by sweat or other biological fluids. The penetration of chromium species either as particulate or soluble forms occurred via three possible routes including transcellular by crossing the cell, intercellular by partitioning into the lipid matrix, and transappendageal by entering hair follicle and sebaceous glands [20, 21]. There are many factors involved in the penetration process including concentration of chromium species, medium (solvent and pH), intrinsic properties of chromium species (molecular volume of chromium species, counter ion, nature of chemical bond and polarity, solubility, and valence), reactivity towards protein, previous penetration or accumulation, skin characteristics (gender and race, age skin, density of sebaceous gland, thickness of skin, and anatomy of skin), and environmental factors (temperature, humidity, and UV radiation) [21, 22]. In a normal skin condition, Cr(VI) ions tend to have higher solubility [23] and percutaneous permeability than Cr(III) ions [24, 25]. However, Cr(III) have higher protein affinity to form metal-protein complex which tends to make it retain in the skin epidermis [26]. After penetrating the skin, Cr(VI) species are reduced by proteins or endogenous antioxidants to form Cr(III) [27] which then react further with any DNA or protein to form Cr(III)-protein complex as the actual allergen (haptens) [28].

In systematic exposure, chromium mostly enter the human body via oral consumption of certain chromium sources such as food or food supplement [29], foodstuff [30], and drinking water [31] or from applications of chromium-based implants [32]. In the digestive system, most of Cr(III) consumed are excreted to feces and some of it (~2%) is absorbed by epithelial cells covering the stomach and enterocytes covering the intestines through passive absorption (diffusion) [33]. This absorption was affected (increased or decreased) by the presence of various ligand such as amino acids, vitamins, carbohydrates, plasma proteins, certain metals, and other chelating agents [34]. After the absorption, Cr(III) complex would be accumulated inside the cells or actively transported to the blood stream by still an

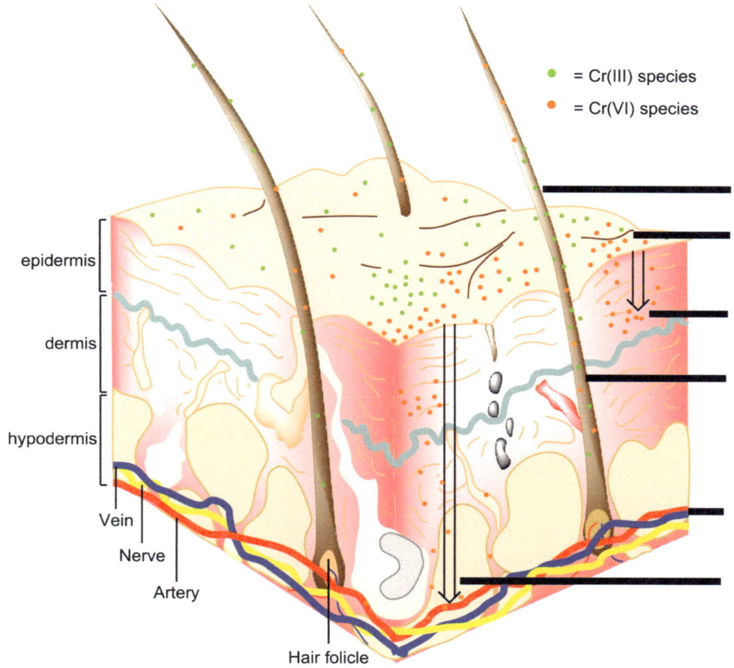

Figure 1.
Several pathways of dermal exposure: (A) deposition of Cr(VI) in the skin surface causes irritation or chemical burn; (B) penetration of Cr(VI) through transcellular and intercellular pathways induces cell apoptosis and contact dermatitis, while Cr(III) tend to suspended in the skin surface due to strong affinity to skin protein; (C) deeper penetration to blood stream causes organ disruptions, (D) Cr species from systemic administration trigger systemic contact dermatitis, (E) deposition of Cr species on hair causes hair discoloration; (F) penetration of Cr species into hair follicle induces rapid hair fall. Figure is drawn using ChemBioDraw Ultra version 14.0 software.

unknown transporter. Cr(III) ions then bound to transferrin (siderophilin) or other plasma proteins in the blood stream and travel to the whole body [33, 34].

3. Toxicities of chromium

3.1 Contact dermatitis

Contact dermatitis is a common skin disease caused by repeated dermal contact with certain allergens (haptens) leading to delayed-type hypersensitivity effect [35]. Many haptens have been identified to cause contact dermatitis such as metals, fragrances, and flavors, preservatives, plastics, rubber, pharmaceutical, cosmetics, woods and plants, textile, etc. [35]. Chromium-induced contact dermatitis is characterized by the presence of certain clinical manifestations in feet and hands. Acute dermatitis is usually indicated by the formation of erythema, oedema, papules, vesicles, and weeping, while chronic dermatitis tends to form scaly, dry, and fissured skin [36]. Various chromium-induced contact dermatitis cases have been reported involving different chromium sources such as cement [37, 38], leather [38–43], tattoo ink [44], cellular phone [45, 46], etc. Concentration threshold for soluble chromium in each chromium-containing product should not exceed 1 ppm to minimize elicitation of contact dermatitis [47].

In general, chromium-induced contact dermatitis is formed through several steps which can be described as the following (**Figure 2**) [48, 49]: initially, after penetrating the skin, Cr(VI) ions are reduced by endogenous antioxidant to form Cr(III) and oxygen reactive species (ROS). Cr(III) as the real allergen is bound to certain proteins to form the hapten, while ROS induces the releasing of interleukin-1β (IL-1β) which then activates antigen-presenting cells (Langerhans cells (LC)). Activated antigen-presenting cells bind with the hapten, mature, travel to the regional lymph nodes, and stay in paracortical T-cell areas. After that, activated antigen presenting cells-hapten complex activates naïve T cells by helping in vigorous blast formation and proliferation to become chromium-specific T cells. Activated chromium-specific T cells then travel through blood stream and recirculate to give hypersensitivity effect detecting a lower concentration of hapten in different parts of the skin.

Treatment of chromium-induced contact dermatitis could be conducted in several approaches including avoiding direct contact to chromium source and topical application of chelating agent and barrier creams to prevent any cutaneous permeation, corticosteroid to relieve inflammation, and antioxidant to reduce oxidative stress [36, 50]. Various antioxidants have been tested in treating chromium-induced contact dermatitis such as *N*-acetylcysteine [51], ascorbic acid [52], pine bark extract (pycnogenol®) [53], and pterostilbene [54]. Two chelating agents, ethylenediaminetetraacetic acid (EDTA) and diethylenetriaminepentaacetic acid (DTPA), also have been examined for preventing contact dermatitis, but both of these ligands showed a low effectiveness as a protecting agent [55, 56]. The developments in

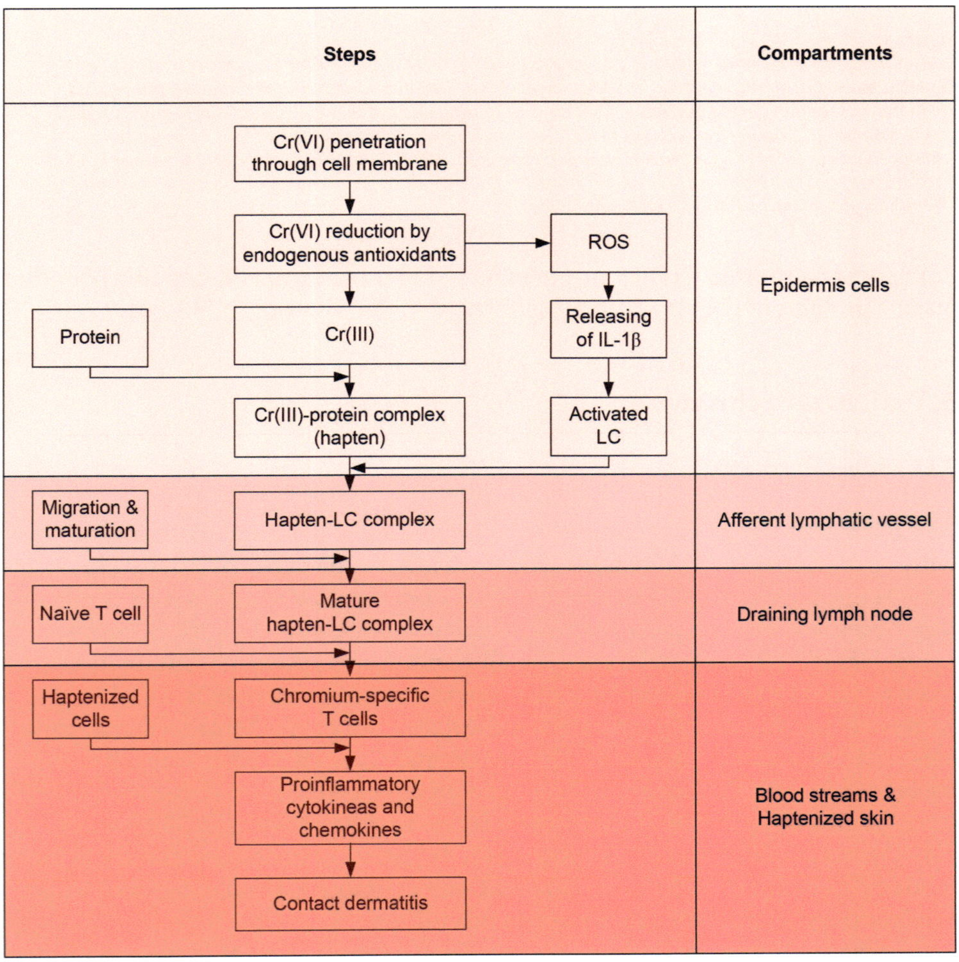

Figure 2.
Simplified elicitation mechanism of chromium-induced contact dermatitis adapted from several references [48, 49].

chromium-induced contact dermatitis are still widely opened. These developments could be focused on finding natural antioxidant and chelating agents that effectively relieve oxidative stress and reduce the reactivity of chromium ions, respectively.

3.2 Systemic contact dermatitis

Systemic contact dermatitis is a kind of dermatitis elicited by systemic administration of certain allergen that previously sensitizes the skin through direct dermal contact [57]. Several groups of allergens have been identified to cause this inflammatory disease such as metals, medications, food, plants and herbals, and certain chemicals [57, 58]. Systemic administration of these allergens is also described in various routes including oral, subconjunctival, intramuscular, pulmonary inhalation, intranasal, intrauterine, endocardial, arthroplastic, intravenous, intraarticular, subcutaneous, intradermal, dental, intratubal, and endovascular [59]. The pathophysiology of this disease still remains unclear, but several mechanisms have been proposed [60–65]. Moreover, the theory suggesting type 3 immune response (antigen-antibody complexes) involved in systemic contact dermatitis (SCD) has not fully been proven [58].

Chromium as a metal allergen has been found to cause systemic contact dermatitis either through oral, dental, or arthroplastic routes [13, 66, 67]. Consumption of Cr(III)-based supplements in the form of chromium picolinate [68] and chromium chloride [69] has been shown to cause SCD. Oral ingestion of potassium dichromate previously used as a homeopathic drug also induces dermatitis as clinical manifestation of SCD [70, 71]. In certain case, SCD is also induced by various metal alloys applied in orthopaedical, cardiac, neurological, and abdominal associated devices [72]. In these cases, chromium-containing alloys such as stainless steel SAE 316 L, cobalt-chromium-molybdenum steel, and Vitalium™ release metal ion [73] are reported to cause SCD [74–79]. These reported SCD cases are identified with the presence of several manifestations including erythroderma [74], and localized/generalized eczema or urticarial [76–78]. These findings suggested that any chromium sources applied in systemic routes could elicit SCD with certain clinical manifestations and degrees of severity.

Treatment of chromium-induced systemic contact dermatitis may be conducted in several approaches including managing diets and lifestyles by gut remediation and avoiding the food and sources that contain chromium; systemic or topical treatment using immune-suppressants such as corticosteroid; phototherapy; and hyposensitization therapy [80, 81]. Sharma developed a guideline for the preparation of low chromate diet that could help in controlling daily chromium consumption from food and ameliorating skin condition [82]. In case of SCD from arthroplastic routes, revision or removal of implant needs to be conducted by considering the time of hypersensitivity incidence after surgery and degree of severity [74]. Revision of implant could be conducted by using less allergenic implant such as titanium-based implant or chromium-based implant coated with certain biocompatible materials such as polytetrafluoroethylene, ZrN multilayers, diamond-like carbon, titanium nitride, graphite-like carbon, and tantalum [74, 79, 83]. Revision or removal of implants may not produce rapid disappearance [76] probably due to the presence of soluble or particulate debris of implant that produce inflammations [84, 85].

3.3 Skin cancer

Chromium, especially Cr(VI), is a potent human carcinogen. In a human cell, carcinogenesis of chromium(VI) (**Figure 3**) occurs through the penetration of chromium(VI) species into the cell via sulphate/phosphate anion transport system, reduction of chromium(VI) by endogenous antioxidant to produce ROS and chromium(III), and alteration of DNA directly by chromium(VI) or indirectly by ROS [86]. The alteration of DNA then produces different types of products including Cr-DNA adducts, DNA–protein and DNA interstrand cross-links, DNA breaks, and DNA base damage [87]. Carcinogenicity of chromium(VI) has induced lung cancer in workers from various chromium-based industries [88–90] and has been associated to incidence of other cancers [91]. However, the meta-analysis study showed that the correlation between exposure of chromium(VI) and the high mortality in skin cancer is not significant [91]. This study indicated that there is no supporting data confirming the chromium species as carcinogen in inducing skin cancer in human.

Several studies, however, showed that chromium species could induce skin cancer in rats and mice either as single carcinogen or cocarcinogen. Oral administrations of drinking water containing sodium dichromate dehydrate to male F344/N rats for 2 years showed that the sample developed various types of skin cancer [92]. Two other studies using hairless SK1-hrBR mice also exhibited that chromium(VI) could act as cocarcinogen in promoting UV-induced skin tumor [93, 94]. Davidson and co-workers [93] showed that oral administration of chromium(VI)-containing drinking water and UV irradiation to hairless mice have synergistic effect in promoting skin tumor. Exposure of chromium(VI) or UV radiation alone did not induce skin tumors [93]. Uddin and co-worker also conducted the same experiment and found that systemic administration of exogenous antioxidant (vitamin E and selenomethionine) did not improve skin condition [94]. It indicated that chromium(VI) cocarcinogenicity may be occurred in different mechanisms without involving ROS [94]. These three studies indicate that acute or chronic oral administration of chromium(VI) species has a great potency in promoting skin cancer in mammals including humans.

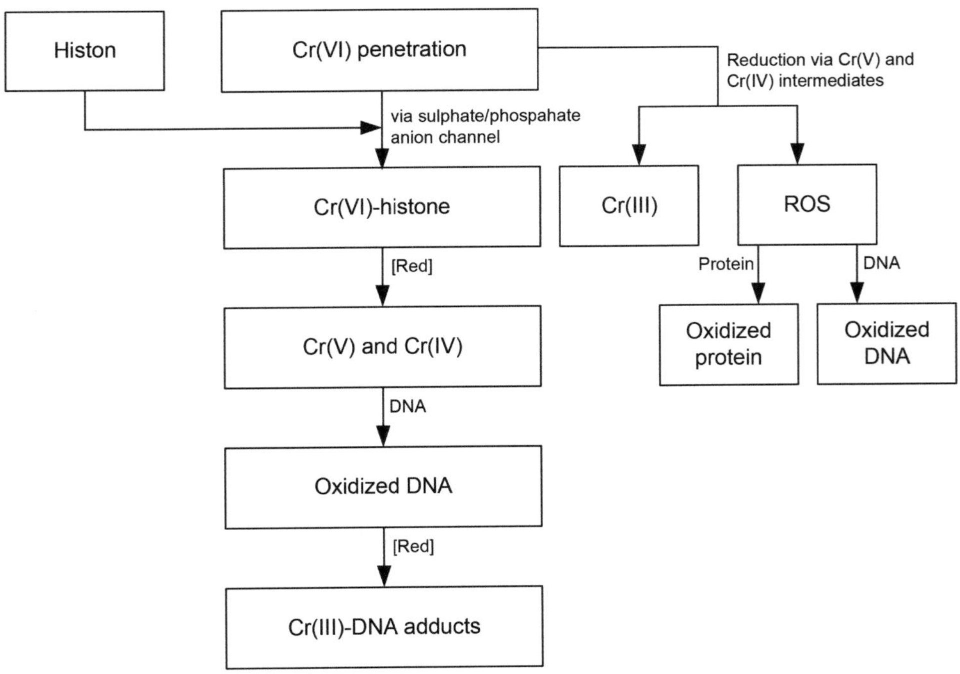

Figure 3.
Simplified mechanism of chromium carcinogenesis adapted from several references [86, 95].

3.4 Irritation and chromium burn

Irritation and chemical burn are caused by dermal exposure of chromium(VI) particle, solution, or mist in large quantities. Solid deposition of chromium(VI) would develop to "chromium ulcers" or "chrome holes" [67], while high concentration of chromium(VI) solution would lead to chromium burn. A mechanism for this ulcer formation is still unclear, but it may be related to the disruption of actin cytoskeleton by chromium(VI) leading to mitochondria-dependent apoptosis in skin fibroblasts cells [96]. Several reports exhibited these irritation and burning effects from different chromium species such as solid CrO_3 [97], chromic acid solution [98–100], hot chromium(III) sulphate solution [101, 102], and chromium acid mist [103].

Management of irritation is conducted by considering preventive and treatment approaches. Prevention of irritation is conducted by using barrier creams, moisturizers, etc., while treatment could be done by using moisturizers and corticosteroids [50]. For chromium burn, treatment is conducted by combining mechanical excision, hemofiltration, and systemic administration of chelating agent and antioxidant [100].

3.5 Hair disorders

Human hair is naturally exposed to a certain amount of chromium [104] that come from various sources [105–107]. Excessive and repeated exposures of chromium in certain environmental condition cause discoloration of blond, dyed-blond, and white hair (to become green) [108–110] and cause rapid hair fall [111, 112]. The mechanisms of these two effects are still unclear. Hair discoloration is probably the result of interaction between chromium ions (and also copper and nickel) and protein in hair (keratin) [113], while rapid hair fall may be related to several mechanisms such as promoting premature end of hair cycle [114] or disruption of hair shaft formation [115].

4. Biological activities of chromium

4.1 Acne vulgaris (antiacne)

Acne vulgaris is a common dermatological condition that affects physical and psychological aspects of patients [116]. Several diseases that show the presence of a certain degree of acne also relate with depression and emotional stress such as type-2 diabetes, rheumatoid arthritis, and polycystic ovarian syndrome (PCOS) [117–119]. Pathophysiology of this disease involves several key mechanisms including excessive sebum production due to hormonal and environmental conditions, alteration of fatty acids composition due to sebum metabolism by *Cutibacterium acnes*, hyperkeratinization within the follicle that clogs up the pore in the form of whitehead or blackhead comedones, inflammation induced by bacterial colonization, and malfunction of locale innate and adaptive immune system [120]. The presence of acne vulgaris is also correlated to the clinical depression in patients [116]. In this case, depression or stress can influence the regulation of sebaceous gland as the main part in sebum production [121]. Catecholamines (epinephrine and norepinephrine) as the main stress hormones also affect the growth of certain *Cutibacterium acnes* strains [122–124]. Catecholamine-treated *C. acnes* strain also can stimulate a limited but significant increase of lipid production in sebaceous gland. However, the increase of intrinsic cytotoxicity or inflammatory potential of *C. acnes* is statistically significant [124].

Several reports exhibited that certain chromium(III) compounds have high activity in improving acne vulgaris. Initially, chromium has been used in the form of high-chromium yeast or chromium GTF (glucose tolerance factor) by consuming 400 μg chromium daily which exhibited comparable improvement in acne conditions [15]. This form of treatment, recently, is considered as a complementary and alternative medicine (CAM) for the treatment of acne vulgaris [125]. Further improvements used different chromium compounds including chromium picolinate [126, 127] and chromium salt such as chromium (III) chloride [128].

In most cases, the usage of chromium compound as antiacne is usually combined with other active compounds such as vitamins, certain minerals, and herbal medicine sources in the form of oral [129–131] or topical [128] formulation to get more effective treatment results. Application of topical formulation containing chromium (III) chloride and magnesium sulphate showed total improvement in acne vulgaris with temporal mild to moderate irritation as a side effect [128]. Oral capsule containing methionine-bound zinc complex, chromium, and vitamins also exhibited 80–100% improvements for mild to moderate acne vulgaris [131]. In another study, a combination of several nutrients with potential antiacne and anti-depressant properties (eicosapentaenoic acid, epigallocatechin-3-gallate, zinc gluconate from green tea extract, selenium, and chromium) may also improve inflammatory acne lesions and mood aspect of patients [132].

Treatment of acne vulgaris in polycystic ovary syndrome (PCOS) showed mix results. A study by Amr and Abdel-Rahim showed that using 200 μg/day oral consumption for 8 weeks has no significant improvement in acne and hirsutism [126]. In a different study, chromium supplementation by women with polycystic ovary syndrome (PCOS) in a randomized, double-blind, placebo-controlled trial exhibited that the treatment gives beneficial effects on acne and hirsutism using 1000 μg/day oral consumption for 6 months [127]. These two studies indicated that the treatment of acne vulgaris in PCOS patients needs greater dose and longer duration.

Action mode of chromium in ameliorating acne vulgaris has not been fully elucidated yet, but there are two mechanisms proposed including (1) by decreasing serum testosterone concentration and (2) lowering the depression of patients. In the first mechanism, chromium can decrease serum testosterone level possibly due to the reduction of testicular steroidogenic enzymes activities [133]. In this case, a lower level of serum testosterone reduces sebum production in sebaceous glands [134]. The second mechanism explains that chromium as an anti-depressant [135–137] may reduce sebum production [138] and affect *C. acnes* growth in sebaceous glands [122–124]. It is clearly showed that these two mechanisms may have direct or indirect synergistic effects [139] in ameliorating skin condition with acne vulgaris.

4.2 Cutaneous ageing (anti-ageing)

Aging is a complex multifactorial process of damage accumulation that causes the deterioration of fitness [140, 141]. Aging has been the main risk factor for several deadly diseases such as cancer, cardiovascular disease, diabetes, and neurodegeneration [141]. In the skin, aging is identified by the presence of folds and wrinkles due to the declining and degradation of collagen [142], loss of elasticity [143], and decreasing of various skin functions [144]. At least, there are seven factors that may produce these clinical manifestations including passage of time, genetics, radiation such as ultraviolet and infrared radiations, lifestyle, chronic debilitating diseases, dysfunction of hormonal system, and gravitational force [145]. Several mechanisms have been proposed to explain the effect of these factors on aging including oxidative stress, telomere shortening, epigenetic dysregulation, DNA damage, genetic mutation, inflammation, mitochondrial dysfunction, and accumulation of glycation end product [146, 147].

Treatment for skin aging can be conducted through three approaches including adjusting lifestyle by routine exercise, calorie restriction, and maintaining mental health; gene therapies; and medications. Among other approaches, medication could be the simplest approach in fighting skin aging such as by using topical or systemic agents [148]. Chromium as dietary supplement (50–200 μg) has been used in preventing skin aging by controlling and regulating blood sugar and lipid levels [17]. Either in topical or systemic applications, chromium is usually combined with different vitamins and minerals to obtain optimum results based on certain parameters such as improving insulin function using chromium picolinate [149], promoting mitochondrial biogenesis and lipid metabolism using oligomannuronate-chromium(III) complexes [150], replacing or removing excess iron production using chromium(III) chloride or chromium picolinate [151, 152], and activating telomerase [153].

Antioxidant activity of chromium may also contribute to its anti-aging properties since oxidative stress has a certain role in the damaging process. Supplementation of chromium(III) in adult male and female with type-2 diabetes mellitus minimized the increase of oxidative stress (thiobarbituric acid reactive substances—TBARS) and increased total antioxidant status [154, 155]. Several combinations have been made by formulating chromium(III) with zinc [156], niacin [157], and vitamin C/E [154] and showed a protective effect against skin damage against oxidative stress.

Antioxidant activity of chromium(III) is correlated to the dose applied as shown in several experiments. Incubation of BALB/3 T3 clone A31 cells and HepG2 cells with chromium(III) chloride concentration higher than 400 μM would induce the formation of oxidative stress, while lower optimized concentration (M = 100–200 μM) would increase superoxide dismutase and catalase antioxidant activities [7]. In vitro study on the effect of chromium(III) and chromium(VI) on catalase activity also showed this dose-dependent activity in which treatment of cell-free catalase using chromium(III) (dose range $1–5 \times 10^{-5}$ mol/L) and chromium(VI) (dose range $1–4 \times 10^{-5}$ mol/L) separately increased the catalase activity [158]. These two studies clearly describe that either chromium(III) or chromium(VI) has a certain optimum concentration to exhibit their beneficial effects.

4.3 Cutaneous wound (rapid wound healing)

Cutaneous wound is the skin defect or skin opening that is caused by external forces [159]. Formation of this wound triggers a set of complex biochemical processes to repair the damage that are called as wound healing or wound repair. In normal condition, there are five consecutive phases occurred in wound healing process including (1) homeostasis phase (immediately) through the migration of thrombocytes and formation of fibrin clot to stop the bleeding; (2) first inflammatory phase (day 1–day 6) by sensing the injury, sending the danger signal, and initializing the inflammation; (3) second inflammatory phase (day 1–day 6) through elimination the pathogens and cleaning the wound; (4) proliferation phase (day 4–day 14) through epithelialization, angiogenesis, granulation tissue formation, and collagen deposition to repair the damage and initialize the tissue remodeling process; and (5) remodeling phase (day 8–year 1) through the deposition of collagen to reach maturation of tissue structure [160–162]. Several internal and external factors have been identified to affect the wound healing process including oxygenation, infection and foreign body, lifestyle, hormonal effect, age, and gender [163].

Several studies have shown that chromium(III), in a certain condition, could improve cutaneous wound either in normal or diabetic Wistar rats using a single dose of a combination of zinc(II) (1.5 mg/kg weight) and chromium(III) (0.02 mg/kg weight) [164] and C57BL6/J mice using chromium(III) chloride (80 μg/kg weight/day) for 21 days [16]. The mechanism of this effect has not been fully elucidated yet, but it may be related to chromium(III) activity in increasing insulin sensitivity, insulin-like growth factor 1 (IGF-1) serum concentration, and protein deposition [16, 165]. In this case, high glucose concentration could inhibit proliferation and differentiation of skin keratinocytes [166] and increase the stiffness of collagen [167] which further inhibits wound healing. In healing acetic acid-induced colitis wound, chromium(III) also acted as an anti-inflammatory agent by inhibiting several inflammatory markers and downregulating pro-inflammatory cytokine genes and antioxidant by suppressing oxidative stress without any significant side effect [168].

In different situations, the use of chromium-based skin clips [169] and orthopedic implant [74] gave an adverse effect by delaying surgery wound healing process. These cases represented a hypersensitivity effect as a manifestation of systemic contact dermatitis. In vitro study using human skin keratinocyte cell line (HaCaT cells) in a medium containing chromium(III) solution (10^{-6} M) showed that chromium(III) ions can decrease wound closure rate and be further decreased when the medium was replaced with another chromium(III) ion-containing medium [170]. Chromium(III) ions also caused downregulation of toll-like receptor-2, -4, and -9 messenger ribonucleic acids (TLR-2, -4, and -9 mRNA), upregulation of matrix metalloproteinase 2 and 13, and upregulation of intercellular adhesion molecule 1 messenger ribonucleic acid (ICAM-1 mRNA) [170].

There's no exact explanation for these opposite effects. However, it may be related to the local concentration of chromium species in wound location. An enhancing effect of wound healing was obtained by applying a relatively small concentration of chromium species via oral administration. In human, for instance, there is only 2% of oral chromium(III) that will be absorbed through stomach and intestine and distributed throughout the body. In the same time, an adverse effect was obtained when local chromium concentration was high due to a particulate and soluble chromium released from the implants.

5. Conclusions

Chromium as versatile heavy metals showed contradictive properties dealing with its dermatologic toxicity and biological activity properties. The main factors that probably correlate to these properties are concentrations and species of chromium. Significant increment of local chromium concentration (more than 1 ppm for chromium[VI] species) either from dermal or systemic administration would increase the risk of dermatologic toxicities, while topical or oral administration of small recommended dietary concentration of chromium (50–200 μg for chromium picolinate) would give several beneficial effects. More studies need to be conducted to know the exact effect of the local concentration of corresponding chromium species in many systems.

Acknowledgements

The financial support from Ministry of Research, Technology and Higher Education of the Republic of Indonesia who has funded the research in calixarene- and resorcinarene-based heavy metal adsorbents is acknowledged.

Conflict of interest

The authors state that there is no conflict of interest.

Author details

Jumina Jumina[1*] and Harizal Harizal[2]

1 Department of Chemistry, Faculty of Mathematics and Natural Sciences, Universitas Gadjah Mada, Yogyakarta, Indonesia

2 Department of Pharmacy, Faculty of Health Sciences, Universitas Esa Unggul, Jakarta, Indonesia

*Address all correspondence to: jumina@ugm.ac.id

IntechOpen

© 2019 The Author(s). Licensee IntechOpen. This chapter is distributed under the terms of the Creative Commons Attribution License (http://creativecommons.org/licenses/by/3.0), which permits unrestricted use, distribution, and reproduction in any medium, provided the original work is properly cited.

References

[1] Lunk H-J. Discovery, properties and applications of chromium and its compounds. ChemTexts. 2015;**1**(6):1-17

[2] Anderson RA. Chromium as an essential nutrient for humans. Regulatory Toxicology and Pharmacology. 1997;**26**(1):S35-S41

[3] Pechova A, Pavlata L. Chromium as an essential nutrient: A review. Veterinární Medicína. 2007;**52**(1):1-18

[4] Wallach S. Clinical and biochemical aspects of chromium deficiency. Journal of the American College of Nutrition. 1985;**4**(1):107-120

[5] Vincent JB. New evidence against chromium as an essential trace element. The Journal of Nutrition. 2017;**147**(12):2212-2219

[6] Katz SA, Salem H. The toxicology of chromium with respect to its chemical speciation: A review. Journal of Applied Toxicology. 1993;**13**(3):217-224

[7] Terpilowska S, Siwicki AK. Pro- and antioxidant activity of chromium(III), iron(III), molybdenum(III) or nickel(II) and their mixtures. Chemico-Biological Interactions. 2019;**298**(25):43-51. DOI: 10.1016/j.cbi.2018.10.028

[8] Kondo K, Takahashi Y, Ishikawa S, Uchihara H, Hirose Y, Yoshizawa K, et al. Microscopic analysis of chromium accumulation in the bronchi and lung of chromate workers. Cancer. 2003;**98**(11):2420-2429

[9] Stearns DM, Belbruno JJ, Wetterhahn KE. A prediction of chromium(III) accumulation in humans from chromium dietary supplements. The FASEB Journal. 1995;**9**(15):1650-1657

[10] Hepburn DDD, Vincent JB. In vivo distribution of chromium from chromium picolinate in rats and implications for the safety of the dietary supplement. Chemical Research in Toxicology. 2002;**15**(2):93-100

[11] Wu LE, Levina A, Harris HH, Cai Z, Lai B, Vogt S, et al. Carcinogenic chromium(VI) compounds formed by intracellular oxidation of chromium(III) dietary supplements by adipocytes. Angewandte Chemie International Edition. 2016;**55**(5):1742-1745

[12] Bregnbak D, Johansen JD, Jellesen MS, Zachariae C, Menné T, Thyssen JP. Chromium allergy and dermatitis: Prevalence and main findings. Contact Dermatitis. 2015;**73**(5):261-280

[13] Yoshihisa Y, Shimizu T. Metal allergy and systemic contact dermatitis: An overview. Dermatology Research and Practice. 2012;**2012**:1-5

[14] Matthews NH, Fitch K, Li W-Q, Morris JS, Christiani DC, Qureshi AA, et al. Exposure to trace elements and risk of skin cancer: A systematic review of epidemiologic studies. Cancer Epidemiology, Biomarkers & Prevention. 2019;**28**(1):3-21

[15] McCarty M. High-chromium yeast for acne? Medical Hypotheses. 1984;**14**(3):307-310

[16] Hsieh Y-T, Hsu T-H, Wang H-C, Chen K-S, Lee W-M. Trivalent chromium restore dexamethasone-induced attenuation effect of insulin-like growth factor-1 and promote skin wound healing in mice. Pakistan Veterinary Journal. 2019;**39**(2):199-204

[17] Janson M. Orthomolecular medicine: The therapeutic use of dietary supplements for anti-aging. Clinical Interventions in Aging. 2006;**1**(3):261-265

[18] Junaid M, Hashmi MZ, Malik RN, Pe D-S. Toxicity and oxidative stress induced by chromium in workers exposed from different occupational settings around the globe: A review. Environmental Science and Pollution Research. 2016;**23**(20):20151-20167

[19] Saha R, Nandi R, Saha B. Sources and toxicity of hexavalent chromium. Journal of Coordination Chemistry. 2011;**64**(10):1782-1806

[20] Liang XW, Xu ZP, Grice J, Zvyagin AV, Roberts MS, Liu X. Penetration of nanoparticles into human skin. Current Pharmaceutical Design. 2013;**19**(35):6353-6366

[21] Filon FL. Penetration of metals through the skin barrier. In: Chen JK, Thyssen JP, editors. Metal Allergy: From Dermatitis to Implant and Device Failure. Cham: Springer; 2018. pp. 67-74

[22] Hostynek JJ. Factors determining percutaneous metal absorption. Food and Chemical Toxicology. 2003;**41**(3):327-345

[23] Rowbotham AL, Levy LS, Shuker LK. Chromium in the environment: An evaluation of exposure of the UK general population and possible adverse health effects. Journal of Toxicology and Environmental Health, Part B. 2000;**3**(3):145-178

[24] Gammelgaard B, Fullerton A, Avnstorp C, Menné T. Permeation of chromium salts through human skin in vitro. Contact Dermatitis. 1992;**27**(5):302-310

[25] Pan T, Wang P, Huang C, Chen C, Fang J. Elucidation of the percutaneous absorption of chromium compounds by functional proteomics. Proteomics. 2009;**9**(22):5120-5131

[26] Rytter M, Haustein U-F. Hapten conjugation in the leucocyte migration inhibition test in allergic chromate eczema. The British Journal of Dermatology. 1982;**106**(2):161-168

[27] Valko M, Morris H, Cronin MTD. Metals, toxicity and oxidative stress. Current Medicinal Chemistry. 2005;**12**(10):1161-1208

[28] Sinigaglia F. The molecular basis of metal recognition by T cells. Journal of Investigative Dermatology. 1994;**102**(4):398-401. DOI: 10.1111/1523-1747.ep12372149

[29] Smart GA, Sherlock JC. Chromium in foods and the diet. Food Additives and Contaminants. 1985;**2**(2):139-147

[30] Hamilton EM, Young SD, Bailey EH, Watts MJ. Chromium speciation in foodstuffs: A review. Food Chemistry. 2018;**250**:105-112. DOI: 10.1016/j.foodchem.2018.01.016

[31] Zhitkovich A. Chromium in drinking water: Sources, metabolism, and cancer risks. Chemical Research in Toxicology. 2011;**24**(10):1617-1629

[32] Basko-plluska JL, Thyssen JP, Schalock PC. Cutaneous and systemic hypersensitivity reactions to metallic implants. Dermatitis. 2011;**22**(2):65-79

[33] Vincent JB, Edwards KC. The absorption and transport of chromium in the body. In: Vincent JB, editor. The Nutritional Biochemistry of Chromium(III) [Internet]. 2nd ed. Amsterdam: Elsevier B.V.; 2019. pp. 129-174. DOI: 10.1016/B978-0-444-64121-2.00004-0

[34] Ducros V. Chromium metabolism. A literature review. Biological Trace Element Research. 1992;**32**(1-3):65-77

[35] Uter W, Werfel T, White IR, Johansen JD. Contact allergy: A review of current problems from a clinical perspective. International Journal of Environmental Research and Public Health. 2018;**15**(6):E1108

[36] Thyssen JP, Menne T. Metal allergy—A review on exposures, penetration, genetics, prevalence, and clinical implications. Chemical Research in Toxicology. 2010;**23**(2):309-318

[37] Wong CC, Gamboni SE, Palmer AM, Nixon RL. Occupational allergic contact dermatitis to chromium from cement: Estimating the size of the problem in Australia. The Australasian Journal of Dermatology. 2015;**56**(4):290-293

[38] Bregnbak D, Thyssen JP, Zachariae C, Johansen JD. Characteristics of chromium-allergic dermatitis patients prior to regulatory intervention for chromium in leather: A questionnaire study. Contact Dermatitis. 2014;**71**(6):338-347

[39] Hansen MB, Menne T, Johansen JD. Cr (III) and Cr (VI) in leather and elicitation of eczema. Contact Dermatitis. 2006;**54**(5):278-282

[40] Hedberg YS, Lidén C, Lindberg M. Chromium dermatitis in a metal worker due to leather gloves and alkaline coolant. Acta Dermato-Venereologica. 2016;**96**(1):104-105

[41] Hedberg YS, Erfani B, Matura M, Lidén C. Chromium (III) release from chromium-tanned leather elicits allergic contact dermatitis: A use test study. Contact Dermatitis. 2018;**78**(5):307-314

[42] Lim JH, Kim HS, Park YM, Lee JY, Kim HO. A case of chromium contact dermatitis due to exposure from a golf glove. Annals of Dermatology. 2010;**22**(1):63-65

[43] Thyssen JP, Jellesen MS, Møller P, Menne' T, Johansen JD. Allergic chromium dermatitis from wearing 'chromium-free' footwear. Contact Dermatitis. 2014;**70**(3):185-187

[44] Dikicier BS, Yaldız M, Çetinkaya R, Erdem T. Tattoo associated allergic contact dermatitis. Australian Journal of Dermatology. 2015;**2**(2):1037

[45] Seishima M, Oyama Z, Oda M. Cellular phone dermatitis with chromate allergy. Dermatology. 2003;**207**(1):48-50

[46] Tan S, Nixon R. Allergic contact dermatitis caused by chromium in a mobile phone. Contact Dermatitis. 2011;**65**(4):246-247

[47] Basketter DA, Angelini G, Ingber A, Kern PS, Menné T. Nickel, chromium and cobalt in consumer products: Revisiting safe levels in the new millennium. Contact Dermatitis. 2003;**49**(1):1-7

[48] Rustemeyer T, van Hoogstraten IMW, von Blomberg BME, Scheper RJ. Mechanisms of allergic contact dermatitis. In: Rustemeyer T, Elsner P, John S-M, Maibach HI, editors. Kanerva's Occupational Dermatology. Berlin, Heidelberg: Springer; 2012. pp. 113-146

[49] Buters J, Biedermann T. Chromium (VI) contact dermatitis: Getting closer to understanding the underlying mechanisms of toxicity and sensitization! Journal of Investigative Dermatology. 2017;**137**(2):274-277. DOI: 10.1016/j.jid.2016.11.015

[50] Saary J, Qureshi R, Palda V, Dekoven J, Pratt M, Skotnicki-grant S, et al. A systematic review of contact dermatitis treatment and prevention. Journal of the American Academy of Dermatology. 2005;**53**(5):845.e1-845.e13

[51] Lee Y-H, Su S-B, Huang C-C, Sheu H-M, Tsai J-C, Lin C-H, et al. N-Acetylcysteine attenuates hexavalent chromium-induced hypersensitivity through inhibition of cell death, ROS-related signaling and cytokine expression. PLoS One. 2014;**9**(9):e108317

[52] Bradberry SM, Vale JA. Therapeutic review: Is ascorbic acid of value in chromium poisoning and chromium dermatitis? Journal of Toxicology. Clinical Toxicology. 1999;**37**(2):195-200

[53] Lee I, Kim S, Shin I, Moon C, Park S, Kim S, et al. Protective effects of pine bark extract on hexavalent chromium-induced dermatotoxicity in rats. Phytherapy Research. 2012;**26**(10):1534-1540

[54] Wang B-J, Chiu H-W, Lee Y-L, Li C-Y, Wang Y-J, Lee Y-H. Pterostilbene attenuates hexavalent chromium-induced allergic contact dermatitis by preventing cell apoptosis and inhibiting IL-1β-related NLRP3 inflammasome activation. Journal of Clinical Medicine. 2018;**7**(12):E489

[55] Allenby CF, Goodwin BFJ. Influence of detergent washing powders on minimal eliciting patch test concentrations of nickel and chromium. Contact Dermatitis. 1983;**9**(6):491-499

[56] Wöhrl S, Kriechbaumer N, Hemmer W, Focke M, Brannath W, Götz M, et al. A cream containing the chelator DTPA (diethylenetriaminepenta-acetic acid) can prevent contact allergic reactions to metals. Contact Dermatitis. 2001;**44**(4):224-228

[57] Winnicki M, Shear NH. A systematic approach to systemic contact dermatitis and symmetric drug-related intertriginous and flexural exanthema (SDRIFE) a closer look at these conditions and an approach to intertriginous eruptions. American Journal of Clinical Dermatology. 2011;**12**(3):171-180

[58] Aquino M, Rosner G. Systemic contact dermatitis. Clinical Reviews in Allergy and Immunology. 2019;**56**(1):9-18

[59] Silvestri DL. Systemic contact dermatitis. Skin & Aging. 2012;(January):22-29

[60] Jensen CS, Lisby S, Larsen JK, Veien NK, Menné T. Characterization of lymphocyte subpopulations and cytokine profiles in peripheral blood of nickel-sensitive individuals with systemic contact dermatitis after oral nickel exposure. Contact Dermatitis. 2004;**50**(1):31-38

[61] Posadas SJ, Pichler WJ. Delayed drug hypersensitivity reactions—New concepts. Clinical and Experimental Allergy. 2007;**37**(7):989-999

[62] Lachapelle J-M. The spectrum of diseases for which patch testing is recommended patients who should be investigated. In: Lachapelle I-M, Maibach HI, Ri I, editors. Patch Testing-Prick Testing: A Practical Guide. New York: Springer-Verlag Berlin Heidelberg GmbH; 2003. p. 7-26

[63] Häusermann P, Harr T, Bircher AJ. Baboon syndrome resulting from systemic drugs: Is there strife between SDRIFE and allergic contact dermatitis syndrome? Contact Dermatitis. 2004;**51**(5-6):297-310

[64] Özkaya E. Current understanding of baboon syndrome. Expert Review of Dermatology. 2009;**4**(2):163-175

[65] Miyahara A, Kawashima H, Okubo Y, Hoshika A. A new proposal for a clinical-oriented subclassification of baboon syndrome and a review of baboon syndrome. Asian Pacific Journal of Allergy and Immunology. 2011;**29**(2):150-160

[66] Hubler WR Jr, Hubler WR Sr. Dermatitis from a chromium dental plate. Contact Dermatitis. 1983;**9**(5):377-387

[67] Shelnutt SR, Goad P, Belsito DV. Dermatological toxicity of hexavalent

chromium. Critical Reviews in Toxicology. 2007;**37**:375-387

[68] Fowler JF. Systemic contact dermatitis caused by oral chromium picolinate. Cutis. 2000;**65**(2):116

[69] Özkaya E, Topkarci Z, Özarmağan G. Systemic allergic dermatitis from chromium in a multivitamin/multimineral tablet. Contact Dermatitis. 2010;**62**(3):184-184

[70] Veien NK, Hattel T, Laurberg G. Chromate-allergic patients challenged orally with potassium dichromate. Contact Dermatitis. 1994;**31**(3):137-139

[71] Van Ulsen J, Stolz E, van Joost T. Chromate dermatitis from a homeopathic drug. Contact Dermatitis. 1988;**18**(1):56-57

[72] Pacheco KA. Allergy to surgical implants. Clinical Reviews in Allergy and Immunology. 2019;**56**(1):72-85

[73] Hedberg YS, Wallinder IO. Metal release from stainless steel in biological environments: A review. Biointerphases. 2016;**11**(1):018901. DOI: 10.1116/1.4934628

[74] Phedy P, Djaja YP, Boedijono DR, Wahyudi M, Silitonga J, Solichin I. Hypersensitivity to orthopaedic implant manifested as erythroderma: Timing of implant removal. International Journal of Surgery Case Reports. 2018;**49**:110-114. DOI: 10.1016/j.ijscr.2018.06.011

[75] Oleffe J, Wilmet J. Generalized dermatitis from an osteosynthesis screw. Contact Dermatitis. 1980;**6**(5):365-365

[76] Rostoker G, Robin J, Binet O, Blamoutier J, Paupe J, Lessana-Leibowitch M, et al. Dermatitis due to orthopaedic implants. A review of the literature and report of three cases. Journal of Bone and Joint Surgery. American Volume. 1987;**69**(9):1408-1412

[77] Balato N, Limbo G, Patruno C, Ayala F. Generalized dermatitis due to an osteosynthesis screw. Contact Dermatitis. 1991;**24**(4):310-310

[78] Gao X, He R, Yan S, Wu L. Case report: Dermatitis associated with chromium following total knee arthroplasty. Journal of Arthroplasty. 2011;**26**(4):665.e13-665.e16. DOI: 10.1016/j.arth.2010.06.002

[79] Thomsen M, Rozak M, Thomas P. Pain in a chromium-allergic patient with total knee arthroplasty: Disappearance of symptoms after revision with a special surface-coated TKA—A case report. Acta Orthopaedica. 2011;**83**(3):386-388

[80] Lampel HP, Silvestri DL. Systemic contact dermatitis: Current challenges and emerging treatments. Current Treatment Options in Allergy. 2014;**1**(4):348-357

[81] Wu G, Xiao X, Feng P, Xie F, Yu Z, Yuan W, et al. Gut remediation: A potential approach to reducing chromium accumulation using lactobacillus plantarum TW1-1. Scientific Reports. 2017;7:1-12. DOI: 10.1038/s41598-017-15216-9

[82] Sharma AD. Low chromate diet in dermatology. Indian Journal of Dermatology. 2009;**54**(3):293-295

[83] Ching HA, Choudhury D, Nine MJ, Osman NAA. Effects of surface coating on reducing friction and wear of orthopaedic implants. Science and Technology of Advanced Materials. 2014;**15**(1):014402

[84] Hallab NJ, Jacobs JJ. Biologic effects of implant debris. Bulletin of the NYU Hospital for Joint Diseases. 2009;**67**(2):182-188

[85] Burton L, Paget D, Binder NB, Bohnert K, Nestor BJ, Sculco TP, et al. Orthopedic wear debris mediated inflammatory osteolysis is mediated in part by NALP3 inflammasome activation. Journal of Orthopaedic Research. 2013;**31**(1):73-80

[86] Wang Y, Su H, Gu Y, Song X, Zhao J. Carcinogenicity of chromium and chemoprevention: A brief update. OncoTargets and Therapy. 2017;**10**:4065-4079

[87] Salnikow K, Zhitkovich A. Genetic and epigenetic mechanisms in metal carcinogenesis and cocarcinogenesis: Nickel, arsenic, and chromium. Chemical Research in Toxicology. 2008;**21**(1):28-44

[88] Langard S, Andersen A, Ravnestad J. Incidence of cancer among ferrochromium and ferrosilicon workers: An extended observation period. British Journal of Industrial Medicine. 1990;**47**(1):14-19

[89] Langrrd S. One hundred years of chromium and cancer: A review of epidemiological evidence and selected case reports. American Journal of Industrial Medicine. 1990;**17**(2):189-215

[90] Birk T, Mundt KA, Dell LD, Luippold RS, Miksche L, Steinmann-steiner-haldenstaett W, et al. Lung cancer mortality in the German chromate industry, 1958 to 1998. Journal of Occupational and Environmental Medicine. 2006;**48**(4):426-433

[91] Deng Y, Wang M, Tian T, Lin S, Xu P, Zhou L, et al. The effect of hexavalent chromium on the incidence and mortality of human cancers: A meta-analysis based on published epidemiological. Frontiers in Oncology. 2019;**9**:24

[92] Program NT. NTP Technical Report on the Toxicology and Carcinogenesis Studies of Sodium Dichromate Dihydrate (CAS NO. 7789-12-0) in F344/N Rats and B6C3F1 Mice. North Carolina; 2008

[93] Davidson T, Kluz T, Burns F, Rossman T, Zhang Q, Uddin A, et al. Exposure to chromium (VI) in the drinking water increases susceptibility to UV-induced skin tumors in hairless mice. Toxicology and Applied Pharmacology. 2004;**196**(3):431-437

[94] Uddin AN, Burns FJ, Rossman TG, Chen H, Kluz T, Costa M. Dietary chromium and nickel enhance UV-carcinogenesis in skin of hairless mice. Toxicology and Applied Pharmacology. 2007;**221**(3):329-338

[95] Lay PA, Levina A. Metal carcinogens. In: Reedijk J, Poeppelmeier K, editors. Comprehensive Inorganic Chemistry II [Internet]. 2nd ed. Amsterdam: Elsevier Ltd.; 2013. pp. 835-856. DOI: 10.1016/B978-0-08-097774-4.00333-8

[96] Rudolf E, Cervinka M, Cerman J, Schroterova L. Hexavalent chromium disrupts the actin cytoskeleton and induces mitochondria-dependent apoptosis in human dermal fibroblasts. Toxicology in Vitro. 2005;**19**(6):713-723

[97] Gibb HJ, Lees PSJ, Pinsky PF, Rooney BC. Clinical findings of irritation among chromium chemical production workers. American Journal of Industrial Medicine. 2000;**38**(2):127-131

[98] Boloorchi A, Sinna R, Benhaim T, Gobel F, Robbe M. Brûlure par acide chromique: Prévention systématique de la toxicité systémique. Annales de Chirurgie Plastique Esthétique. 2007;**52**:621-623

[99] Terrill PJ, Gowar JP. Chromic acid burns; beware, be aggressive, be watchful. British Journal of Plastic Surgery. 1990;**43**(6):699-701

[100] Lin C-C, Wu M-L, Yang C-C, Ger J, Tsai W-J, Deng J-F. Acute severe chromium poisoning after dermal exposure to hexavalent chromium. Journal of the Chinese Medical Association. 2009;**72**(4):219-221. DOI: 10.1016/S1726-4901(09)70059-0

[101] Laitung JK, Earley M. The role of surgery in chromic acid burns: Our experience with two patients. Burns. 1984;**10**(5):378-380

[102] Wang X-W, Davies JWL, Sirvent RLZ, Robinson WA. Chromic acid burns and acute chromium poisoning. Burns. 1985;**11**(3):181-184

[103] Ogawa M, Nakajima Y, Endo Y. Four cases of chemical burns thought to be caused by exposure to chromic acid mist. Journal of Occupational Health. 2007;**49**(5):402-404

[104] Mikulewicz M, Chojnacka K, Gedrange T, Górecki H. Reference values of elements in human hair: A systematic review. Environmental Toxicology and Pharmacology. 2013;**36**(3):1077-1086. DOI: 10.1016/j.etap.2013.09.012

[105] Bartlett RJ. Chromium cycling in soils and water: Links, gaps, and methods. Environmental Health Perspectives. 1991;**92**:17-24

[106] Testa SM. Sources of chromium contamination in soil and groundwater. In: Guertin J, Jacobs JA, Avakian CP, editors. Chromium(VI) Handbook. Florida: CRC Press; 2005. pp. 143-164

[107] Nocoń K, Kozłowska WR, Widziewicz K. Research on chromium and arsenic speciation in atmospheric particulate matter: Short review. E3S Web of Conferences. 2018;**28**:01026

[108] Horev L. Exogenous factors in hair disorders. Exogenous Dermatology. 2004;**3**(5):237-245

[109] Melnik BC, Plewig G, Daldrup T, Borchard F, Pfeiffer B, Zahn H. Green hair: Guidelines for diagnosis and therapy. Journal of the American Academy of Dermatology. 1986;**15**(5):1065-1068

[110] Bhat GR, Lukenbach ER, Kennedy RR, Parreira RM. The green hair problem: A preliminary investigation. Journal of the Society of Cosmetic Chemists. 1979;**30**(1):1-8

[111] Sarwar N, Imran M, Rashid M, Ishaque W, Asif M, Matloob A, et al. Chemosphere phytoremediation strategies for soils contaminated with heavy metals: Modifications and future perspectives. Chemosphere [Internet]. 2017;**171**:710-721. DOI: 10.1016/j.chemosphere.2016.12.116

[112] Salem HM, Eweida EA, Farag A. Heavy metals drinking water and their environmental impact on human health. In: Proceedings of International Conference for Environmental Hazards Mitigation (ICEHM 2000); Giza. 2000. p. 542-556

[113] Smart KE, Kilburn M, Schroeder M, Martin BG, Hawes C, Marsh JM, et al. Copper and calcium uptake in colored hair. Journal of Cosmetic Science. 2009;**60**:337-345

[114] Pierard GE. Toxic effects of metals from the environment on hair growth and structure. Journal of Cutaneous Pathology. 1979;**6**(4):237-242

[115] Trüeb RM. Systematic approach to hair loss in women. JDDG: Journal der Deutschen Dermatologischen Gesellschaft. 2010;**8**(4):284-297

[116] Uhlenhake E, Yentzer BA, Feldman SR. Acne vulgaris and depression: A retrospective examination. Journal of Cosmetic Dermatology. 2010;**9**(1):59-63

[117] Marques AR, Silva C, Colmonero S, Andrade P. Diabetes mellitus and polycystic ovary syndrome: Beyond a dermatological problem. Diabetes Case Reports. 2016;**1**(3):10000113

[118] Sadeeqa S, Mustafa T, Latif S. Polycystic ovarian syndrome—Related depression in adolescent girls: A review. Journal of Pharmacy & Bioallied Sciences. 2018;**10**(2):55-59

[119] Gåfvels C, Hägerström M, Rane K, Wajngot A, Wändell PE. Depression and anxiety after 2 years of follow-up in patients diagnosed with diabetes or rheumatoid arthritis. Health Psychology Open. 2016;**3**(2):1-12

[120] Moradi Tuchayi S, Makrantonaki E, Ganceviciene R, Dessinioti C, Feldman SR, Zouboulis CC. Acne vulgaris. Nature Reviews Disease Primers. 2015;**1**:15029. DOI: 10.1038/nrdp.2015.29

[121] Zouboulis CC, Böhm M. Neuroendocrine regulation of sebocytes—A pathogenetic link between stress and acne. Experimental Dermatology. 2004;**13**(Suppl 4):31-35

[122] Roberts A, Matthews JB, Socransky SS, Freestone PPE, Williams PH, Chapple ILC. Stress and the periodontal diseases: Effects of catecholamines on the growth of periodontal bacteria in vitro. Oral Microbiology and Immunology. 2002;**17**(5):296-303

[123] Boyanova L. Anaerobe stress hormone epinephrine (adrenaline) and norepinephrine (noradrenaline) effects on the anaerobic bacteria. Anaerobe [Internet]. 2017;**44**:13-19. DOI: 10.1016/j.anaerobe.2017.01.003

[124] Borrel V, Thomas P, Catovic C, Racine P-J, Konto-Ghiorghi Y, Lefeuvre L, et al. Acne and stress: Impact of catecholamines on cutibacterium acnes. Frontiers in Medicine. 2019;**6**(July):155

[125] Fox L, Csongradi C, Aucamp M, Plessis J, Gerber M. Treatment modalities for acne. Molecules. 2016;**21**(8):E1063

[126] Amr N, Abdel-Rahim HE. The effect of chromium supplementation on polycystic ovary syndrome in adolescents. Journal of Pediatric and Adolescent Gynecology. 2015;**28**(2):114-118. DOI: 10.1016/j.jpag.2014.05.005

[127] Jamilian M, Bahmani F, Siavashani MA, Mazloomi M, Asemi Z, Esmaillzadeh A. The effects of chromium supplementation on endocrine profiles, biomarkers of inflammation, and oxidative stress in women with polycystic ovary syndrome: A randomized, double-blind, placebo-controlled trial. Biological Trace Element Research. 2016;**172**(1):72-78

[128] Russell KO. Method of administration of chromium and magnesium sulfate for treatment of acne. United States of America; US 9,642.877 B1. 2017

[129] Goodless DR. Composition and method for treatment of acne. Unites States of America; WO 2005/049060 A1. 2005

[130] Fitzjarrell EA. Method and composition for treating acne. United States of America; US00575.9559A. 1998

[131] Sardana K, Garg VK. An observational study of methionine-bound zinc with antioxidants for mild to moderate acne vulgaris. Dermatologic Therapy. 2010;**23**(4):411-418

[132] Rubin MG, Kim K, Logan AC. Acne vulgaris, mental health and omega-3 fatty acids: A report of cases. Lipids in Health and Disease. 2008;**7**(36):1-5

[133] Yousofvand N, Zarei F, Ghanbari A. Exogenous testosterone, finasteride and castration effects on testosterone, insulin, zinc and chromium in adult male rats. Iranian Biomedical Journal. 2013;**17**(1):49-53

[134] Makrantonaki E, Ganceviciene R, Zouboulis C. An update on the role of the sebaceous gland in the pathogenesis of acne. Dermatoendocrinology. 2011;**3**(1):41-49

[135] Davidson JRT, Abraham K, Connor KM, McLeod MN. Effectiveness of chromium in atypical depression: A placebo-controlled trial. Biological Psychiatry. 2003;**53**(3):261-264

[136] Mcleod MN, Golden RN. Chromium treatment of depression. The International Journal of Neuropsychopharmacology. 2000;**3**(4):311-314

[137] Docherty J, Sack D, Roffman M, Finch M, Komorowski J. A double-blind, placebo-controlled, exploratory trial of chromium picolinate in atypical depression: Effect on carbohydrate craving. Journal of Psychiatric Practice. 2005;**11**(5):302-314

[138] Szöllősi AG, Oláh A, Bíró T, Tóth BI. Recent advances in the endocrinology of the sebaceous gland. Dermatoendocrinology. 2018;**9**(1):e1361576

[139] Hendrick V, Gitlin M, Altshuler L, Korenman S. Antidepressant medications, mood and male fertility. Psychoneuroendocrinology. 2000;**25**(1):37-51

[140] Partridge L. The new biology of ageing. Philosophical Transactions of the Royal Society B. 2010;**365**(1537):147-154

[141] Niccoli T, Partridge L. Ageing as a risk factor for disease. Current Biology. 2012;**22**(17):R741-R752. DOI: 10.1016/j.cub.2012.07.024

[142] Varani J, Dame MK, Rittie L, Fligiel SEG, Kang S, Fisher GJ, et al. Decreased collagen production in chronologically aged skin: Roles of age-dependent alteration in fibroblast function and defective mechanical stimulation. The American Journal of Pathology. 2006;**168**(6):1861-1868

[143] Takema Y, Yorimoto Y, Kawai M, Imokawa G. Age-related changes in the elastic properties and thickness of human facial skin. The British Journal of Dermatology. 1994;**131**(5):641-648

[144] Hall GK, Phillips TJ. Skin and hormone therapy. Clinical Obstetrics and Gynecology. 2004;**47**(2):437-449

[145] Piérard GE. The quandary of climacteric skin ageing. Dermatology. 1996;**193**:273-274

[146] Mchugh D, Gil J. Senescence and aging: Causes, consequences, and therapeutic avenues. The Journal of Cell Biology. 2018;**217**(1):65-77

[147] Zhang S, Duan E. Fighting against skin aging: The way from bench to bedside. Cell Transplantation. 2018;**27**(5):729-738

[148] Ganceviciene R, Liakou AI, Theodoridis A, Makrantonaki E, Zouboulis CC. Skin anti-aging strategies. Dermatoendocrinology. 2012;**4**(3): 308-319

[149] McCarty MF. Homologous physiological effects of phenformin and chromium picolinate. Medical Hypotheses. 1993;**41**(4):316-324

[150] Hao J, Hao C, Zhang L, Liu X, Zhou X, Dun Y, et al. OM2, a novel oligomannuronate-chromium (III) complex, promotes mitochondrial biogenesis and lipid metabolism in 3T3-L1 adipocytes via the

AMPK-PGC1α pathway. PLoS One. 2015;**10**(7):e0131930

[151] Loubet PE. Topical mixture and method for dermal application to remove excess iron and other heavy metals from cellular tissue. United States of America; US 2013/0344166A1. 2013

[152] Russell KO. Method and composition for inhibiting aged skin. United States of America; US 2018/0318184 A1. 2018

[153] Tsoukalas D, Fragkiadaki P, Docea AO, Alegakis AK, Sarandi E, Thanasoula M, et al. Discovery of potent telomerase activators: Unfolding new therapeutic and anti-aging perspectives. Molecular Medicine Reports. 2019;**20**(4):3701-3708

[154] Lai M-H. Antioxidant effects and insulin resistance improvement of chromium combined with vitamin C and E supplementation for type-2 diabetes mellitus. Journal of Clinical Biochemistry and Nutrition. 2008;**43**(3):191-198

[155] Cheng H-H, Lai M-H, Hou W-C, Huang C-L. Antioxidant effects of chromium supplementation with type 2 diabetes mellitus and euglycemic subjects. Journal of Agricultural and Food Chemistry. 2004;**52**(5):1385-1389

[156] Anderson RA, Roussel A-M, Zouari N, Mahjoub S, Matheau J-M, Kerkeni A. Potential antioxidant effects of zinc and chromium supplementation in people with type 2 diabetes mellitus. Journal of the American College of Nutrition. 2001;**20**(3):212-218

[157] Yanardag R, Peksel A, Yesilyaprak B, Doger MM, Arisan-Atac I. Effects of a combination of niacin and chromium (III)-chloride on the skin and lungs of hyperlipemic rats. Biological Trace Element Research. 2005;**103**(3):249-260

[158] Chen L, Zhang J, Zhu Y, Zhang Y. Interaction of chromium(III) or chromium(VI) with catalase and its effect on the structure and function of catalase: An in vitro study. Food Chemistry. 2018;**244**:378-385. DOI: 10.1016/j.foodchem.2017.10.062

[159] Panasci K. Burns and wounds. In: Paz JC, West MP, editors. Acute Care Handbook for Physical Therapists [Internet]. 4th ed. Missouri: Elsevier Inc.; 2014. pp. 283-311. DOI: 10.1016/B978-1-4557-2896-1.00012-3

[160] Lindenbaum E. Compositions and methods for stimulating wound healing. United States of America; US 10,292,997 B2. 2019

[161] Broughton G, Janis J, Attinger C. Wound healing: An overview. Plastic and Reconstructive Surgery. 2006;117(Suppl 7):1e.S-32e.S

[162] Cañedo-dorantes L, Cañedo-ayala M. Skin acute wound healing: A comprehensive review. International Journal of Inflammation. 2019;**2019**:ID 3706315

[163] Guo S, DiPietro LA. Factors affecting wound healing. Journal of Dental Research. 2010;**89**(3):219-229

[164] de Moraes SP, Chaves FR, Banci S, Rover PA, Georgetti F, Neto JA d R. Zinco e cromo na cicatrização de feridas em ratos normais e diabéticos. Revista do Colégio Brasileiro de Cirurgiões. 2000;**27**(6):394-399

[165] Peng Z, Qiao W, Wang Z, Dai Q, He J, Guo C, et al. Chromium improves protein deposition through regulating the mRNA levels of IGF-1 , IGF-1R, and Ub in rat skeletal muscle cells. Biological Trace Element Research. 2010;**137**(2):226-234

[166] Spravchikov N, Sizyakov G, Gartsbein M, Accili D, Tennenbaum T,

Wertheimer E. Glucose effects on skin keratinocytes: Implications for diabetes skin complications. Diabetes. 2001;**50**(7):1627-1635

[167] Snedeker JG, Gautieri A. The role of collagen crosslinks in ageing and diabetes—The good, the bad, and the ugly. Muscles, Ligaments and Tendons Journal. 2014;**4**(3):303-308

[168] Odukanmi OA, Salami AT, Koda K, Morakinyo OL, Olaleye SB. Trivalent chromium promotes healing of experimental colitis in mice by suppression of inflammation and oxidative stress. Journal of Bioscience and Medicine. 2017;**5**(8):108-126

[169] Lhotka CG, Szekeres T, Fritzer-Szekeres M, Schwarz G, Steffan I, Maschke M, et al. Are allergic reactions to skin clips associated with delayed wound healing? American Journal of Surgery. 1998;**176**(4):320-323

[170] Perfetto B, Stellavato A, Melito A, De Gregorio V, Cammarota M, Giuliano M. A time-lapse approach to examine chromium and nickel effects on wound healing in vitro. Journal of Immunotoxicology. 2012;**9**(4):392-400

Section 3

Removal and Remediation Techniques

Chapter 8

Polyaniline-Based Nanocomposites for Environmental Remediation

Thabiso C. Maponya, Mpitloane J. Hato, Thabang R. Somo, Kabelo E. Ramohlola, Mogwasha D. Makhafola, Gobeng R. Monama, Arjun Maity, Kwena D. Modibane and Lebogang M. Katata-Seru

Abstract

With growth in civilisation and industrialisation, there is an increase in the release of toxic heavy metal ions and dyes into water system, which is of public concern. As a result, appropriate treatment methods have to be implemented in order to mitigate and prevent water pollution. The discovery of nanotechnology has led to the development and utilisation of various nanoadsorbent for the removal of pollutants from water. PANI nanostructures and nanocomposites are noble adsorbents that have gained popularity in addressing water pollution issues and have been reported in literature. In this chapter, the main focus is on the synthesis of PANI nanocomposites and nanostructures and their application as efficient adsorbents for water treatment. Detailed discussions on different synthetic routes and characterisation have been dedicated to applications of these materials and are compared for the adsorptive removal of heavy metal ions and dyes from water.

Keywords: conducting polymers, polyaniline, composites, water remediation

1. Introduction

The existence of various toxic pollutants in natural water systems originating from different activities such as textile industry, mining operation and steel manufacturing is still a global challenge [1]. Hence, it is vital to develop efficient technologies to remove these pollutants from wastewater prior release into the environment. Among several water treatment processes (coagulation, membrane filtration, chemical reduction and precipitation, reverse osmosis), adsorption technology is deemed an efficient technology to treat wastewater due to exceptional advantages such as low cost, high removal efficiency and regeneration [1]. For many years, there has been an ongoing research on the development of high affinity materials towards water pollutants, which can effectively adsorb and remove them from industrial wastewater [2]. Conducting polymers (CPs) including polypyrrole (PPy), polythiophene (PTh), polyacetylene (PA) and polyaniline (PANI) have attained great attention for water treatment, owing to their intriguing properties such as ease synthesis, tunable structure and the presence of ideal functional groups [3]. Hence, CPs have been widely applied in materials such as rechargeable batteries, chemical sensors, electrochromic devices, surface coating for corrosion protection and water treatment [4]. Among these conducting polymers, PANI has been widely studied due to its low cost, ease of synthesis, good environmental stability, unique doping/de-doping property and relatively high conductivity [5]. The discovery of PANI dates back to about 180 years ago to the experiments made by Runge [6]. From his work reported in 1834, he discovered that a dark green PANI changed to black when the mixture of copper(II) chloride and aniline nitrate is heated on a porcelain plate to 100°C [6, 7]. Anciently, PANI was known as 'aniline black', after forming an undesirable black powder deposit on the anode during oxidation of aniline [8, 9] and is the most stable CP that can be easily protonated (with an acid) to increase conductivity or deprotonated (with a base) to reduce its conductivity [10]. In 1862, Letheby prepared it through oxidation of aniline under mild conditions [9, 11]. Attempt to control the synthesis conditions of polyaniline grew until in the 1910s when Green and Woodhead managed successfully to control the conditions, which led to the discovery of its four oxidation states [8]. This was followed by Jozefowic's group in the 1960s and 1970s for better understanding of the material [6, 8]. After this, the study of polyaniline with other (intrinsic conducting polymers) ICPs increased tremendously worldwide and were studied for different applications.

2. Synthesis of polyanilines

Polyaniline have different chemical structures which is attributed to the oxidation state of the polymer backbone, hence it exists in different states [12, 13]. The general representation of the polyaniline structure can be described by the following structural formula (**Figure 1**):

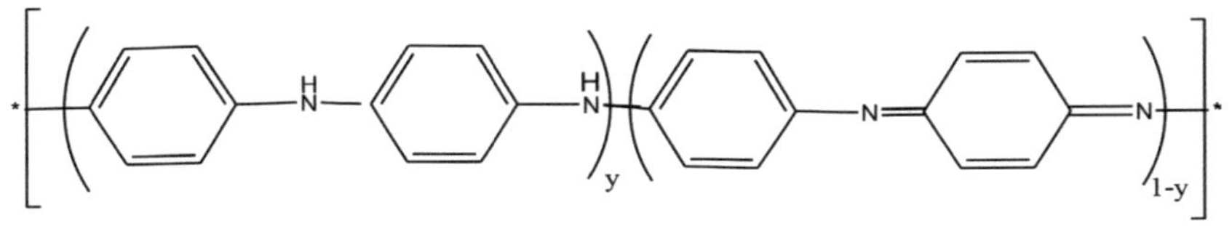

Figure 1.
The general structure of polyaniline.

were y = 1, 0.5 and 0 correspond to fully reduced polyaniline (leucoemeraldine), the half oxidised polyaniline (emeraldine) and fully oxidised polyaniline (pernigraniline), respectively [13]. Upon doping, PANI can be interconverted from one oxidation state to another [13]. The interconversions can be clearly deduced as presented in **Figure 2**.

There are two general methods which are employed to synthesise conducting polymers through electrochemical oxidation of the monomers or chemical oxidation of the monomers, namely: electrochemical polymerisation and chemical polymerisation.

2.1 Electrochemical polymerisation

Electrochemical polymerisation or method can be carried out by employing one of the three techniques: (i) applying a constant current (galvanostatic), or (ii) applying a constant potential (potentiostatic), and lastly (iii) by applying a potential scanning/cycling to the aqueous solution of aniline [14]. Polymerisation process is performed in strongly acidic aqueous electrolyte using a radical polymerisation mechanism which allows a formation of anilinium radical cation

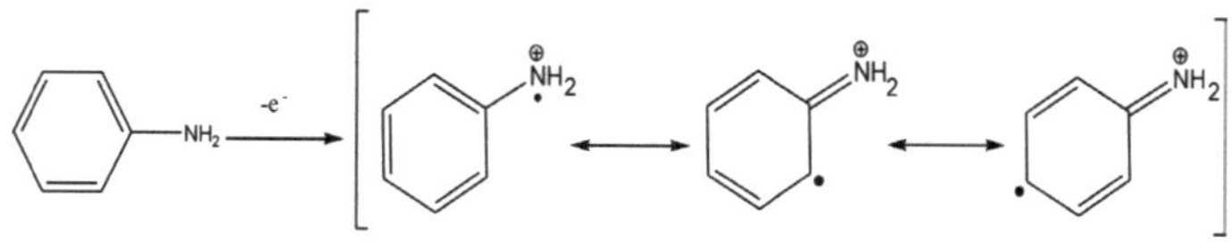

Figure 2.
Oxidation of aniline monomer during polymerisation of aniline [21, 22].

by aniline oxidation on the electrode[15, 16]. Electrochemical polymerisation strongly depends on the following factors:

a. low pH which is needed for preparation of conductive polymeric materials,

b. the dopant anion incorporated into polymer to determine the morphology, conductivity, rate of polymerisation growth and influences degradation process and

c. inert electrode such as Pt, Au and graphite [17].

The electrochemical process is more advantageous since film properties such as thickness and conductivity can be controlled by the synthesis parameters, including the current density, substrate, pH, nature and concentration of electrolyte [18].

2.2 Chemical polymerisation

Like electrochemical polymerisation, chemical polymerisation is also carried out in an acidic medium such as hydrochloric acid (HCl) and formic acid which helps in yielding primary polymer [19]. For this process to occur, a dopant or oxidant is required. The dopant reagents used during this process are the oxidising agents such as ammonium persulfate (APS), ferric chloride ($FeCl_3$), hydrogen peroxide (H_2O_2) and ceric nitrate ($Ce(NO_3)_3$). The principal function of the oxidant is to withdraw a proton from an aniline molecule, without forming a strong coordination bond either with the substrate intermediate or with the final product [20].

The general mechanism involved during polymerisation of aniline proceeds dominantly via radical mechanisms. Radical mechanisms can be subdivided into initiation, chain propagation and termination steps, which results in stable intermediate resonance structures. The three different stages of polymerisation are illustrated from **Figures 2–4**.

2.2.1 Step 1: initiation step (oxidation of aniline monomer)

The initial step of aniline oxidative polymerisation is the generation of the aniline cation radical in the oxidation of aniline with an oxidant as shown in **Figure 2** [21, 22]. The aniline cation radical undergoes resonance to attain the most stable and reactive radical cation which is free from steric hindrances [23]. This step is the slowest step in the reaction, hence it's deemed as the rate determining step in aniline polymerisation [24].

2.2.2 Step 2: radical coupling and re-aromatisation

Head to tail coupling of the *N*- and *para*- radical cations takes place (**Figure 3**), yielding a dicationic dimer species. This dimer further undergoes the process of re-aromatisation which causes it to revert to its neutral state, yielding an

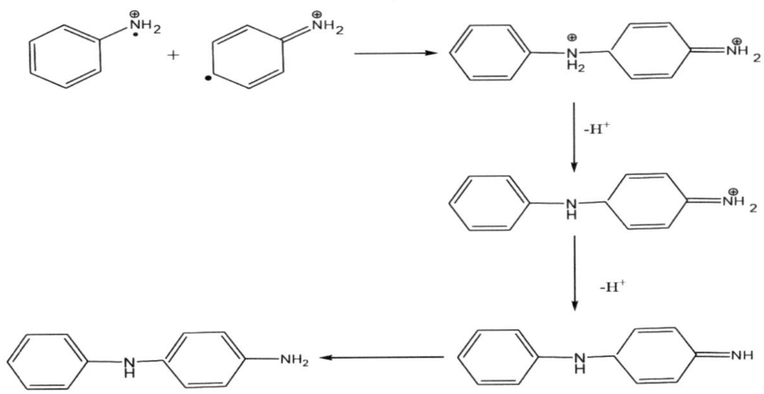

Figure 3.
Formation of a dimer [25, 27].

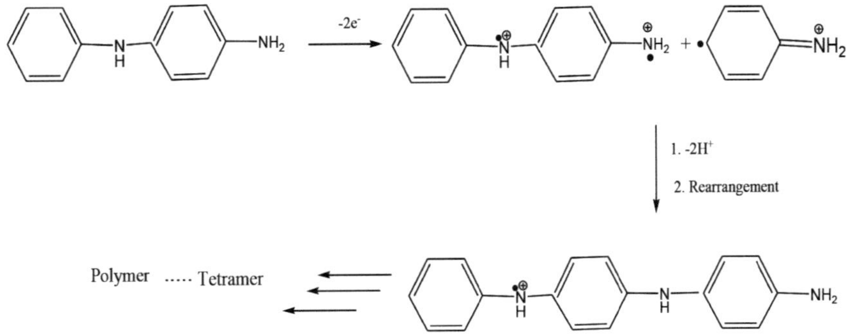

Figure 4.
Formation of a trimer and polymer formation [27].

intermediate referred to as *p*-aminodiphenylamine (PADPA) [25, 26]. These processes are also accompanied by the elimination of two protons.

2.2.3 Step 3: chain propagation

The dimers are immediately oxidised and then react with a stable aniline cation radical via an electrophilic aromatic substitution, followed by deprotonation and rearrangements to afford the trimer as seen in **Figure 4** [27]. The trimer further undergoes oxidation and reacts with aniline cation radical to form a tetramer and so on.

3. Synthesis and characterisation of polyaniline nanocomposites

Different nanocomposites based on polyaniline have been reported for waste-water remediation in order to enhance the removal efficiency of polyaniline. Depending on the method of synthesis, a variety of PANI based nanostructures and nanocomposites can be developed. For example, Ren et al. [28] reported PANI/PAN (polyacrylonitrile) nanocomposite synthesised via in–situ polymerisation of ANI monomer using DBSA and APS to obtain a nanofibrous structures as depicted in **Figure 5a**. Rachna and co-workers [29] reported zinc ferrite-PANI nanocomposite prepared following similar preparation method and monomer using toluene as a solvent and $CuSO_4$. The SEM image (**Figure 5b**) showed that the nanocomposite

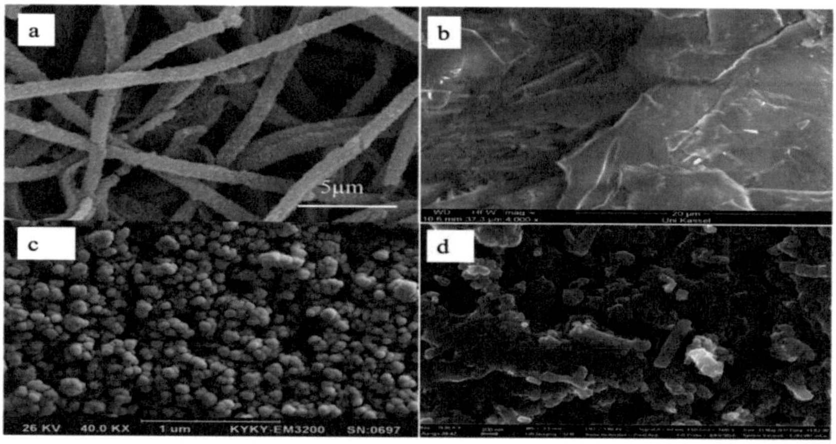

Figure 5.
SEM images of different structures of PANI nanocomposites (a) fibres [28], (b) smooth surface [29], (c) spherical [30] and (d) tubular [2].

had a smooth surface. Tanzifi et al. [30] prepared PANI/carboxymethyl cellulose/titanium dioxide PANI/CMC/TiO_2 nanocomposite in acidic medium using the same polymerisation route and APS as an oxidising agent. The nanocomposite structure formed uniform spherical particles as indicated in **Figure 5c**. PANI@$Ni(OH)_2$ nanocomposite was prepared by Bhaumik et al. [2] via in–situ polymerisation using APS and hydrazine hydrate. The SEM image showed a tubular structure as represented in **Figure 5d**.

Hallajiqomi et al. [31] synthesised PANI/PVP nanocomposite via in–situ polymerisation using KIO_3 and PVP as oxidising agent and surfactant, respectively. The nanocomposite showed irregular structural morphology. PANI/reduced graphene oxide (RGO) nanocomposite was synthesised by Li et al. [32] following the same polymerisation route using APS as an oxidising agent. The nanocomposite exhibited a lamellar structure. In another study, Harijan and Chandra [33] reported a PANI-GO nanocomposite synthesised by similar preparation method and the nanocomposite was composed of sheet-like morphology. Wang et al. [34] synthesised PANI/α-ZrP with plate–like structures via in–situ polymerisation using the same oxidising agent. In another study, Abdolahi et al. [35] synthesised uniform PANI nanofibers through interfacial polymerisation with different sizes ranging from micro- to nanometers. Gold-polyaniline (AuPANI) nanocomposite was prepared by simple interfacial polymerisation, performed in an immiscible water/toluene biphasic system using tetrachloroaurate, as an oxidant [36]. The TEM images of AuPANI nanocomposite showed rod-like Au nanoparticles embedded in a PANI matrix (see **Figure 6a–d**). Dhachanamoorthi et al. [37] prepared PANI-iron oxide (Fe_3O_4) ternary nanocomposites with improved crystallinity upon addition of Fe_3O_4 by mechanical mixing approach. Similar method was used for the synthesis of PANI-zinc oxide (ZnO) nanocomposites with enhanced electrical conductivity and homogeneous distribution on ZnO nanoparticles in the polymer matrix [38]. Basavaiah et al. [39] prepared polyaniline nanorods and magnetite nanoparticles via self-assembly route.

Table 1 shows some of the PANI based nanocomposites reported for the removal of pollutants from wastewater. The most commonly used method is the in–situ chemical polymerisation, which can result in various morphological structures. The structure of the nanocomposite is strongly affected by the type of oxidant, surfactant or stabiliser, the precursor and the ratio of the precursor to that of ANI.

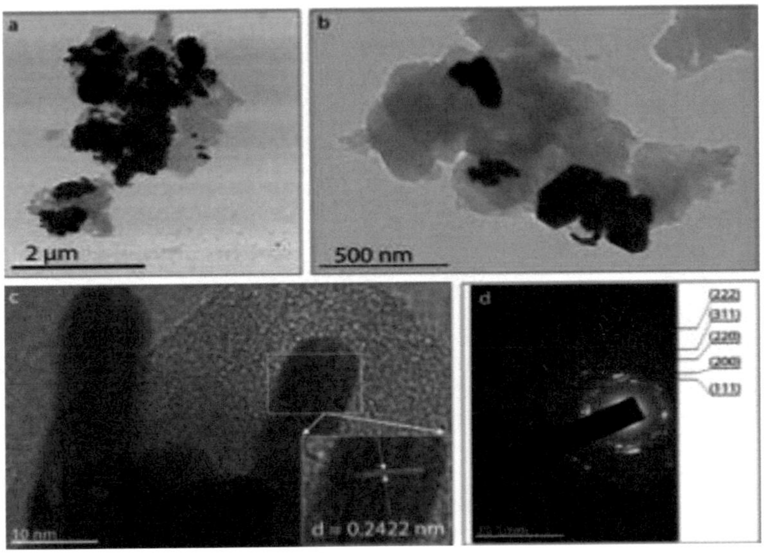

Figure 6.
TEM images of AuPANI nanocomposite with different magnifications (a, b, c) showing rod-like Au nanoparticles embedded in a PANI synthesised by interfacial polymerisation and (d) corresponding selected area electron diffraction (SAED) [36].

Adsorbent	Synthesis method, oxidant and/or other reagents	Morphology	Refs.
PANI-PPy	a, $FeCl_3$	Fibres	[40]
PANI nanoadsorbent	a, APS, stabiliser	Spherical particles	[41]
PANI/tin(II) molybdophosphate	In-situ method	Spherical particles	[42]
PANI/Fe^0	a, $FeCl_3$	Fibrous structure	[43]
PANI/ZrO_2	Direct mixing, APS	Irregular rougher surface	[44]
PANI–ZnO	a, APS	Flaky structure	[45]
PANI/SiO_2	a, APS, NaCMCNa	Uniform spherical particles	[46]
Fe_3O_4/G/PANI	a, APS	Mixture of sheets and spherical particles	[47]
PANI/zeolite	a, APS	Mixture of sheets and tubular particles	[48]
PANI/MnO_2/TiO_2	a, $KMnO_2$	Aggregated spherical particles	[49]
PPy-PANI/Fe_3O_4	a, $FeCl_3$	roughly spherical particles	[3]

a = in-situ polymerisation, NaCMCNa = sodium carboxymethyl cellulose sodium.

Table 1.
Some of PANI nanocomposites and their reported structures.

4. Polyaniline and its composites for wastewater treatment

The application of PANI for wastewater treatment has been widely studied owing to its exceptional structure that comprises abundant amine and imine functional groups [50, 51]. The nitrogen atoms on these functional groups have lone pairs of electrons to facilitate chelation and adsorption of pollutants [28, 52]. However, PANI has disadvantages such as poor mechanical properties and processability as well as low solubility. These limitations emanate from its high conjugation and strong electrostatic interaction between chains, which decreases its performance and limit its commercial application [51, 53]. Composites formation offers potential in addressing the above shortfalls. Polyaniline composites can be regarded as a material consisting of PANI matrix and one or more components such as semiconductors, metal nanoparticles, organic compounds, inorganic compounds as well as biological and natural products in order to improve polymer backbone properties or extend its functionalities [51, 54]. In order to prepare the composite materials successfully, several methods like physical mixing, sol–gel technique, *in–situ* chemical polymerisation, emulsion technology, sonochemical process and irradiation technique are employed [55]. Since then, various PANI composites have been reported for the removal of pollutants from wastewater using membrane technology [56, 57], chemical reduction [50], photocatalytic degradation [58, 59] and adsorption technology [60, 61]. Among these methods, adsorption has been identified as a prestigious technology, due to its flexibility and simplicity of design, initial cost, ease operation and insensitivity to toxic pollutants [62]. Different polyaniline-based composites for adsorption of pollutants have been widely studied. Samani et al. [60] reported PANI/PEG (polyethylene glycol) composite for the removal of hexavalent chromium ions (Cr(VI)), which had the maximum adsorption capacity (q_{max}) of 68.97 mg/g. Debnath et al. [63] reported a PANI/lignocellulose composite with q_{max} of 1672.5 mg/g for Congo red (CR) removal. PANI/chitosan was studied by Janaki et al. [64] for the removal of dyes (CR, Coomassie Brilliant Blue (CBB), and Remazol Brilliant Blue R (RBBR)). They obtained the maximum capacities to be 322.58, 357.14 and 303.03 for CR, CBB and RBBR, respectively. PANI/silica (SiO_2) gel was reported by Karthik et al. [52] with q_{max} = 63.41 mg/g for the removal of Cr(VI). However, most of these PANI composites have lower removal efficiency and adsorption capacity, owing to their irregular structure resulting from agglomeration, poor water dispersion and reduced surface area [35]. Numerous researches have been focussed on the development of nanostructured PANI composites with improved surface area and maximum adsorption capacity [35, 53, 65, 66].

5. Application of polyaniline nanocomposites as adsorbents for water treatment

In recent years, polyaniline nanocomposites have been used as adsorbents for the removal of various pollutants from wastewater [30]. More studies have been focussed on the adsorption of organic dyes and heavy metals ions due to their good interaction with PANI nanocomposites functional groups [52]. In the study of interaction between PANI nanocomposites and pollutants, various adsorption parameters such as pH, contact time, adsorbent dose, temperature, nature of the adsorbent and concentration of the pollutants are investigated [38]. From these parameters, the efficiency and adsorption capacity of nanocomposites can be determined to confirm the potential of the PANI nanocomposites as adsorbents for water purification.

5.1 Adsorption of heavy metals

The general sources of heavy metals are weathering of rocks due to their abundance in nature and mining industries as a result of mineral process of metal ores [61]. Various heavy metals known to pollute water include nickel, cadmium, lead, mercury, chromium, arsenic and copper. The water pollution by these toxic metals is a global concern owing to their acute toxicity and enduring accumulation [67].

Heavy metals are problematic since they are mutagenic, carcinogenic, are not biodegradable and can form various species [68].

5.1.1 PANI nanocomposites for adsorption of heavy metals ions

A PANI/RGO nanocomposite was reported by Li *et al.* [32] for the removal of Hg(II) ions from aqueous solution. It was shown that the PANI/RGO had high equilibrium adsorption capacity in comparison to PANI (**Figure 7a**). The obtained monolayer maximum adsorption capacity was 1000 mg/g at pH 4, 33°C and 400 mg/L for 200 mg adsorbent dose. Bhaumik *et al.* [38] reported a PANI/Fe^0 nanocomposite for the removal of arsenic (As), which had Langmuir maximum adsorption capacity of 232.5 and 227.7 mg/g for both As(III) and As(V) at pH 7, 25°C and 1 mg/L for 10 mg of PANI/Fe^0 nanocomposite. The obtained pH effects results (**Figure 7b**) showed higher removal efficiency by the nanocomposite in comparison to the neat PANI. Harijan and Chandra [33] reported a PANI/GO for the removal Cr(VI) from aqueous solution. It was demonstrated that the nanocomposite had high Langmuir maximum capacity of 192 mg/g at pH 6.5, 30°C and 100 mg/L for 25 mg of PANI/GO in comparison to the neat PANI (**Figure 7c**). **Table 2** shows some of the PANI nanocomposites reported for removal of various heavy metal ions under different experimental conditions.

5.2 Adsorption of organic dyes

Organic pollutants are generally materials that comprise of aromatic rings in their structure. Numerous organic pollutants including dyes, chlorinated, aliphatic and phenolic compounds are carcinogenic and mutagenic [70]. Dyes are of major concern due to their wide application in textile, paper, pigment and plastic industries. Their presence in water systems results in water decolouration, which can

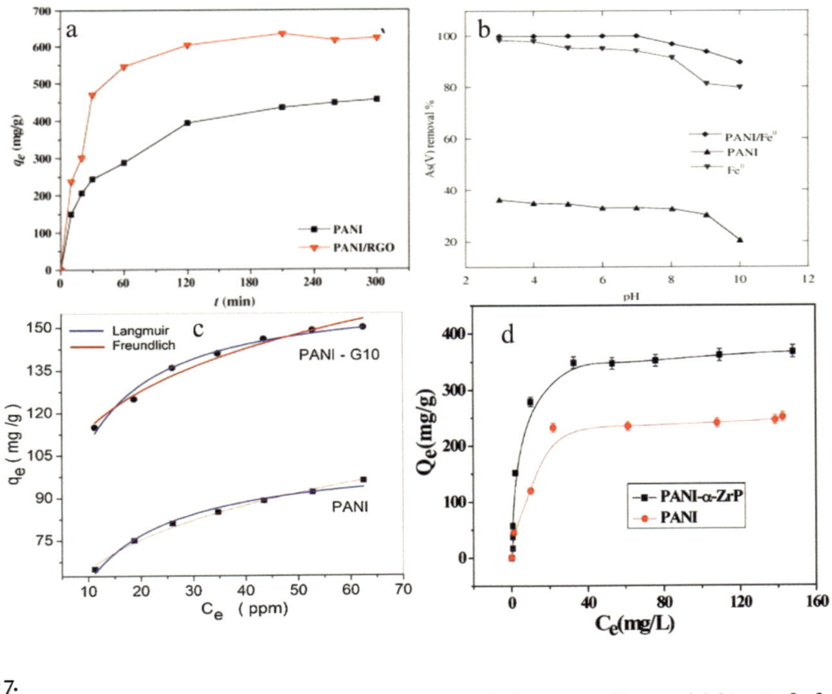

Figure 7.
Comparison of PANI and PANI nanocomposites for the removal of various pollutants (a) kinetics [32], (b) pH effect [43], (c) isotherms [33] and (d) equilibrium adsorption capacity [34].

Adsorbent	Pollutant	q_{max} (mg/g)	pH	Conc. (mg/L)	Adsorbent dose (mg)	Refs.
PANI/PAN	Cr(VI)	67.03	2	5	10	[28]
PAMpDA@Fe_3O_4	Co(II)	116.3	6	50	50	[69]
PPy-PANI/Fe_3O_4	Cr(VI)	303.0	2	100	50	[3]
Fe_3O_4/GO/PANI	Cr(VI)	153.4	6.5	100	50	[47]
PANI@$Ni(OH)_2$	Cr(VI)	625.0	4	100	10	[2]
PANI/zeolite	Cr(VI)	—	2	50	200	[48]
PANI/PVP	Mn(II)	50.30	7	100	250	[31]

Table 2.
Some of the PANI nanocomposites for heavy metals adsorption.

negatively affect aquatic life by influencing the photosynthetic process [71]. Some of PANI based nanocomposites for dye removal are given in **Table 3**.

5.2.1 PANI nanocomposites for adsorption of dyes

Wang et al. [34] reported PANI/α-ZrP for the removal of methyl orange (MO) cationic dye. It was demonstrated by **Figure 7d** that the nanocomposite (5 mg) had high removal efficiency capacity at pH 4 and 25°C for 100 mg/L MO solution. The monolayer maximum adsorption capacity was obtained to be 377 mg/g. Tanzifi et al. [46] reported PANI/SiO_2 nanocomposite for the removal of amido black 10B. The obtained Langmuir maximum adsorption capacity was 42.24 mg/g at pH 2, 25°C and 30 mg/L for a 100 mg adsorbent dose. Gharbani [37] reported PANI/tin(II)molybdophosphate for the removal of malachite green (MG). It was demonstrated that the removal efficiency was 93% for 50 mg/L of MG at pH 10 and the adsorption process followed Freundlich isotherm model. In another study, Ballav et al. [72] synthesised PANI coated ligno-cellulose composite (PLC) via in-situ polymerisation of aniline monomer for the removal of Reactive Black 5 (RB-5) from aqueous solutions. The authors reported that the equilibrium adsorption isotherm studies revealed that the Langmuir isotherm provided the best fit with monolayer adsorption capacity of 312 mg/g. The Maity's research group also reported the use of PANI-coated lignin-based adsorbent for the uptake of reactive dye eosin yellow (EY) from aqueous solution [73]. The adsorption capability of the adsorbent was found to be more effective than the unmodified adsorbent at lower pH.

Adsorbent	Pollutant	q_{max} (mg/g)	pH	Conc. (mg/L)	Adsorbent dose (mg)	Refs.
PANI/MWCNTs	CR	147	2	50	7	[74]
PANI/PA 6	MO	48.8	—	10	30	[54]
$ZnFe_2O_4$-PANI	Rhodamine B RHB	1000	2	10	500	[29]
PANI/CMC/TiO_2	CR	94.28	2.6	82	140	[30]
PANI/Fe^0	CR	99.6	7	100	1000	[75]
PPy–PANI NFs	CR	222.22	4	200	1000	[35]
Starch/PANI	Reactive Black 5	811.30	5	10	60	[76]

Table 3.
Some of the PANI based nanocomposites for dye adsorption.

6. Conclusions

This chapter encompasses the work done on PANI nanocomposites/nanostructures for the removal of heavy metal ions and dyes from aqueous solution. Adsorption technology was widely studied due to its flexibility in design and operational simplicity. The use of PANI nanocomposites as adsorbents has been investigated for adsorption of pollutants owing to their inherent properties such as high surface area, environmental stability, easy preparation and good water dispersion. The literature showed that a variety of morphological structures can be obtained, which depend strongly on the method of synthesis, and thus have various effects on the adsorption of the pollutant. For example, PANI based nanocomposites and nanostructures are preferable prepared via in-situ polymerisation. It was demonstrated that the pH, contact time, temperature, adsorbent dosage, concentration, nature of adsorbent and the presence of competing ions have an influence on the removal efficiency of nanocomposites. Hence, optimisation of adsorption parameters has been studied for the removal of heavy metals such Cr(VI), Hg(II), As(V), Co(II) and dyes such as CR, MB, MO, MG and RB5 from wastewater using various PANI nanocomposites and nanostructures. Overall, these nanocomposites display improved removal efficiency towards heavy metals and dyes adsorption. Hence, data in this chapter provides insight into PANI based materials for potential use as economically valuable adsorbents for the removal of dyes and heavy metals ions from wastewater.

Acknowledgements

KDM and MJH greatly acknowledge the financial support from the National Research Foundation (NRF) (Grant Nos. 117727 and 118113), Sasol Inzalo Foundation and University of Limpopo, South Africa.

Conflict of interest

We declare that there is no conflict of interest.

Author details

Thabiso C. Maponya[1,2], Mpitloane J. Hato[1,3]*, Thabang R. Somo[1], Kabelo E. Ramohlola[1], Mogwasha D. Makhafola[1], Gobeng R. Monama[1], Arjun Maity[2], Kwena D. Modibane[1]* and Lebogang M. Katata-Seru[4]

1 Department of Chemistry, School of Physical and Mineral Sciences, University of Limpopo (Turfloop), Sovenga, Polokwane, South Africa

2 DST/CSIR Innovation Centre, National Centre for Nanostructured Materials, CSIR Materials Science and Manufacturing, Pretoria, South Africa

3 Department of Environmental Sciences, College of Agriculture and Environmental Sciences, University of South Africa (UNISA), Florida Science Campus, Johannesburg, South Africa

4 Department of Chemistry, School of Physical and Chemical Sciences, North West University, Mmabatho, South Africa

*Address all correspondence to: mpitloane.hato@ul.ac.za and kwena.modibane@ul.ac.za

IntechOpen

© 2019 The Author(s). Licensee IntechOpen. This chapter is distributed under the terms of the Creative Commons Attribution License (http://creativecommons.org/licenses/by/3.0), which permits unrestricted use, distribution, and reproduction in any medium, provided the original work is properly cited.

References

[1] Bhaumik M, Maity A, Srinivasu VV, Onyango MS. Removal of hexavalent chromium from aqueous solution using polypyrrole-polyaniline nanofibers. Chemical Engineering Journal. 2012;**181**(182):323-333. DOI: 10.1016/j.cej.2011.11.088

[2] Bhaumik M, Kumar V, Maity A. Synergetic enhancement of Cr(VI) removal from aqueous solutions using polyaniline@Ni$(OH)_2$ nanocomposites adsorbent. Journal of Environmental Chemical Engineering. 2018;**6**: 2514-2527. DOI: 10.1016/j.jece.2018.03.047

[3] Kera NK, Bhaumik M, Pillay P, Ray SS, Maity A. Selective removal of toxic Cr(VI) from aqueous solution by adsorption combined with reduction at a magnetic nanocomposite surface. Journal of Colloid and Interface Science. 2017;**503**:214-228. DOI: 10.1016/j.jcis.2017.05.018

[4] Krishnani KK, Srinives S, Mohapatra BC, Boddu VM, Hao J, Meng X, et al. Hexavalent chromium removal mechanism using conducting polymers. Journal of Hazardous Materials. 2013;**252**(253):99-106. DOI: 10.1016/j.jhazmat.2013.01.079

[5] Thambidurai S, Pandiselvi K. Polyaniline/natural polymer composites and nanocomposites. In: Polyaniline Blends, Composites, and Nanocomposites. 2018. pp. 235-256. DOI: 10.1016/B978-0-12-809551-5.00009-6

[6] Rasmussen SC. The early history of polyaniline: Discovery and origins. Substantia. 2017;**1**:99-109. DOI: 10.13128/substantia-30

[7] Ciri G. Recent advances in polyaniline research: Polymerization mechanisms, structural aspects, properties and applications. Synthetic Metals. 2013;**177**:1-47. DOI: 10.1016/j.synthmet.2013.06.004

[8] Kang ET, Neoh KG, Tan KL. Polyaniline: A polymer with many interesting intrinsic redox states. Progress in Polymer Science. 1998;**23**:277-324. DOI: 10.1016/s0079-6700(97)00030-0

[9] Li D, Huang J, Kaner RB. Polyaniline nanofibers: A unique polymer nanostructure for versatile applications. Accounts of Chemical Research. 2009;**42**:135-145. DOI: 10.1021/ar800080n

[10] Parel MMP, Gillado AV, Herrera MU. Morphology and electrical conductivity of polyaniline coating on acetate film. Surfaces and Interfaces. 2018;**10**:74-77. DOI: 10.1016/j.surfin.2017.12.00

[11] Trojanowicz M. Application of conducting polymers in chemical analysis. Microchimica Acta. 2003;**143**:75-91. DOI: 10.1007/s00604-003-0066-5. Review article

[12] Sattari S, Reyhani A, Khanlari MR, Khabazian M, Heydari H. Synthesis of polyaniline-multi walled carbon nanotubes composite on the glass and silicon substrates and methane gas sensing behavior of them at room temperature. Journal of Industrial and Engineering Chemistry. 2014;**20**:1761-1764. DOI: 10.1016/j.jiec.2013.08.029

[13] Bhadra S, Khastgir D, Singha NK, Lee JH. Progress in preparation, processing and applications of polyaniline. Progress in Polymer Science. 2009;**34**:783-810. DOI: 10.1016/j.progpolymsci.2009.04.003

[14] Liu X, Zhang L, Li Y, Bian L, Su Z, Zhang L. Electropolymerization of aniline in aqueous solutions at pH 2 to

12. Journal of Materials Science. DOI: 10.1007/s10853-005-0854-x

[15] Gurunathan K, Murugan AV, Marimuthu R, Mulik UP, Amalnerkar DP. Electrochemically synthesised conducting polymeric materials for applications towards technology in electronics, optoelectronics and energy storage devices. Materials Chemistry and Physics. 1999;**61**:173-191. DOI: 10.1016/SO254-0584(99)00081-4

[16] Gvozdenovi MM, Jugovi BZ, Stevanovi JS, Grgur BN. Electrochemical synthesis of electroconducting polymers. Hemijska Industrija. 2014;**68**:673-684. DOI: 10.2298/HEMIND131122008G

[17] Jaymand M. Recent progress in chemical modification of polyaniline. Dedicated to professor Dr. Ali Akbar Entezami. Progress in Polymer Science. 2013;**38**:1287-1306. DOI: 10.1016/j.progpolymsci.2013.05.015

[18] Toshima N, Hara S. Direct synthesis of conducting polymers from simple monomers. Progress in Polymer Science. 1995;**20**:155-183. DOI: 10.1016/0079-6700(94)00029-2

[19] Campos TLA, Kersting DF, Ferreira CA. Chemical synthesis of polyaniline using sulphanilic acid as dopant agent into the reactional medium. Surface and Coatings Technology. 1999;**122**:3-5. DOI: 10.1016/S0257-8972(99)00399-0

[20] Ciri G, Konyushenko EN, Trchov M. Chemical oxidative polymerization of anilinium sulfate versus aniline: Theory and experiment. Synthetic Metals. *2008*;**158**:200-211. DOI: 10.1016/j.synthmet.2008.01.005

[21] Mazur M, Michota-kami A, Bukowska J. Surface-catalyzed growth of poly(2-methoxyaniline) on gold. Electrochimica Acta. 2007;**52**:5669-5676. DOI: 10.1016/j.electacta.2006.10.043

[22] Nicolas-debarnot D, Poncin-Epaillard F. Polyaniline as a new sensitive layer for gas sensors. Analytica Chimica Acta. 2003;**475**:1-15. DOI: 10.1016/S0003-2670(02)01229-1

[23] Cordeiro MA, Goncalves D, Bulhoes LO, Cordeiro JM. Synthesis and characterisation of poly-o-toulidine: Kinetics and structural aspects. Materials Science. 2005;**8**:5-10

[24] Sapurina IY, Stejskal J. The effect of pH on the oxidative polymerization of aniline and the morphology and properties of products. Russian Chemical Reviews. 2010;**79**:1123-1143. DOI: 10.1070/RC2010v079n12ABEH004140

[25] Tang HJ, Wang AT, Lin SY, Huang KY, Yang CC, Yeh JM, et al. Polymerization of aniline under various concentrations of APS and HCl. Polymer Journal. 2011;**43**:667-675. DOI: 10.1038/pj.2011.43

[26] Ahmed SM. Mechanistic investigation of the oxidative polymerization of aniline hydrochloride in different media. Polymer Degradation and Stability. 2004;**85**:605-614. DOI: 10.1016/j.polymdegradstab.2004.01.003

[27] Song E, Choi J. Conducting polyaniline nanowire and its applications in chemiresistive sensing. Nanomaterials. 2013;**3**:498-523. DOI: 10.3390/nano3030498

[28] Ren J, Huang X, Wang N, Lu K, Zhang X, Li W. Preparation of polyaniline-coated polyacrylonitrile fiber mats and their application to Cr(VI) removal. Synthetic Metals. 2016;**222**:255-266. DOI: 10.1016/j.synthmet.2016.10.027

[29] Rachna K, Agarwal A, Singh NB. Preparation and characterization of zinc ferrite—polyaniline nanocomposite for removal of rhodamine B dye from aqueous solution. Environmental

Nanotechnology Monitoring and Management. 2018;**9**:154-163. DOI: 10.1016/j.enmm.2018.03.001

[30] Tanzifi M, Yaraki MT, Karami M, Karimi S, Kiadehi AD, Karimipour K, et al. Modelling of dye adsorption from aqueous solution on polyaniline/carboxymethyl cellulose/TiO_2 nanocomposites. Journal of Colloid and Interface Science. 2018;**519**:154-173. DOI: 10.1016/j.jcis.2018.02.059

[31] Hallajiqomi M, Eisazadeh H. Adsorption of manganese ion using polyaniline and its nanocomposite: Kinetics and isotherm studies. Journal of Industrial and Engineering Chemistry. 2017;**55**:191-197. DOI: 10.1016/j.jiec.2017.06.045

[32] Li R, Liu L, Yang F. Preparation of polyaniline/reduced graphene oxide nanocomposite and its application in adsorption of aqueous Hg(II). Chemical Engineering Journal. 2013;**229**:460-468. DOI: 10.1016/j.cej.2013.05.089

[33] Harijan DKL, Chandra V. Polyaniline functionalized graphene sheets for treatment of toxic hexavalent chromium. Journal of Environmental Chemical Engineering. 2016;**4**:3006-3012. DOI: 10.1016/j.jece.2016.06.014

[34] Wang L, Wu X, Xu W, Huang X, Liu J, Xu A. Stable organic–inorganic hybrid of polyaniline/α-zirconium phosphate for efficient removal of organic pollutants in water environment. ACS Applied Materials and Interfaces. 2012;**4**:2686-2692. DOI: 10.1021/am300335e

[35] Abdolahi A, Hamzah E, Ibrahim Z, Hashim S. Synthesis of uniform polyaniline nanofibers through interfacial polymerization. Materials. 2012;**5**:1487-1494. DOI: 10.3390/ma5081487

[36] Bogdanovic U, Pasti I, Ciric-Marjanovic G, Mitric M, Ahrenkiel SP, Vodnik V. Interfacial synthesis of gold-polyaniline nanocomposite and its electrocatalytic application. ACS Applied Materials & Interfaces. 2015;**7**:28393-28403. DOI: 10.1021/acsami.5b09145

[37] Dhachanamoorthi K, Chandra L, Suresh P, Perumal K. Facile preparation and characterization of polyaniline-iron oxide ternary polymer nanocomposites by using "mechanical mixing" approach. Mechanics: Materials Science and Engineering. 2017;**9**. DOI: 10.2412/mmse.41.37.672

[38] Ansari SP, Mohammad F. Studies on nanocomposites of polyaniline and zinc oxide nanoparticles with supporting matrix of polycarbonate. ISRN Materials Science. 2012;**2012**:1-7. DOI: 10.5402/2012/129869

[39] Basavaiah K, Prasad MD, Rao AVP. Simultaneous synthesis of polyaniline nanorods and magnetite nanoparticles via self-assembly method. Journal of Experimental Nanoscience. 2014;**9**:491-500. DOI: 10.1080/17458080.2012.670276

[40] Bhaumik M, McCrindle R, Maity A. Efficient removal of Congo red from aqueous solutions by adsorption onto interconnected polypyrrole-polyaniline nanofibres. Chemical Engineering Journal. 2013;**28**:506-515. DOI: 10.1016/j.cej.2013.05.026

[41] Tanzi M, Hosseini SH, Kiadehi AD, Olazar M, Karimipour K, Rezaiemehr R, et al. Artificial neural network optimization for methyl orange adsorption onto polyaniline nano-adsorbent: Kinetic, isotherm and thermodynamic studies. Journal of Molecular Liquids. 2017;**244**:189-200. DOI: 10.1016/j.molliq.2017.08.122

[42] Gharbani P. Synthesis of polyaniline tin(II) molybdophosphate nanocomposite and application of it in the removal of dyes from aqueous

solutions. Journal of Molecular Liquids. 2017;**242**:229-234. DOI: 10.1016/j.molliq.2017.07.017

[43] Bhaumik M, Noubactep C, Gupta VK, McCrindle RI, Maity A. Polyaniline/Fe^0 composite nanofibers: An excellent adsorbent for the removal of arsenic from aqueous solutions. Chemical Engineering Journal. 2015;**271**:135-146. DOI: 10.1016/j.cej.2015.02.079

[44] Agarwal S, Tyagi I, Kumar V, Golbaz F. Synthesis and characteristics of polyaniline/zirconium oxide conductive nanocomposite for dye adsorption application. Journal of Molecular Liquids. 2016;**218**:494-498. DOI: 10.1016/j.molliq.2016.02.040

[45] Mostafaei A, Zolriasatein A. Synthesis and characterization of conducting polyaniline nanocomposites containing ZnO nanorods. Progress in Natural Science: Materials International. 2012;**22**:273-280. DOI: 10.1016/j.pnsc.2012.07.002

[46] Tanzifi M, Yaraki MT, Kiadehi AD, Hosseini SH, Olazar M, Bharti AK, et al. Adsorption of amido black 10B from aqueous solution using polyaniline/SiO_2 nanocomposite: Experimental investigation and artificial neutral network modeling. Journal of Colloid and Interface Science. 2018;**510**:246-261. DOI: 10.1016/j.jcis.2017.09.055

[47] Harijan DKL, Chandra V. Magnetite/graphene/polyaniline composite for removal of aqueous hexavalent chromium. Journal of Applied Polymer Science. 2016;**133**:1-7. DOI: 10.1002/app.44002

[48] Shyaa AA, Hasan OA, Abbas AM. Synthesis and characterization of polyaniline/zeolite nanocomposite for the removal of chromium(VI) from aqueous solution. Journal of Saudi Chemical Society. 2015;**19**:101-107. DOI: 10.1016/j.jscs.2012.01.001

[49] Vellaichamy V, Periakaruppan P, Nagulan B. Reduction of Cr^{6+} from wastewater using a novel in situ-synthesized PANI/MnO_2/TiO_2 nanocomposite: Renewable, selective, stable, and synergistic catalysis. ACS Sustainable and Chemical Engineering. 2017;**5**:9313-9324. DOI: 10.1021/acssuschemeng.7b02324

[50] Olad A, Nabavi R. Application of polyaniline for the reduction of toxic Cr(VI) in water. Journal of Hazardous Materials. 2007;**147**:845-851. DOI: 10.1016/j.jhazmat.2007.01.083

[51] Tiburcio G, Santos I, Ortega MM, Castillo JM, Castro T, Romero MA, et al. Selective adsorption of metallic complex using polyaniline or polypyrrole. Materials Chemistry and Physics. 2016;**182**:39-48. DOI: 10.1016/j.matchemphys.2016.07.003

[52] Karthik R, Meenakshi S. Removal of hexavalent chromium ions using polyaniline/silica gel composite. Journal of Water Processing and Engineering. 2014;**1**:37-45. DOI: 10.1016/j.jwpe.2014.03.001

[53] Ayad M, El-Hefnawy G, Zaghlol S. Facile synthesis of polyaniline nanoparticles; its adsorption behavior. Chemical Engineering Journal. 2013;**217**:460-465. DOI: 10.1016/j.cej.2012.11.099

[54] Xia Y, Li T, Chen J, Cai C. Polyaniline (skin)/polyamide 6 (core) composite fiber: Preparation, characterization and application as a dye adsorbent. Synthetic Metals. 2013;**175**:63-169. DOI: 10.1016/j.synthmet.2013.05.012

[55] Karim MR, Lim KT, Lee MS, Kim K, Yeum JH. Sulfonated polyaniline-titanium dioxide nanocomposites synthesized by one-pot UV-curable polymerization method. Synthetic Metals. 2009;**159**:209-213. DOI: 10.1016/j.synthmet.2008.09.001

[56] Ghaemi N, Zereshki S, Heidari S. Removal of lead ions from water using PES-based nanocomposite membrane incorporated with polyaniline modified GO nanoparticles: Performance optimization by central composite design. Process Safety and Environmental Protection. 2017;**111**:475-490. DOI: 10.1016/j.psep.2017.08.011

[57] Tan S, Laforgue A, Bélanger D. Characterization of a cation-exchange/polyaniline composite membrane. Langmuir. 2003;**19**:744-751. DOI: 10.1021/la0263054

[58] Soji DV, Bani ND, Lazarevi MJ, Cetojevi DD, Fin NL, Comor MI. Photocatalytic decomposition of selected biologically active compounds in environmental waters using TiO_2/polyaniline nanocomposites: Kinetics, toxicity and intermediates assessment. Environmental Pollution. 2018;**239**: 457-465. DOI: 10.1016/j.envpol.2018.04.039

[59] Miao J, Xie A, Li S, Huang F, Cao J, Shen Y. A novel reducing graphene/polyaniline/cuprous oxide composite hydrogel with unexpected photocatalytic activity for the degradation of Congo red. Applied Surface Science. 2016;**360**: 594-600. DOI: 10.1016/j.apsusc.2015.11.005

[60] Samani MR, Borghei SM, Olad A, Chaichi MJ. Removal of chromium from aqueous solution using polyaniline-polyethylene glycol composite. Journal of Hazardous Materials. 2010;**184**: 248-254. DOI: 10.1016/j.jhazmat.2010.08.029

[61] Kumar PA, Chakraborty S, Ray M. Removal and recovery of chromium from wastewater using short chain polyaniline synthesized on jute fiber. Chemical Engineering Journal. 2008;**141**:130-140. DOI: 10.1016/j.cej.2007.11.004

[62] Zare EN, Motahari A, Sillanpää M. Nanoadsorbents based on conducting polymer nanocomposites with main focus on polyaniline and its derivatives for removal of heavy metal ions/dyes: A review. Environmental Research. 2018;**162**:173-195. DOI: 10.1016/j.envres.2017.12.025

[63] Debnath S, Ballav N, Maity A, Pillay K. Development of a polyaniline-lignocellulose composite for optimal adsorption of Congo red. International Journal of Biological and Macromolecules. 2015;**75**:199-209. DOI: 10.1016/j.ijbiomac.2015.01.011

[64] Janaki V, Oh BT, Shanthi K, Lee KJ, Ramasamy AK, Kamala-Kannan S. Polyaniline/chitosan composite: An eco-friendly polymer for enhanced removal of dyes from aqueous solution. Synthetic Metals. 2012;**162**:974-980. DOI: 10.1016/j.synthmet.2012.04.015

[65] Wang J, Zhang K, Zhao L. Sono-assisted synthesis of nanostructured polyaniline for adsorption of aqueous Cr(VI): Effect of protonic acids. Chemical Engineering Journal. 2014;**239**:123-131. DOI: 10.1016/j.cej.2013.11.006

[66] Duruibe OJ, Ogwuegbu MOC, Egwurugwu JN. Heavy metal pollution and human biotoxic effects. International Journal of Physical Sciences. 2007;**2**:112-118

[67] Sounthararajah DP, Loganathan P, Kandasamy J, Vigneswaran S. Adsorptive removal of heavy metals from water using sodium titanate nanofibres loaded onto GAC in fixed-bed columns. Journal of Hazardous Materials. 2015;**287**:306-316. DOI: 10.1016/j.jhazmat.2015.01.067

[68] Choppala G, Kunhikrishnan A, Seshadri B, Hee J, Bush R, Bolan N. Comparative sorption of chromium species as influenced by pH, surface charge and organic matter content

in contaminated soils. Journal of Geochemical Exploration. 2018;**184**:255-260. DOI: 10.1016/j.gexplo.2016.07.012

[69] Zare EN, Lakouraj MM, Ramezani A. Effective adsorption of heavy metal cations by superparamagnetic poly(aniline-co-m-phenylenediamine)@ Fe_3O_4 nanocomposite. Advances in Polymer Technology. 2015;**34**:1-11. DOI: 10.1002/adv.21501

[70] Zhou Y, Zhang L, Cheng Z. Removal of organic pollutants from aqueous solution using agricultural wastes: A review. Journal of Molecular Liquids. 2015;**212**:739-762. DOI: 10.1016/j.molliq.2015.10.023

[71] Srinivasan A, Viraraghavan T. Decolorization of dye wastewaters by biosorbents: A review. Journal of Environmental Management. 2010;**1**:1915-1929. DOI: 10.1016/j.jenvman.2010.05.003

[72] Ballav N, Debnath S, Pillay K, Maity A. Efficient removal of reactive black from aqueous solution using polyaniline coated ligno-cellulose composite as a potential adsorbent. Journal of Molecular Liquids. 2015;**209**:387-396. DOI: 10.1016/j.molliq.2015.05.051

[73] Debnath S, Ballav V, Maity A, Pillay K. Efficient removal of reactive black from aqueous solution using polyaniline coated ligno-cellulose composite as a potential adsorbent. International Journal of Biological Macromolecules. 2015;**72**:732-739. DOI: 10.1016/j.ijbiomac.2014.09.018

[74] Shariri R, Mahmoodi NO. Synthesis and characterization of polypyrrole, polyaniline nanoparticles and their nanocomposite for removal of azo dyes; sunset yellow and Congo red. Journal of Cleaner Production. 2018;**179**:235-245. DOI: 10.1016/j.jclepro.2018.01.03

[75] Bhaumik M, McCrindle RI, Maity A. Enhanced adsorptive degradation of Congo red in aqueous solutions using polyaniline/Fe^0 composite nanofibers. Chemical Engineering Journal. 2015;**260**:716-729. DOI: 10.1016/j.cej.2014.09.014

[76] Janaki V, Vijayaraghavan K, Oh B, Lee K, Muthuchelian K, Ramasamya AK, et al. Starch/polyaniline nanocomposite for enhanced removal of reactive dyes from synthetic effluent. Carbohydrates Polymers. 2012;**90**:1437-1444. DOI: 10.1016/j.carbpol.2012.07.012

Chapter 9

Low Dimensional Nanostructures: Measurement and Remediation Technologies Applied to Trace Heavy Metals in Water

María Luisa García-Betancourt, Sandra I. Ramírez Jiménez, Apsahara González-Hodges, Zandra E. Nuñez Salazar, Ismailia Leilani Escalante-García and Jeannete Ramírez Aparicio

Abstract

A nanostructure is a system in which at least one external dimension is in the nanoscale, it means a length range smaller than 100 nm. Nanostructures can be natural or synthetic and determine the physicochemical properties of bulk materials. Due to their high surface area and surface reactivity, they can be an efficient alternative to remove contaminants from the environment, including heavy metals from water. Heavy metals like mercury (Hg), cadmium (Cd), arsenic (As), lead (Pb), and chromium (Cr) are highly poisonous and hazardous to human health due to their non-biodegradability and highly toxic properties, even at trace levels. Thus, efficient, low-cost, and environmentally friendly methodologies of removal are needed. These needs for removal require fast detection, quantification, and remediation to have heavy metal-free water. Nanostructures emerged as a powerful tool capable to detect, quantify, and remove these contaminants. This book chapter summarizes some examples of nanostructures that have been used on the detection, quantification, and remediation of heavy metals in water.

Keywords: nanostructures, remediation, water nanobioremediation, immobilized biological substrates, heavy metals

1. Introduction

Pollution due to heavy metals possesses a serious issue not only to human health but also to the environment and urban infrastructure. Heavy metals can be found in wastewater, groundwater, lakes, and streams, but also in soils or sediments. Heavy metals come from natural sources, but they can also originate from different anthropogenic activities (**Figure 1**). Human exposure to them has risen dramatically as a result of an exponential increase in their use in several industrial, agricultural, domestic, and technological applications [1]. Even though metals such as cobalt (Co), copper (Cu), iron (Fe), manganese (Mn), selenium (Se), or zinc (Zn)

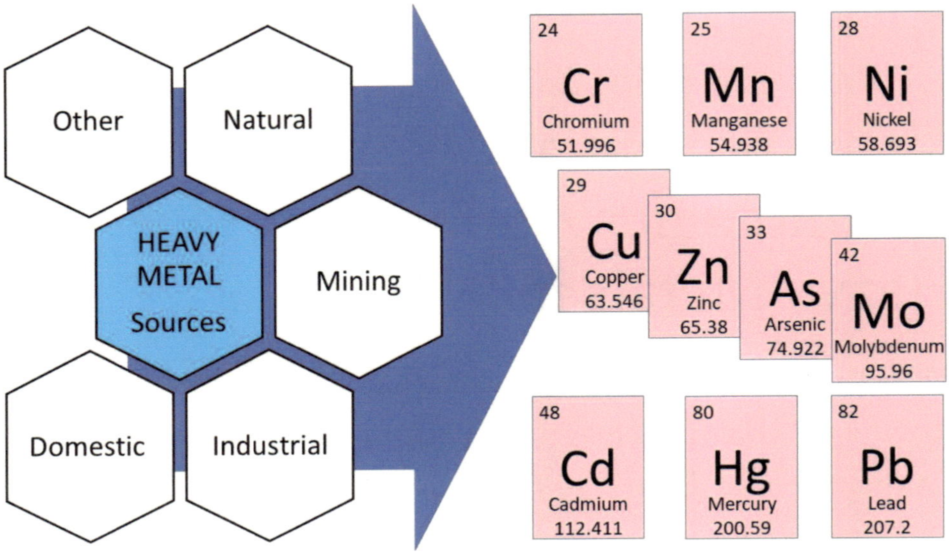

Figure 1.
The origin of some of the heavy metals released to the environment.

play essential biochemical roles in the cells of living beings, in high concentrations or as a result of long-term exposure, they are associated with cellular and tissue damage leading to adverse effects and diseases [2]. Notably, arsenic (As), cadmium (Cd), chromium (Cr), lead (Pb), and mercury (Hg) are ranked among the metals of great public health significance [1].

In order to minimize the damage to public health and other living organisms due to heavy metals, remediation appeared to rectify and redeem the environment after anthropogenic disturbance [3]. Remediation aims to implement and realize efficient actions for the eradication of diverse forms of pollution in three main activities: (1) remove formation damage, (2) testing before each remediation, and (3) routine treatment applications (maintenance) [4].

Remediation is just one of the multiple promising applications of nanotechnology. The utilization of nanotechnology concerning the environment classifies as follows: (1) the design and production of sustainable materials, (2) remediation using nanostructures, and (3) nano-based sensors [5]. Nanotechnology may contribute to any of the remediation facets, listed above, providing high selectivity (speciation and clean-up) and sensitivity (preconcentration) [6] to achieve trace levels in water, usually 100 microgram per gram.

The nanomaterials mainly used for removal are metal and iron oxides as they facilitate the precipitation and reduction of heavy metals to less harmful ionic species [7]. They are followed by carbonaceous materials such as fibers, carbon nanotubes, graphene, and graphene oxide. The porosity and surface area of carbonaceous materials allow them to easily form membranes for an efficient metal removal; besides, they serve as support of nanometals for easy recovery [8–10]. Other nanostructures based on silicon, such as silica and zeolite, are widely used due to its versatility and availability. The reactive and high surface area of porous silica allows the functionalization for multiple adsorptions of heavy metals, and zeolite allows selective separation of cations by ion exchange and adsorption from water [11, 12].

From this brief review, three main facts can be listed: (1) the importance of removing toxic heavy metals, (2) the role of nanotechnology on remediation, and (3) the use of diverse nanostructures for the removal of heavy metal from water.

In this context, the need for rapid and efficient removal of pollutants focused on toxic heavy metals with high selectivity and sensitivity from water is addressed in this work [1, 5, 6], considering the advantages offered by nanostructures. This book chapter summarizes some examples of nanostructures employed for wastewater treatment considering the three aspects listed above, as well as ways to reach heavy-metal concentrations lower than trace-level limit, after removal.

2. Nanotechnology and nanostructures classification

Nanotechnology is a multidisciplinary science focused on the design, processing, manipulation, and application of new materials at the nanoscale; it means between 1 and 100 nanometers (nm). It also comprises the modification of some properties of bulk materials. Nanostructures are nanometric materials with a defined morphology. They form a self-assembling of atoms, molecules, or macromolecules.

Nowadays, thanks to the application of nanostructures, nanotechnology can contribute to the solution or mitigation of part of the worldwide problems involving environmental pollution [13]. The main strategies focus on different categories: (1) sustainable design and synthesis of nanostructures; (2) new generation technology using functionalized nanostructures to remove contaminants from water, air, and soil; (3) sensor design for the estimation and quantification of pollutants; (4) study of toxicity and evaluation of environmental and health implications associated with the use of nanostructures for remediation; (5) emerging solutions [5, 14, 15].

Nanostructures are materials or structures with at least one dimension in the nanometer scale (1–100 nm) [16]. A zero-dimension (0D) material has a nanometer scale in each of the three directions (x, y, z), while a one-dimension (1D) material has a nanometer-scale only in two directions. A two-dimension (2D) structure contains a nanometer-scale only in one direction. Three-dimensional (3D) nanostructures are included in this classification even though their dimensions are higher than 100 nm, but their construction is a hierarchical architecture that grows in all directions using 0D, 1D, and 2D nanostructures. **Figure 2** presents a scheme

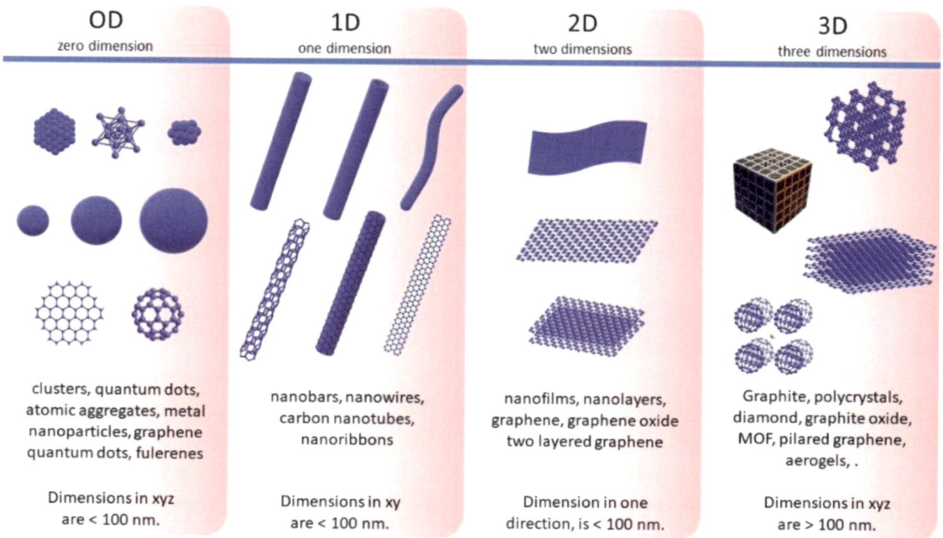

Figure 2.
Schematic for the classification of low dimensional nanostructures: zero dimension (0D), one dimension (1D), two dimension (2D), and three dimension (3D). Examples for each dimension are mentioned in the text.

classification of some examples of nanostructures: 0D nanostructures are quantum dots, nanoparticles, fullerenes, clusters, metal nanoparticles, or graphene quantum dots, among others. One-dimensional structures are carbon nanotubes, single layer graphene nanoribbons, graphite nanoribbons, or nanobars. Two-dimensional nanostructures include nanofilms, graphene, graphene oxide, and two layered graphene. Three-dimensional nanomaterials include graphite, metallic-organic frameworks, aerogels, and composites.

3. Measurement methods using nanostructures

All the elements from the periodic table are suitable to form nanostructures in all dimensions, individually or combined, increasing in this way the variety of properties, behavior, and functionality. As a result, new opportunities are now available with nanotechnology to enhance the efficiency of existing methods for detection, quantification, and remediation of pollutants. However, environmental remediation is still a challenge and some important facts should be considered, such as selectivity, efficiency, reusability, eco-friendly, and low cost. It is important to notice that nanomaterials are non-biodegradable and so they tend to accumulate in organisms, leading to serious health problems when ingested by humans [17]. Thus, using nanotechnology either as a remediation strategy or measurement method needs to consider the recuperation and reutilization of nanomaterials as mandatory steps.

Some authors have conducted relevant investigations to know the levels of heavy metals in water. Florence Lagarde and Jaffrezic-Renault proposed complementary methods for the monitoring and treatment of metal-contaminated water [18], which Agostino et al. resumed in two main sections: fast and conventional [19]. The first consists of the development of in-situ, low cost, non-specific early warning systems to operate online. The second one is focused on accurate detection and quantification using conventional methods that are expensive, more time consuming, and require manual operation.

Detection and quantification of heavy metals are crucial tasks in the characterization of different water types: surface water, groundwater, tap water, and wastewater. Nowadays, the monitoring of toxic effects becomes mandatory since water resources are being scarce, and the development of monitoring tools is essential for an easy, fast, and accurate detection of heavy metals in a few steps. Sensing is an integral part of both fast and conventional measurement methods for the quantification of heavy metals proposed in nowadays research [20]. Nanotechnology used in sensing improves not only the efficiency but also the selectivity and reduces the time needed in each step of the methodology of quantification of heavy metals in water. Sensing using nanotechnology can be implemented in-situ for a faster detection of heavy metals in water with a low-cost technique, followed by the quantification with a sensitive and accurate analytical method on-lab such as atomic absorption spectroscopy, X-ray fluorescence, electrochemical techniques, and among others that are described below.

Some of the 0D nanostructures used for the detection and quantification of heavy metals in water are the metal nanoparticles. These nanoparticles can be made of gold (Au), silver (Ag), titanium oxide (TiO_2), iron oxide (Fe_3O_4), zinc sulfide (ZnS), or cadmium sulfide (CdS) due to the facility that these materials offer to modify their surface, and the easiness they offer to control their reproducibility and flexibility in an aqueous medium [21]. The heavy metals are detected by a spectroscopic, electrochemical, or optical technique. The methodologies adapted to these strategies using nanostructures have some advantages and disadvantages as can be expected [22]. The spectroscopic methods allow not only the detection but also the quantification step, and they are usually very sensitive to multiple heavy metal ions. Some examples of spectroscopic techniques are ICP-MS (inductively coupled plasma mass spectrometry), AAS (atomic absorption spectroscopy), HR-SPS (high-resolution surface plasmon resonance spectroscopy), and SERS (surface-enhanced Raman spectroscopy). SERS is useful for sensing applications since the surface plasmon resonance of metal nanoparticles can be identified with a detection limit of 10 ppt in a pH buffered solution, and of 70 ppt in a real sample when compared with AAS, whose detection limit is around 0.2 ppb [22]. ICP-MS quantifies a wide variety of elements in one-single analysis step with high accuracy, sensitivity, and selectivity. Low detection limits for metal ions can be achieved by using ICP-MS, normally in the ppt range [23, 24]. Such techniques are typically used for quantification but not for sensing. The main disadvantage of the spectroscopic methodologies is the high cost of the analytical instrumentation, the time-consuming procedures, and the use of sophisticated equipment that requires trained personnel [22, 25]. On the other hand, electrochemical methods are low cost, low time-consuming, and easy to handle. They can be conducted under any of the following techniques: potentiometric, amperometric, voltammetric, impedimetric, or electrochemiluminescence. They have a high sensitivity and low detection limits (from 0.25 ng/L to 0.12 μg/L) [26–28]. The main advantage of the electrochemical methods is the possibility of sensing or monitoring of pollutants with in-line systems coupled with water streams. However, the design and miniaturization of the electrodes needed to build portable sensors are one of their major challenges [22, 28]. The use of optical methods for the detection of heavy metals is also an economical and fast option. Optical procedures usually utilized for this purpose are absorption, reflection, or luminescence spectrometry. The main limitations could be that some non-optical indicators interfere with some metal ions [22].

Table 1 summarizes some examples of detection and quantification of heavy metals using nanostructures, reported in recent literature, where the spectroscopic, electrochemical, and optical methodologies have been applied.

For a fast and in situ detection, some authors propose an innovative and cheap methodology based on an online operation aided by a smartphone. The strategy uses the wavelength shifting of the resonant plasmon. It means that nanoparticles suffer a color change, from red to purple, when cadmium (Cd^{2+}), mercury (Hg^{2+}), or copper (Cu^{2+}) ions are present on a solution [29–31]. The nanosensor uses a gold nanoparticle-based assay platform that records, processes signals within the range of 1 ng/mL–32 ng/mL of Hg^{2+}, and proposes a concentration of 0.28 ng/mL as the limit of detection [29]. The *in situ* detection of Cd^{2+} with the nanogold-based detection system controlled by the smartphone is fast and reliable as it displays a linear range for the Cd^{2+} concentration, between 2 and 20 μg/L with a detection limit of 1.12 μg/L [30]. This range is below the permissible limit of Cd^{2+} ion in water and smaller than those based on non-portable and other nano-based technologies [32]. The main advantage for these types of systems is the portability, as well as the immobilization of metal ions to gold nanoparticles that allows the tuning of sensitivity. For example, the aptamer is a particular recognition of Cd^{2+} [33], with a low recognition of Pb^{2+}, and acceptable detection even though the specimen contains other metal ions in high concentration [30].

The quenching of quantum dot (QD) was studied for cadmium (Cd^{2+}), mercury (Hg^{2+}), and lead (Pb^{2+}) to understand the colorimetric/visual sensors. The composite formed by Mn-doped ZnS QD and graphene oxide detects Pb^{2+} down to 0.4 ppb. Further, the interaction between the metal ion and the passivating ligands attached to the QD also contributes partway to the specific detection combined with the quenching of QD [34]. The green approach is also present on the fast detection, which is vital since the production of nanostructures requires to minimize the use

Ion	Nanostructure and sizes	D	Main methodology	Ref.
Cd^{2+} Hg^{2+} Cu^{2+}	Spherical gold nanoparticles ~15 nm	0D	Smartphone-based colorimetric reader system	Xiao et al., Gan eta al., and Cao et al. [29–31]
Hg^{2+} Pb^{2+} Cd^{2+}	Mn–doped ZnS quantum Dots ~5 nm	0D	Detection through the quenching of QD emission	Devaiah Chonamada et al. [34]
Hg^{2+}	Spherical silver nanoparticles (10–30) nm	0D	Green approach using aqueous extract of *Vigna mungo* beans for reduction	Choudhary et al. [35]
Ag^{+} Cu^{2+} Hg^{2+}	Core-shell nanoparticles. ZnS coated up conversion nanoparticles 60 nm in diameter and 20 nm in the shell	0D	Detection of residuals in water using ion metals	Chu et al. [36]
Pb^{2+} Cd^{2+} Cu^{2+} Zn^{2+}	Rhodium/ antimony co-doped TiO_2 nanorod and titanate nanotube ~20 nm	2D in 2D	Quantification and removal of metals and degradation of organic pollutants. The adsorbed metal enhanced the photocatalytic degradation of organic pollutants	Dhandole et al. [37]
Pb^{2+} Cd^{2+}	Functionalized Fe3O4/ NaP zeolite nanocomposite ~30 nm	0D in 3D	Quantification of metals and removal of bacteria	Zendehdel et al. [38]
Pb^{2+} Cd^{2+}	SnS-decorated Bi_2O_3 nanosheets (3–4) nm	2D in 3D	Electrochemical detection of Cd(II) and Pb(II) in real samples from lake and tap water	Jin et al. [39]

Table 1.
New methodologies using nanostructures for sensing and quantification of heavy metals.

of solvent, energy, time, and the release of odd substances to the environment. An extract of *Vigna mungo* beans was used as a reducing agent of nanoparticles to minimize the use of traditional chemical compounds and solvents; the detection limit for Hg^{2+} ions was near 0.13 μM [35].

In previous investigations, two processes were combined to enhance the implementation of nanostructures on heavy metal detection [36–38]. The main advantage of this approach is that after the detection and quantification steps, the removed metal agglomerates in a nanometric size scale that is useful for further remediation. After quantification and removal of any heavy metal, it is possible to convert core-shell nanoparticles, such Ag^{+}, Cu^{2+}, and Hg^{2+} coated on ZnS and use them for the detection of another kind of contaminants. The ZnS nanoparticle can remove up to 3.98 μmol of Ag^{+} [36]. The metal adsorbed on TiO_2 nanostructures enhanced the photocatalytic degradation of organic pollutants. The degraded pollutants were orange (II) dye and bisphenol-A. Metals increased photodegradation after 5 hours in a batch experiment (70% of dye and 80% of Bisphenol A) due to the photodeposition of metal ions on the TiO_2 nanoparticle surface [37]. Functionalized Fe_3O_4/NaP zeolite nanocomposite removed metal, more than 95%, after removed bacteria (*Bacillus subtilis*) [38]. A different methodology was employed for *in situ* measurements by electrochemical detection using SnS-decorated Bi_2O_3 nanosheets. The composite exhibited high sensitivity and efficient detection for the removal of heavy metals with high toxicity, which limits of detection was 1.50 nM for Cd^{2+} and 1.40 nM for Pb^{2+} [39].

Quantification of metallic ions is essential during the different steps of the water cycle that goes through various biological processes. The main quantification techniques are atomic absorption spectrometry, inductively coupled plasma mass spectrometry, anodic stripping voltammetry, X-ray fluorescence spectrometry, and microprobes. These techniques require sample pretreatment procedures, analyte pre-concentration steps, and expensive instrumentation [40]. Due to the presence of dangerous heavy metal ions in water, an instantaneous measurement and quantification are desirable. The use of nanotechnology, along with the investigation of novel approaches, will allow the quantification of heavy metals in an inexpensive, rapid, and simple way. For example, Stenberg, Massad-Ivanirb, and Ester Segal presented a nanostructured porous silicon (Si) biosensor for the detection and quantification of copper ion in real water samples. The monitoring is based on the Laccase relative activity. Lacasse is an enzyme, a multi-copper oxidase, immobilized within the oxidized nanostructured porous silicon. The Laccase-based biosensor exhibited a lower detection limit of 1.30 μM, smaller than other values [41].

Quantification of the metal ion content, as well as the chemical speciation of the different chemical forms of a specific metal in water, becomes essential to estimate its toxicity and persistence in living organisms. The environmental health effects can be understood by the study of natural chemistry, quantification, and speciation of chemical species of metal ions. The chemical form of any metal depends strongly on the chemical conditions in which it is exposed, and the toxicity not only relays on its ionic form but also on the chemical species that formed. For example, the methylmercury ion (CH_3Hg^+) is the toxic specie of the inorganic mercury II (Hg^{2+}). Speciation analysis requires complex, expensive, and time-consuming pre-separation analyses, which are incompatible with *in situ* measurements. Guerrini et al. [42] presented a Surface-Enhanced Raman Scattering (SERS) methodology for the chemical speciation of Hg^{2+} and CH_3Hg^+. In this work, the SERS consisted of an active platform of closely spaced spherical gold nanoparticles anchored on polystyrene microparticles. The ion receptor, mercaptopyridine (MPY), forms strong bonds between the gold atoms and the mercaptan group, and coordinates a nitrogen group with both species Hg^{2+} and CH_3Hg^+. The coordination with nitrogen is determined by the changes in vibrational SERS of MPY, which could give insights on the qualitative and quantitative modifications correlated with Hg^{2+} or CH_3Hg^+.

4. Nanoremediation

Tague defines remediation as "all measures taken for treatment of damaged wells for restoring an optimal performance" [4]. Nowadays, the environment requires those actions for the protection of human health and all living systems on Earth. Environmental remediation consists of the eradication, removal, or transformation of contaminants from natural resources [43]. Although remediation is a complex task, nanoremediation has emerged as an optimal alternative for the removal of pollutants from different waters (groundwater, surface water, and residual water), soil, air, and sediments [43].

Water is by far, one of the most contaminated resources in the planet; that is why the remediation and removal of contaminants are an urgent need together with easy and fast monitoring tools. The available treatments used for removal of heavy metals from water are classified as follows: chemical precipitation, membrane filtration, ion exchange, reverse osmosis, and adsorption [44]. The adsorption using nanomaterials has been of great interest since several nanostructured adsorbents have demonstrated a high performance [44–46]. Adsorption on nanostructured materials is complicated, but some authors have proposed possible mechanisms that depend mainly on the nature of the surface area. The fundamental mechanisms are based on physical adsorption (physisorption), chemical adsorption (chemisorption), electrostatic attraction, and sorption-precipitation [9, 47, 48]. **Figure 3** shows a schematic representation for the adsorption mechanisms of heavy metal on porous nanomaterials. Lu et al. reported that biochar, a 3D network, is a material rich in cations and surface interaction sites for lead adsorption. Electrostatic cation exchange or metal exchange reactions mechanisms may occur when calcium (Ca^{2+}), magnesium (Mg^{2+}), potassium (K^{+}), and sodium (Na^{+}) ions released from biochar in the adsorption of Pb ions, but the electrostatic interaction and surface complexation with pi-cationic and functional groups interaction, may also happen in the adsorption of Pb ions [48].

4.1 Carbonaceous nanostructures

The removal of heavy metals with carbon nanotubes (CNT) varies because the adsorption mechanisms depend on the affinity order of metals ions and the surface functionalization of CNT. Some examples of adsorption have been summarized by Ihsanullah et al. [47]. The mechanism is based on the release of protons (H^{+}) from the surface when oxidation of CNT was achieved. After that, the attachment of divalent metals occurs on the surface. This process depends on the concentration of the metal ion. A higher number of ions increase the pH value due to the increment of H^{+} in the solution. It was found that the adsorption of divalent ions, such as Cd^{2+} and Zn^{2+}, is dominated by a combination of physisorption (Van deer Waals forces) and sorption-precipitation when the pH is increasing in the solution. It means that the acid treatment and surface modification have a significant effect on the adsorption capacity of CNT. Some examples of carbon materials used for metal removal under different metal concentrations in the ppm range are listed in **Table 2** [49–52]. Those experiments were performed without temperature or pH

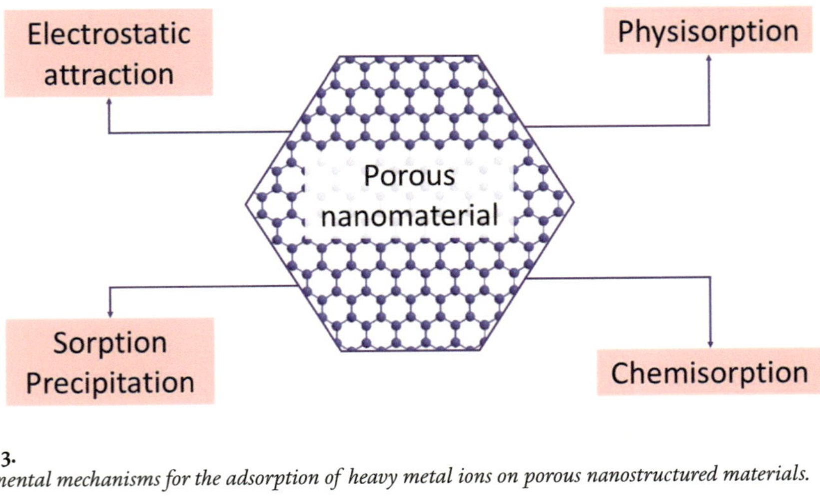

Figure 3.
Fundamental mechanisms for the adsorption of heavy metal ions on porous nanostructured materials.

Material	Copper	Lead	Other metals	Reference
Activated carbon	—	30 ppm – 83% 50 ppm – 74%	**Chrome** 30 ppm – 50.6% 50 ppm – 48.2% **Nickel** 30 ppm – 90% 50 ppm – 87.8% **Cadmium** 30 ppm – 86% 50 ppm – 84%	Karnib et al. [49]
Activated carbon (coconut waste)	126 ppm 73% pH 5	709 ppm – 100% pH 4	**Nickel** 996 ppm – 92% pH 4	Kadirvelu et al. [50]
Graphite oxide	—	30 ppm – 85% 50 ppm – 84%	**Chrome** 30 ppm – 63% **Nickel** 30 ppm – 89.9%	Sheet et al. [51]
Carbon nanotubes	100 ppm – 41%	40 ppm – 83%	**Chrome** 0.5 ppm – 66% **Nickel** 15 ppm -75.4% **Cadmium** 6 ppm - 87%	Mubarak et al. [52]
N-doped carbon nanotubes growth on red volcanic rock	10 ppm – 82% 20 ppm – 84% 40 ppm – 51%	10 ppm – 91% 20 ppm – 90% 40 ppm – 93%	—	Gonzalez Hodges [53]
N-doped carbon nanotubes growth on black volcanic rock	10 ppm – 90% 20 ppm – 95% 40 ppm - 71%	10 ppm – 91% 20 ppm – 81% 40 ppm – 98%	—	Gonzalez Hodges [53]
N-doped carbon nanotubes growth on yellow volcanic rock	10 ppm – 92% 20 ppm – 83% 40 ppm – 63%	10 ppm – 100% 20 ppm – 89% 40 ppm - 99%	—	Gonzalez Hodges [53]
Red volcanic rock	5 ppm – 15%	5 ppm – 14%	—	Mabel et al. [54]

Table 2.
Heavy metal removal percentages using different carbon-based materials.

variation, except for the analysis that uses activated carbon from coconut waste, where the pH was adjusted.

Table 2 presents higher percentages of copper and lead removal when using nitrogen-doped carbon nanotubes grown on volcanic rock [53] than other carbon materials or red volcanic rock [54] without any treatment. The increased efficiency of these carbon nanotubes grown on volcanic rock can be correlated with the presence of nitrogen on their graphitic lattice. Doping carbon nanostructures with heteroatoms, such as nitrogen, increases the adsorption capacity without any acid treatment [4]. Nitrogen has a similar atom diameter but an extra electron as carbon, favoring the atom replacement, and the electron-donating properties [55]. N-doping occurs in different forms, such as pyridine, pyrrole-like, and quaternary-N salts, but may include nitrogen oxide groups and amines [56]. In particular, quaternary nitrogen atoms are considered to increase the adsorption capacity by electrostatic interaction, as the surface carbon nanostructures doped in this way are more negatively charged than the undoped ones [55, 57]. Besides, nitrogenated carbon nanostructures act as reducing agents for toxic Cr^{6+} converting it into harmless Cr^{3+} [55].

4.2 Silica-based nanostructures

The high surface area found on porous materials, such as the silica-based ones, is attributed to the 3D network that displays a hierarchical organization that can be found at different scales (nano, micro, and milli). The nanostructured organization is mainly attributed to the form and size of the pore, which allows the accommodation of other nanostructures, as well as the functionalization of the surface. The methodology to produce nanostructured porous silica is based on the liquid-crystal template mechanism. This allows getting pores around 2–10 nm in size [58]. A surfactant at high concentration is set at a specific pH and temperature values to form a liquid-crystalline phase. Then, the crystalline form may be organized by spherical or rod-shaped micelles arranged on a periodic 3D structure, which serves as a template to be filled with tetraethoxysilane (TEOS) or tetramethylorthosilica (TMOS). Finally, the surfactant is removed to form a mesoporous material [59].

Surface functionalization is a useful tool for the realization of a well-defined set of functions which improves the adsorption properties of silica. It can be carried out by chemical bonds or physical interactions. Silica surface functionalization is of interest due to the multiple options that it offers for heavy metals removal. For example, the thiol-functionalization of mesoporous silica has been proved to exhibit a high affinity for thiophilic heavy-metal ions like Cu^{2+}, Zn^{2+}, Cr^{3+}, and Ni^{2+} but focused on highly toxic metals such Hg^{2+}. At the same time, the amino functionalization increases the efficiency of the adsorption of Hg^{2+} [59, 60]. For some specific types of silica materials such as SBA-15 (Santa Barbara Amorphous-15), the imidazole functionalization improves the selective affinity adsorption of Pd^{2+} and Pt^{2+}, despite the high concentration of Ni^{2+}, Cu^{2+}, and Cd^{2+} in the mixture [61]. Specific morphologies, such as PMO-like (periodic mesoporous organosilica) or spherical mesostructured, increased the surface area and the particular site of functionalization, consequently, affinity, and selectivity are improved [62, 63].

4.3 Zeolites

Zeolites are known as microporous materials with a crystalline structure composed by tetrahedral building blocks of alumina (AlO_4) and silica (SiO_4) that give rise to a three-dimensional network linked by oxygen atoms [64]. Zeolites are an effective, economical, and eco-friendly option to remove heavy metals and organic contaminants from wastewater [64–66]. There are more than 40 naturally occurring zeolites each with different physicochemical properties. Among their most interesting characteristics are their high chemical stability, molecular sieve, adsorption, and ion exchange capacity [67].

The Si/Al ratio in the structure of a zeolite determines its adsorption and ion exchange capacities, which are directly related to the amount of aluminum contained in the natural zeolite [66], so the isomorphic replacement of Si^{4+} by Al^{3+} (aluminum-rich zeolites) will have a more negative charge on the framework, providing a higher affinity for polar molecules. The negative charge is balanced with interchangeable cations (generally sodium, potassium, or calcium), therefore these cations are used for cation exchange processes in solutions containing lead, chromium, or mercury, among others ions [68]. The adsorption capacity of zeolites also depends on the charge density and the diameter of the hydrated ion, which is why zeolites have a great potential to remove heavy metal ions from wastewater [64].

Natural zeolites such as clinoptilolite, mordenite, and chabazite were investigated for the removal of heavy metals [64, 65]. The natural zeolite clinoptilolite has attracted attention because of its abundance. The selectivity of the sodium form of clinoptilolite, extracted from natural deposits, was found to be

$Pb^{2+} > Cd^{2+} > Cs^{+} > Cu^{2+} > Co^{2+} > Cr^{3+} > Zn^{2+} > Ni^{2+} > Hg^{2+}$. This behavior confirmed that natural zeolites have a great ability to selectively remove metallic ions from wastewater. In particular, sodium loaded zeolites resulted the most effective because sodium acts as an exchangeable ion with heavy metals [69]. Further, clinoptilolite exhibited a higher capacity than carbonaceous materials such as carbon nanotubes (CNTs) for lead removal (up to 15.5 times higher) [70, 71].

Nano zeolite [72] and modified zeolites [73–75] demonstrated a higher removal capacity of lead than that reported by natural zeolites, even much higher than functionalized reduced graphene oxide (RGO) [71]. The silica nano-zeolite X [72] presented an adsorption capacity of 909.09 mg/g of Pb^{2+} being 5.5 times higher compared to that reported for natural clinoptilolite (166 mg/g) [69]. The main characteristic of nano zeolites and modified zeolites is a larger surface area and pore size, which make them more efficient and facilitate the adsorption of heavy metals and organic molecules, compared to natural zeolites [76]. The efficiency in the removal of heavy metals and other cations will strongly depend on the high surface area of the zeolites, and in their efficiency in removing the metal ions present in wastewater. The surface area in porous materials is determined by the specific surface area (BET) by standard multipoint techniques of nitrogen adsorption. There are several studies where the largest surface area is for nano zeolites (692 m^2/g) [76], followed by modified silica natural zeolites (702 m^2/g) [72]. A smaller surface area is found for the sodium form of clinoptilolite (70.4 m^2/g) [77], and the smallest area is for the simple natural zeolite clinoptilolite (15.36702 m^2/g) [78].

4.4 Iron oxides

Iron oxide nanostructures are gaining attention for metal removal from water due to their high surface area, excellent adsorption capacities, innocuousness with the environment, and easiness of separation as one can make good use of their magnetic properties. Despite their advantageous characteristics, their use in real scenarios has not been proved. It is essential to explore their removal mechanisms, not only for iron oxides but also for zerovalent iron nanomaterials.

Zerovalent iron nanoparticles (nZVI) are considered a strong reducing agent that is bringing degradation to less harmful substances from a wide range of organic and inorganic pollutants. Sorption of co-precipitation of heavy metals on the surface forms an iron oxide or hydroxide shell when nZVI is exposed to air [79]. The most cited example is the transformation of Cr^{6+}, which precipitates on nZVI with corrosion products. However, this property usually depends on the surface functionalization of the nZVI. Functionalization is important since iron nanoparticles form aggregates very easily and suffer oxidation under acidic and oxygenated environments. The best approach is to coat the nanoparticle surface with Fe_3O_4 or some polymer to reduce the contact with the environmental oxygen, but maintaining the reactivity [79].

Previous studies on the adsorption of Cr^{6+} on the surface of Fe_3O_4 show the formation of a different crystalline structure by chemical adsorption. For chromite ($FeCr_2O_4$), the Cr^{6+} is reduced to Cr^{3+}, followed by the precipitation of Cr^{3+} onto the F_3O_4 nanoparticles. Low desorption is indicative of adsorption because desorption is due to physical adsorption, mainly by electrostatic interactions. This adsorption process predominates on γ-Fe_2O_3, as previously demonstrated by X Ray Photoelectron Spectroscopy (XPS). Another evidence of the physical adsorption is the non-modification of the crystallographic structure, an indication that the removal process was not due to a chemical interaction [80, 81]. The positively charged surface on the iron oxide nanoparticles determines the attraction for negatively charged pollutants; the positive charge of iron oxides depends on the

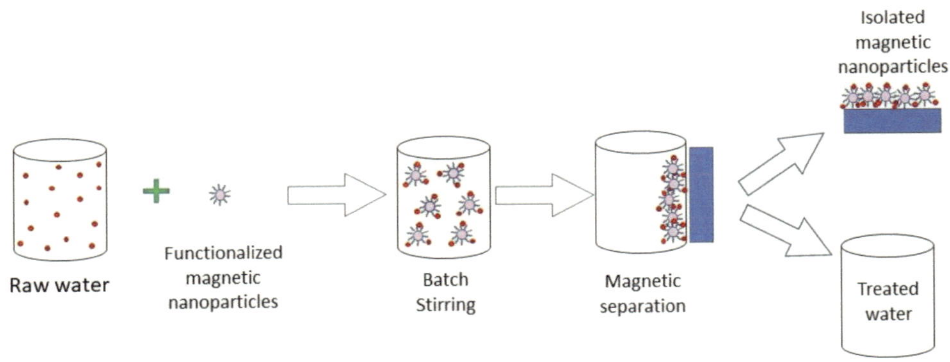

Figure 4.
The different steps on removal of contaminants from water using magnetic nanoparticles.

polarization of oxygen atoms on the surface, which can be modified by pH. The successful removal of pollutants such as Cr^{6+} and As^{5+} is highly pH-dependent [82].

4.4.1 Recovery and reuse of nanostructures using iron oxides

Some of the problems faced when iron nanostructures are used in water remediation are related to their capture, recovery, and reuse. Sometimes complicated steps are required. Many authors have proposed to take advantage of the magnetic nanoparticles as an alternative to capture contaminants and nanostructures that could remain in the environment. Goon et al. [45] studied the capture and quantification of cupric ions at trace level using a composite formed by polyethyleneimine (PEI-) coated with Fe_3O_4 nanoparticles. They captured trace levels (~2 ppb) of Cu^{2+}. The PEI is amine-rich, so it captures the Cu ions easily, while the magnetite nanoparticles allow the magnetic separation of the material from water. Hu and coworkers [44] used graphene oxide coped Fe_3O_4 nanoparticles for highly efficient removal of Pb^{+2}.

The particularity of the composite with graphene oxide is that adsorption capacity improved at pH 7, the natural pH value found in a faucet. They also observed that the system graphene oxide/Fe_3O_4 could be recyclable because it maintains an 80% adsorption capacity after 10 adsorption-desorption cycles. This process can be generalized to the removal, or capture, of any contaminant by the interaction with a specific functionalized nanoparticle followed by its recovery from water (see the scheme in **Figure 4**). Tang and Lo [79] consider that magnetic separation could be a low-cost and a convenient method over the use of a membrane-separation filtration method because the separation of tiny magnetic nanoparticles with the adsorbed heavy metal is easier. The magnetic separation usually occurs with the help of a magnetic field or with a hand-held magnet [79].

5. Remediation based on nanobiotechnology

Natural resources, such as biological systems as microorganisms and plants can also be mixed with nanostructured nanoparticles, usually by cross-linking bonding using a bifunctional reagent; nanoparticles should be inert and biocompatible materials. Even though this methodology is fast, simple, and exhibits an electron transfer, the main disadvantage is the formation of covalent bonds between the functional groups at the outer membranes of the biological living system [18]. Biological systems can be used to directly originate nanomaterials for heavy metal remediation, but they can also be the active coating on nanostructured materials for similar purposes. Even more, the utilization of microbes for intracellular/extracellular synthesis of nanoparticles with different chemical composition, size/shapes, and controlled monodispersity can be a novel, economically viable, and eco-friendly strategy that can reduce toxic chemicals in the conventional protocol [83]. Studies were conducted on some bacteria to produce an iron sulfide compound, which acts as an adsorbent for several toxic metal ions [84]. Nanoparticles obtained from the plant *Noaea mucronata* were used for the accumulation of heavy metals, such as Pb, Cu, Cd, Zn, Fe, and Ni in groundwater, streams, and rivers [85]. The study conducted on plant species such as *Centaurea virgata*, *Scariola orientalis*, *Noaea mucronata*, *Chenopodium album*, *Cydonia oblonga*, *Reseda lutea*, and *Salix excelsa* revealed that these plants are very good heavy metal accumulators. Specifically, *Noaea mucronata* is a suitable accumulator for Pb to a level higher than 1000 ppm [86].

Biological substrates like bacteria, fungi, algae, yeast, and plant derivatives can be immobilized on nanomaterials or nanoparticles and play an essential role in the retention of metal ions. They offer several advantages, namely biodegradability, natural abundance, low cost, simple production, high surface to volume ratio, and various active sites such as carboxyl, hydroxy, amino, sulfate, or phosphate groups [8]. Besides, immobilized biological substrates have been employed as living and non-living cells. The use of dead bacteria offers the possibility to develop continuous flow systems on different solid supports [87]. The main advantage is that dead microorganisms avoid the risk of contamination of water with bacteria. Some examples were summarized by Escudero et al. [87] in an extensive literature review of biological substrates, which includes composites with biological materials and nanostructures as a green alternative in trace elemental preconcentration and speciation analysis. Such nano-based technology has been proved successful on the laboratory scale, but only a few have been used for small-scale testing or commercialization [88]. Specific studies are summarized in the following lines to exemplify the advantages of the biological substrate-based methods in the removal of heavy metals from water. According to the information reported by Escudero and collaborators, the most employed biological substrates are plant-derivatives, bacteria, and fungi, and they are mainly used for water treatments [87].

Heat inactivated *Fusarium verticillioides* filamentous fungi has been immobilized on nano-silica particles for biosorption of calcium (Ca^{2+}) and magnesium (Mg^{2+}) cations, in helping the preconcentration technique of solid-phase extraction to reduce the hardness of aqueous solutions. Maximum capacities were found to be 1000.0 μmol/g for magnesium and 1333.3 μmol/g for calcium [89]. Lead (Pb) has also been removed using a chromatographic column filled with biomass of this same fungal species immobilized on TiO_2 nanoparticles and using hydrochloric acid (HCl) as eluent [90]. Dead coliform bacteria have been immobilized on nanoparticles of titanium oxide (TiO_2) for Pb preconcentration which was then analyzed using a flow injection analysis system coupled to a flame atomic absorption spectrometer [91]. The biosorption of aluminum (Al^{3+}) and cadmium (Cd^{2+}) ions over an exopolysaccharide obtained from the bacterium *Lactobacillus rhamnosus* was possible due to the presence of hidroxyl (–OH) and carboxyl (–COOH) groups that facilitated a complex formation with the target analytes [92]. The well recognized bacteria *Escherichia coli* was immobilized on multiwalled carbon nanotubes to help in the determination of trace elements such as Cd, Co, Cu, and Ni by flame atomic absorption spectrometry [93]; while the metallothionein of the cyanobacterium genera Synechococcus decorated graphene oxide nanosheets for the selective adsorption of Cd from different waters [94]. Different mechanisms are involved in the extraction of metal ions by biological substrates. They include ion exchange, microprecipitation, complexation, and oxide-reduction processes whose adequate election depends basically of the target contaminant and the type of sample to be treated.

6. Perspectives

There is still a need to develop smarter nanomaterials for remediation purposes on different environments, particularly at trace levels. Significant facts can be listed as follows:

- Laboratory experiences on the removal of heavy metals based on the use of nanomaterials must be extended toward real environments potentializing their advantages, but also having in mind their potential risks to human health and to the ecosystems, areas which are poorly understood and might lead to unintended consequences.

- The mobility, bioavailability, toxicity, and persistence of a wide variety of manufactured nanoparticles that are already in use, need to be studied [95]. This will provide qualitative and quantitative information for a better understanding of their potential risks, beyond their use in the heavy metal removal. For this issue, full-scale ecosystem-wide studies can be carried out by machine learning programs [96, 97].

- More studies focused on profitable strategies applied to the recovery and reuse of nanostructures need to be achieved. They will also allow the proposal of safe handling and disposal guidelines for the already used nanoparticles not only in environmental remediation protocols but in any other area that is already releasing them into the environment.

- Accessible and low-cost pollution sensors based on an electrochemical or an optical response that can be easily implemented in an electronic device such as a smartphone, are urgently needed. These technologies are much less expensive that the traditional spectroscopic techniques and can transform every device into a mobile sensor, with probably thousands displayed in huge territorial areas.

- Strategies for the maintenance or the improvement of water quality become mandatory in times when water access is becoming more and more difficult. In this context, the presence of different nanomaterials as an efficient, low-cost, and environmentally friendly alternative for the removal of heavy metals from waters to keep healthily environments [98] is a research and educational area that is just emerging.

7. Conclusions

The increasing pressure that the exacerbated demand on primary goods and commodities is placing to our environment by the development of larger communities will lead us to catastrophic and irreversible damages if a change of actions is not taken. Toxic and dangerous pollutants that are discharged in increasing quantities mean negative impacts to our natural resources, among which water is one of the most vulnerable as it is an absolutely necessary requisite for any form of life on

Earth. Palliative solutions have been proposed in different scenarios with variable degrees of success. In this sense, we have presented a summary of some of the alternatives that have been explored to help in the remediation of water reservoirs, based on the application of a variety of nanomaterials coupled with traditional analytical techniques, and on the other hand supported with some innovations in terms of the design, reuse or efficiency.

One of the multiple promising applications of nanotechnology is remediation. The current work mentioned some examples of applied nanotechnology such as testing before and after a remediation strategy is applied. The presented low dimensional (0D, 1D, 2D, and 3D) nanomaterials were coupled with traditional analytical techniques, and on the other hand, supported with some innovations in terms of the design, reuse, or efficiency. Most of the examples showed that concentration values lower than trace levels can be reached, which is an important fact concerning highly toxic metals.

The combination of nanotechnology and remediation opens an avenue of multiple options for environmental improvement strategies. It is an efficient alternative for the sensing, quantification, and removal of contaminants from water, specifically the cleaning of heavy metals, ideally to quantities lower than trace levels, as was exemplified in this chapter. The main advantages of the nanomaterials used for this purpose reside in their high sensitivity and selectivity that can be achieved at a reduced cost and with a lower time-consumption than other non-processed materials (for example volcanic rock shown in **Table 2**). Also, the nanoscale size means an impressive increment of the surface area of nanomaterials that can be translated into a higher reactivity. The reactivity allows the tuning of selectivity or combined selectivity; it means that any molecule with a specific affinity to any metal can be anchored to the nanostructure surface. Besides, nanostructured materials could be adapted for subsequent remediations and magnetic removal, reducing secondary contamination.

The combination of living systems, nanotechnology, and remediations extends the possibilities. Some microorganisms (dead or dead) are natural accumulators at trace levels and combined with nanostructures make a synergistic effect with improved sorption capacity; such combination allows the production of sustainable materials. Toxicity also concerns with this field, due to one of the main concerns about the extensive use of nanomaterials resides on the fact that the details on the nanostructure interaction with the environment and with different living organisms are still unknown. It means that systematic and extensive studies are required to aid to fill the current voids of data or information. Then, the toxicity is still a limitation on the use of nanomaterials in real samples. Other limitations such as the industrial manufacture tackle multiple difficulties due to do not readily exist regulation neither guidance on information requirements nor safety assessment. Apart from regulation, the right on intellectual property delays the industrial application.

It is true that one of the main concerns about the extensive use of nanomaterials resides on the fact that the details on the nanostructure interaction with the environment and with different living organisms are still unknown. It means that systematic and extensive studies are required to aid to fill the current voids of data or information. Another important issue is the recuperation of the nanomaterials once used. In some cases, the magnetic removal represents a practical option for recovery and reuse, but additional innovative strategies are needed. Overall, the application of nanomaterials to the removal of heavy metals from water is still a vast research opportunity area to be covered by current and future generations of scientists.

Conflict of interest

The authors declare no conflict of interest.

Author details

María Luisa García-Betancourt[1*], Sandra I. Ramírez Jiménez[1], Apsahara González-Hodges[1], Zandra E. Nuñez Salazar[1], Ismailia Leilani Escalante-García[2] and Jeannete Ramírez Aparicio[3]

1 Center of Chemical Research, IICBA, Autonomous University of the State of Morelos, Cuernavaca, Mexico

2 Department of Chemical Sciences, Autonomous University of Zacatecas, Zacatecas, Mexico

3 CONACYT - National Institute of Electricity and Clean Energies, Cuernavaca, Mexico

*Address all correspondence to: mluisa.garcia@uaem.mx

IntechOpen

© 2020 The Author(s). Licensee IntechOpen. This chapter is distributed under the terms of the Creative Commons Attribution License (http://creativecommons.org/licenses/by/3.0), which permits unrestricted use, distribution, and reproduction in any medium, provided the original work is properly cited. CC BY

References

[1] Bradl H, Kim C, Kramar U, StÜben D. In: Heavy Metals in the Environment: Origin, Interaction and Remediation. UK: Elsevier; 2005. p. 28-164. DOI: 10.1016/s1573-4285(05)80021-3

[2] Tchounwou PB, Yedjou CG, Patlolla AK, et al. Heavy metal toxicity and the environment. EXS. 2012;**101**:133-164. DOI: 10.1007/978-3-7643-8340-4_6

[3] Elliott M, Burdon D, Hemingway KL, et al. Estuarine, coastal and marine ecosystem restoration: Confusing management and science - a revision of concepts. Estuarine, Coastal and Shelf Science. 2007;**74**:349-366. DOI: 10.1016/j.ecss.2007.05.034

[4] Civan F. Formation damage abatement—Guidelines, methodology, preventive maintenance, and remediation treatments. In: Reservoir Formation Damage. Amsterdam: Elsevier; 2016. DOI: 10.1016/b978-0-12-801898-9.00023-0

[5] Tratnyek PG, Johnson RL. Nanotechnologies for environmental cleanup. Nano Today. 2006;**1**:44-48. DOI: 10.1016/S1748-0132(06)70048-2

[6] Smichowski P, Londonio A. A retrospective and prospective of the use of bio- and nanomaterials for preconcentration, speciation, and determination of trace elements: A review spanning 25 years of research. Analytical and Bioanalytical Chemistry. 2020:1-4. DOI: 10.1007/s00216-020-02536-5

[7] Khin MM, Nair AS, Babu VJ, et al. A review on nanomaterials for environmental remediation. Energy and Environmental Science. 2012;**5**: 8075-8109. DOI: 10.1039/c2ee21818f

[8] Zhao G, Li J, Ren X, et al. Few-layered graphene oxide nanosheets as superior sorbents for heavy metal ion pollution management. Environmental Science and Technology. 2011;**45**:10454-10462. DOI: 10.1021/es203439v

[9] Lee LZ, Zaini MAA, Tang SH. Porous nanomaterials for heavy metal removal. In: Handbook of Ecomaterials. Switzerland: Springer; Vol. 1. 2019. pp. 469-494. DOI: 10.1007/978-3-319-68255-6_27

[10] Qu X, Alvarez PJJ, Li Q. Applications of nanotechnology in water and wastewater treatment. Water Research. 2013;**47**:3931-3946. DOI: 10.1016/j.watres.2012.09.058

[11] Mohan D, Pittman CU. Arsenic removal from water/wastewater using adsorbents – A critical review. Journal of Hazardous Materials. 2007;**142**:1-53. DOI: 10.1016/j.jhazmat.2007.01.006

[12] Hassan AM, Wan Ibrahim WA, Bakar MB, et al. New effective 3-aminopropyltrimethoxysilane functionalized magnetic sporopollenin-based silica coated graphene oxide adsorbent for removal of Pb(II) from aqueous environment. Journal of Environmental Management. 2020;**253**:109658. Epub ahead of print. DOI: 10.1016/j.jenvman.2019.109658

[13] Theron J, Walker JA, Cloete TE. Nanotechnology and water treatment: Applications and emerging opportunities. Critical Reviews in Microbiology. 2008;**34**:43-69. DOI: 10.1080/10408410701710442

[14] Patil SS, Shedbalkar UU, Truskewycz A, et al. Nanoparticles for environmental clean-up: A review of potential risks and emerging solutions. Environmental Technology and Innovation. 2015;**5**:10-21. DOI: 10.1016/j.eti.2015.11.001

[15] Diallo MS, Fromer NA, Jhon MS. Nanotechnology for

sustainable development: Retrospective and outlook. In: Nanotechnology for Sustainable Development. 1st ed. Vol. 2050. 2014. pp. 1-16. DOI: 10.1007/978-3-319-05041-6_1

[16] Azmi MA, Shad KF. Role of nanostructure molecules in enhancing the bioavailability of oral drugs. In: Nanostructures for Novel Therapy: Synthesis, Characterization and Applications. Amsterdam: Elsevier; 2017. Epub ahead of print. DOI: 10.1016/B978-0-323-46142-9.00014-1

[17] Lu F, Astruc D. Nanomaterials for removal of toxic elements from water. Coordination Chemistry Reviews. 2018;**356**:147-164. DOI: 10.1016/j.ccr.2017.11.003

[18] Lagarde F, Jaffrezic-Renault N. Cell-based electrochemical biosensors for water quality assessment. Analytical and Bioanalytical Chemistry. 2011;**4**:947. DOI: 10.1007/s00216-011-4816-7

[19] Agostino V, Massaglia G, Gerosa M, et al. Environmental electroactive consortia as reusable biosensing element for freshwater toxicity monitoring. New Biotechnology. 2020;**55**:36-45. DOI: 10.1016/j.nbt.2019.09.005

[20] Ahmad L, Arshid M, Aaliya B, et al. Detection and removal of heavy metal ions: A review. Environmental Chemistry Letters. 2019;**17**:1495-1521. DOI: 10.1007/s10311-019-00891-z

[21] Kumar A, Gupta K. RNA-mediated fluorescent colloidal CdSe nanostructures in aqueous medium-analysis of Cd2+ induced folding of RNA associated with morphological transformation (0D to 1D), change in photophysics and selective Hg2+ sensing. Journal of Materials Chemistry A. 2017;**5**:6146-6163. DOI: 10.1039/c6ta10795h

[22] Dalmieda J, Kruse P. Metal cation detection in drinking water. Sensors. 2019:5134. Epub ahead of print;**19**. DOI: 10.3390/s19235134

[23] Zeiner M, Rezić I, Steffan I. Analytical methods for the determination of heavy metals in the textile industry. Journal of Chemists and Chemical Engineers. 2007;**56**:587-595

[24] Helaluddin ABM, Khalid RS, Alaama M, et al. Main analytical techniques used for elemental analysis in various matrices. Tropical Journal of Pharmaceutical Research. 2016;**15**: 427-434. DOI: 10.4314/tjpr.v15i2.29

[25] Scherzer RD, Gasparov LV, Stagon SP. Enhanced morphological stability of silver nanoparticles supported on rough substrates at high temperatures. Micro & Nano Letters. 2018;**13**:498-501. DOI: 10.1049/mnl.2017.0526

[26] Aragay G, Merkoçi A. Nanomaterials application in electrochemical detection of heavy metals. Electrochimica Acta. 2012;**84**:49-61. DOI: 10.1016/j.electacta.2012.04.044

[27] Bansod BK, Kumar T, Thakur R, et al. A review on various electrochemical techniques for heavy metal ions detection with different sensing platforms. Biosensors and Bioelectronics. 2017;**94**:443-455. DOI: 10.1016/j.bios.2017.03.031

[28] Hernandez-Vargas G, Sosa-Hernández JE, Saldarriaga-Hernandez S, et al. Electrochemical biosensors: A solution to pollution detection with reference to environmental contaminants. Biosensors. 2018;**8**:29. DOI: 10.3390/bios8020029

[29] Xiao W, Xiao M, Fu Q, et al. A portable smart-phone readout device for the detection of mercury contamination based on an aptamer-assay nanosensor. Sensors. 2016;**16**:1871. DOI: 10.3390/s16111871

[30] Gan Y, Liang T, Hu Q, et al. In-situ detection of cadmium with aptamer functionalized gold nanoparticles based on smartphone-based colorimetric system. Talanta. 2020;**208**:120231. DOI: 10.1016/j.talanta.2019.120231

[31] Cao Y, Liu Y, Li F, et al. Portable colorimetric detection of copper ion in drinking water via red beet pigment and smartphone. Microchemical Journal. 2019;**150**:104176. DOI: 10.1016/j.microc.2019.104176

[32] Awual MR. A facile composite material for enhanced cadmium(II) ion capturing from wastewater. Journal of Environmental Chemical Engineering. 2019;**7**:103378. DOI: 10.1016/j.jece.2019.103378

[33] Huang CC, Chang HT. Aptamer-based fluorescence sensor for rapid detection of potassium ions in urine. Chemical Communications. 2008;**12**:1461-1463. DOI: 10.1039/b718752a

[34] Devaiah Chonamada T, Sharma B, Nagesh J, et al. Origin of luminescence-based detection of metal ions by Mn–doped ZnS quantum dots. ChemistrySelect. 2019;**4**:13551-13557. DOI: 10.1002/slct.201903769

[35] Choudhary MK, Garg S, Kaur A, et al. Green biomimetic silver nanoparticles as invigorated colorimetric probe for Hg2+ ions: A cleaner approach towards recognition of heavy metal ions in aqueous media. Materials Chemistry and Physics. 2020;**240**:122164. DOI: 10.1016/j.matchemphys.2019.122164

[36] Chu Z, Wang W, Zhang C, et al. Monitoring and removal of trace heavy metal ions via fluorescence resonance energy transfer mechanism : In case of silver ions. Chemical Engineering Journal. 2019;**375**:121927. DOI: 10.1016/j.cej.2019.121927

[37] Dhandole LK, Kim SG, Bae HS, et al. Simultaneous and synergistic effect of heavy metal adsorption on the enhanced photocatalytic performance of a visible-light-driven RS-TONR/TNT composite. Environmental Research. 2020;**180**:108651. DOI: 10.1016/j.envres.2019.108651

[38] Zendehdel M, Ramezani M, Shoshtari-Yeganeh B, et al. Simultaneous removal of Pb(II), Cd(II) and bacteria from aqueous solution using amino-functionalized Fe3O4/NaP zeolite nanocomposite. Environmental Technology. 2019;**40**:3689-3704. DOI: 10.1080/09593330.2018.1485750

[39] Jin W, Fu Y, Hu M, et al. Highly efficient SnS-decorated Bi2O3 nanosheets for simultaneous electrochemical detection and removal of Cd(II) and Pb(II). Journal of Electroanalytical Chemistry. 2020;**856**:113744. DOI: 10.1016/j.jelechem.2019.113744

[40] Forzani ES, Zhang H. Detection of heavy metal ions in drinking water using a high-resolution differential surface Plasmon resonance sensor. Environmental Science & Technology. 2005;**39**:1257-1262. DOI: 10.1021/es049234z

[41] Shtenberg G, Massad-ivanir N, Segal E. Detection of trace heavy metal ions in water by detection of trace heavy metal ions in water by nanostructured porous Si biosensors. Analyst. 2015;**140**:4507-4514. DOI: 10.1039/c5an00248f

[42] Guerrini L, Rodriguez-Loureiro I, Correa-Duarte MA, et al. Chemical speciation of heavy metals by surface- enhanced Raman scattering spectroscopy: Identification and quantification of inorganic- and methyl-mercury in water. Nanoscale. 2014;**6**:8368-8375. DOI: 10.1039/C4NR01464B

[43] Ingle AP, Seabra AB, Duran N, et al. Nanoremediation: A new and emerging technology for the removal of toxic contaminant from environment. In: Microbial Biodegradation and Bioremediation. Amsterdam: Elsevier; 2014:233-250. DOI: 10.1016/b978-0-12-800021-2.00009-1

[44] Hu Z, Qin S, Huang Z, et al. Recyclable graphene oxide-covalently encapsulated magnetic composite for highly efficient Pb(II) removal. Journal of Environmental Chemical Engineering. 2017;**5**:4630-4638. DOI: 10.1016/j.jece.2017.09.003

[45] Goon IY, Zhang C, Lim M, et al. Controlled fabrication of Polyethylenimine-functionalized magnetic nanoparticles for the sequestration and quantification of free Cu2+. Langmuir. 2010;**56**:12247-12252. DOI: 10.1021/la101196r

[46] Pyrzynska K. Carbon nanostructures for separation, preconcentration and speciation of metal ions. TrAC, Trends in Analytical Chemistry. 2010;**29**:718-727. DOI: 10.1016/j.trac.2010.03.013

[47] Ihsanullah AA, Al-Amer AM, et al. Heavy metal removal from aqueous solution by advanced carbon nanotubes: Critical review of adsorption applications. Separation and Purification Technology. 2016;**157**:141-161. DOI: 10.1016/j.seppur.2015.11.039

[48] Lu H, Zhang W, Yang Y, et al. Relative distribution of Pb 2+ sorption mechanisms by sludge-derived biochar. Water Research. 2012;**46**:854-862. DOI: 10.1016/j.watres.2011.11.058

[49] Karnib M, Kabbani A, Holail H, et al. Heavy metals removal using activated carbon, silica and silica activated carbon composite. Energy Procedia. 2014;**50**:113-120. DOI: 10.1016/j.egypro.2014.06.014

[50] Kadirvelu K, Thamaraiselvi K, Namasivayam C. Removal of heavy metals from industrial wastewaters by adsorption onto activated carbon prepared from an agricultural solid waste. Bioresource Technology. 2001;**76**:63-65. DOI: 10.1016/S0960-8524(00)00072-9

[51] Sheet I, Kabbani A, Holail H. Removal of heavy metals using nanostructured graphite oxide, silica nanoparticles and silica/graphite oxide composite. Energy Procedia. 2014;**50**:130-138. DOI: 10.1016/j.egypro.2014.06.016

[52] Mubarak NM, Sahu JN, Abdullah EC, et al. Removal of heavy metals from wastewater using carbon nanotubes. Separation and Purification Reviews. 2014;**43**:311-338. DOI: 10.1080/15422119.2013.821996

[53] Gonzalez Hodges AN. Nanotubos de carbono multicapa dopados con nitrógeno crecidos sobre roca volcánica y su aplicación en la remoción de metales pesados. México: Tesis de Licenciatura; 2019. pp. 1-105

[54] Mabel VM, Laura MC, Mónica SC, et al. Tratamiento Terciario De Aguas Residuales Por Filtración E Intercambio Iónico. Available from: https://www.researchgate.net/profile/Mabel_Mier/publication/237211478_Tratamiento_Terciario_De_Aguas_Residuales_Por_Filtracion_E_Intercambio_Ionico/links/546494c90cf2c0c6aec58c6d.pdf

[55] Zhang T, Wei S, Waterhouse GIN, et al. Chromium (VI) adsorption and reduction by humic acid coated nitrogen-doped magnetic porous carbon. Powder Technology. 2020;**360**:55-64. DOI: 10.1016/j.powtec.2019.09.091

[56] Byrne TM, Gu X, Hou P, et al. Quaternary nitrogen activated carbons for removal of perchlorate with electrochemical regeneration. Carbon. 2014;**73**:1-12. DOI: 10.1016/j.carbon.2014.02.020

[57] Chu B, Amano Y, Machida M. Preparation of bean dreg derived N-doped activated carbon with high adsorption for Cr(VI). Colloids and Surfaces A: Physicochemical and Engineering Aspects. 2019;**586**:124262. DOI: 10.1016/j.colsurfa.2019.124262

[58] Hoffmann F, Cornelius M, Morell J, et al. Silica-based mesoporous organic-inorganic hybrid materials. Angewandte Chemie, International Edition. 2006;**45**:3216-3251. DOI: 10.1002/anie.200503075

[59] Zhou Y, Quan G, Wu Q, et al. Mesoporous silica nanoparticles for drug and gene delivery. Acta Pharmaceutica Sinica B. 2018;**8**:165-177. DOI: 10.1016/j.apsb.2018.01.007

[60] Nooney RI, Kalyanaraman M, Kennedy G, et al. Heavy metal remediation using functionalized mesoporous silicas with controlled macrostructure. Langmuir. 2001;**17**:528-533. DOI: 10.1021/la000720j

[61] Kang T, Park Y, Choi K, et al. Ordered mesoporous silica (SBA-15) derivatized with imidazole-containing functionalities as a selective adsorbent of precious metal ions. Journal of Materials Chemistry. 2004;**14**:1043-1049. DOI: 10.1039/b315829b

[62] Zhang L, Zhang W, Shi J, et al. A new thioether functionalized organic-inorganic mesoporous composite as a highly selective and capacious Hg2+ adsorbent. Chemical Communications. 2003;**2**:210-211. DOI: 10.1039/b210457a

[63] Walcarius A, Sayen S, Gérardin C, et al. Dipeptide-functionalized mesoporous silica spheres. Colloids and Surfaces A: Physicochemical and Engineering Aspects. 2004;**234**:145-151. DOI: 10.1016/j.colsurfa.2003.11.010

[64] Erdem E, Karapinar N, Donat R. The removal of heavy metal cations by natural zeolites. Journal of Colloid and Interface Science. 2004;**280**:309-314. DOI: 10.1016/j.jcis.2004.08.028

[65] Vaca Mier M, López Callejas R, Gehr R, et al. Heavy metal removal with mexican clinoptilolite: Multi-component ionic exchange. Water Research. 2001;**35**:373-378. DOI: 10.1016/S0043-1354(00)00270-0

[66] Egashira R, Tanabe S, Habaki H. Adsorption of heavy metals in mine wastewater by Mongolian natural zeolite. Procedia Engineering. 2012;**42**:49-57. DOI: 10.1016/j.proeng.2012.07.394

[67] Ramírez-Aparicio J, Sánchez-Martínez A, Ramírez-Bon R. Photodecolorization of rhodamine under sunlight irradiation driven by chabazite. Solar Energy. 2016;**129**:45-53. DOI: 10.1016/j.solener.2016.01.050

[68] Shaheen SM, Derbalah AS, Moghanm FS. Removal of heavy metals from aqueous solution by zeolite in competitive sorption system. International Journal of Environmental Science and Development. 2012;**3**:362-367. DOI: 10.7763/ijesd.2012.v3.248

[69] Zamzow MJ, Eichbaum BR. Separation science and technology removal of heavy metals and other Cations from wastewater using zeolites. Separation Science and Technology. 1990;**25**:1555-1569. DOI: 10.1080/01496399008050409

[70] Santhosh C, Velmurugan V, Jacob G, et al. Role of nanomaterials

in water treatment applications: A review. Chemical Engineering Journal. 2016;**306**:1116-1137. DOI: 10.1016/j.cej.2016.08.053

[71] Online VA, Kam CS, Leung TL, et al. RSC advances form of carbon and surface functionalization. Separation and Purification Technology. 2018;**47**:18355-18362. DOI: 10.1039/c8ra02264j

[72] Rasouli M, Yaghobi N, Hafezi M, et al. Adsorption of divalent lead ions from aqueous solution using low silica nano-zeolite X. Journal of Industrial and Engineering Chemistry. 2012;**18**:1970-1976. DOI: 10.1016/j.jiec.2012.05.014

[73] Oliveira JA, Cunha FA, Ruotolo LAM. Synthesis of zeolite from sugarcane bagasse fly ash and its application as a low-cost adsorbent to remove heavy metals. Journal of Cleaner Production. 2019;**229**:956-963. DOI: 10.1016/j.jclepro.2019.05.069

[74] Liu H, Peng S, Shu L, et al. Magnetic zeolite NaA: Synthesis, characterization based on metakaolin and its application for the removal of Cu2+, Pb2+. Chemosphere. 2013;**91**:1539-1546. DOI: 10.1016/j.chemosphere.2012.12.038

[75] Tran HN, Lien Nghiem TH, Duong Vu TT, et al. Dye-doped silica-based nanoparticles for bioapplications. Advances in Natural Sciences: Nanoscience and Nanotechnology. 2013;**4**:043001. DOI: 10.1088/2043-6262/4/4/043001

[76] Pham T-H, Byeong-Kyu L. Aqueous Phase Adsorption : Theory, Simulations, and Experiments. United Kingdom: Taylor & Francis; 2019. pp. 181-210

[77] García-Mendieta A, Solache-Ríos M, Olguín MT. Evaluation of the sorption properties of a Mexican clinoptilolite-rich tuff for iron, manganese and iron–manganese systems. Microporous and Mesoporous Materials. 2009;**118**:489-495. DOI: 10.1016/j.micromeso.2008.09.028

[78] Bekta N, Kara S. Recent advances in functionalized micro and mesoporous carbon materials: Synthesis and applications. Chemical Society Reviews. 2004;**39**:189-200. DOI: 10.1016/j.seppur.2003.12.001

[79] Tang SCN, Lo IMC. Magnetic nanoparticles: Essential factors for sustainable environmental applications. Water Research. 2013;**47**:2613-2632. DOI: 10.1016/j.watres.2013.02.039

[80] Hu J, Lo IMC, Chen G. Comparative study of various magnetic nanoparticles for Cr(VI) removal. Separation and Purification Technology. 2007;**56**: 249-256. DOI: 10.1016/j.seppur.2007.02.009

[81] Hu J, Lo IMC, Chen G. Performance and mechanism of chromate (VI) adsorption by δ-FeOOH-coated maghemite (γ-Fe2O3) nanoparticles. Separation and Purification Technology. 2007;**58**:76-82. DOI: 10.1016/j.seppur.2007.07.023

[82] Hu J, Lo IMC, Chen G. Removal of Cr(VI) by magnetite nanoparticle. Water Science and Technology. 2004;**12**:139-146

[83] Juwarkar AA, Singh SK, Mudhoo A. A comprehensive overview of elements in bioremediation. Reviews in Environmental Science and Biotechnology. 2010;**9**:215-288. DOI: 10.1007/s11157-010-9215-6

[84] Mahmoud ME, Yakout AA, Abdel-Aal H, et al. Enhanced biosorptive removal of cadmium from aqueous solutions by silicon dioxide nano-powder, heat inactivated and immobilized Aspergillus ustus. Desalination. 2011;**279**:291-297. DOI: 10.1016/j.desal.2011.06.023

[85] Mohsenzadeh F, Rad AC. Bioremediation of heavy metal pollution by nano-particles of Noaea Mucronata. International Journal of Bioscience, Biochemistry and Bioinformatics. 2012;**2**:85. DOI: 10.7763/ijbbb.2012.v2.77

[86] Rizwan M, Singh M, Mitra CK, et al. Ecofriendly application of nanomaterials: Nanobioremediation. Journal of Nanoparticles. 2014;**2014**:7. DOI: 10.1155/2014/431787

[87] Escudero LB, Maniero MÁ, Agostini E, et al. Biological substrates: Green alternatives in trace elemental preconcentration and speciation analysis. TrAC, Trends in Analytical Chemistry. 2016;**80**:531-546. DOI: 10.1016/j.trac.2016.04.002

[88] Tripathi S, Sanjeevi R, Anuradha J, et al. Nano-bioremediation: Nanotechnology and bioremediation. In: Biostimulation Remediation Technologies for Groundwater Contaminants. Singapore: Springer; 2018. pp. 202-219. DOI: 10.4018/978-1-5225-4162-2.ch012

[89] Mahmoud ME, Yakout AA, Abdel-Aal H, et al. Immobilization of Fusarium verticillioides fungus on nano-silica (NSi-Fus): A novel and efficient biosorbent for water treatment and solid phase extraction of Mg(II) and Ca(II). Bioresource Technology. Singapore: Springer; 2013;**134**:324-330. DOI: 10.1016/j.trac.2016.04.002

[90] Bakircioglu Y, Bakircioglu D, Akman S. Biosorption of lead by filamentous fungal biomass-loaded TiO2 nanoparticles. Journal of Hazardous Materials. 2010;**178**:1015-1020. DOI: 10.1016/j.jhazmat.2010.02.040

[91] Bakircioglu D, Ucar G, Kurtulus YB. Coliform bacteria immobilized on titanium dioxide nanoparticles as a biosorbent for trace lead preconcentration followed by atomic absorption spectrometric determination. Microchimica Acta. 2011;**174**:367-374. DOI: 10.1007/s00604-011-0630-3

[92] Polak-Berecka M, Szwajgier D, Waśko A. Biosorption of Al+3 and Cd+2 by an exopolysaccharide from *Lactobacillus rhamnosus*. Journal of Food Science. 2014;**79**:T2404-T2408. DOI: 10.1111/1750-3841.12674

[93] Aydemir N, Tokman N, Akarsubasi AT, et al. Determination of some trace elements by flame atomic absorption spectrometry after preconcentration and separation by Escherichia coli immobilized on multiwalled carbon nanotubes. Microchimica Acta. 2011;**175**:185. DOI: 10.1007/s00604-011-0668-2

[94] Yang T, Liu LH, Liu JW, et al. Cyanobacterium metallothionein decorated graphene oxide nanosheets for highly selective adsorption of ultra-trace cadmium. Journal of Materials Chemistry. 2012;**22**:21909-21916. DOI: 10.1039/c2jm34712a

[95] Nowack B. Pollution prevention and treatment using nanotechnology. Nanotechnology. 2010;**2**:1-15. DOI: 10.1002/9783527628155.nanotech010

[96] Odedele TO, Ibrahim HD. Modelling toxicity behaviour of engineered nanomaterials using computational intelligence approach. In: Procedings of the SPE International Conference and Exhibition on Health, Safety, Security, Environment, and Social Responsibility; 16-18 April 2018; United Arab Emirates. Abu Dabi; ADNOC and SPE; 2018. p. 190498

[97] Cohen Y, Rallo R, Liu R, et al. In silico analysis of nanomaterials hazard and risk. Accounts of Chemical

Research. 2013;**46**:802-812. DOI: 10.1021/ar300049e

[98] Chen X, Yu L, Zou S, et al. Zeolite cotton in tube: A simple robust household water treatment filter for heavy metal removal. Scientific Reports. 2020;**10**:1-9. DOI: 10.1038/s41598-020-61776-8

Chapter 10

Superabsorbent Hydrogels for Heavy Metal Removal

Tanzil Ur Rehman, Luqman Ali Shah, Noor Saeed Khattak, Abbas Khan, Noor Rehman and Sultan Alam

Abstract

The superabsorbent hydrogels (SAHs) are 3D polymer networks having hydrophilic nature, which can swell, absorb, and hold incredible amount of water in aqueous medium showing better sorption ability. The sorption ability enables SAH to absorb organic pollutants, dyes, and heavy metal ions (HMI) from wastewater. Therefore, SAHs have recently got considerable interest from the researchers to be used for wastewater treatment. In order to know the swelling property and to understand the wastewater treatment in general and heavy metal ion removal from industrial effluent in particular, this chapter describes the removal of heavy metal ions from wastewater in details. Thus this chapter will enable us to understand the theoretical and experimental part regarding the removal of heavy metal ions by SAH from wastewater. It also highlights the parameters of adsorption process such as effect of initial concentration of heavy metal ions, effect of external stimuli (pH), effect of temperature on the removal of heavy metal ions, and dosage studies. The synthesis of SAH and its use for removal of heavy metal ions from wastewater as well as recycling, selectivity, and effectiveness are also discussed in detail.

Keywords: superabsorbent hydrogels, swelling ability, sensitivity, heavy metal ions, wastewater treatment

1. Introduction

Recent advancement in industrialization, industrial waste, and effluents has direct impact on water bodies and causes water pollution. This phenomenon not only affects the plants and animals but also has alarming impact on drinking water for human beings and causes severe health issues [1]. The largest water pollution sources are organic pollutants, dyes, and inorganic heavy metals [2]. The industries which are declared as the main source of these water pollutants include pharmaceuticals [3], textiles [4], dyeing and metallurgical industries [5], etc. However, the removal of these organic pollutants and heavy metal ions from the polluted water is a challenging task for researchers, as most of the industries lack the technology and facilities to treat the wastewater before entering into the main streams. For this purpose various chemical, physical, and biological methods were used for wastewater treatment which includes filtration [6], advanced oxidation [7, 8], flocculation and coagulation [9], catalysis [10–13], photo and chemical degradation [14, 15], and adsorption [16]. Due to low cost and easy operation, adsorption is the most appropriate and reasonable choice for the removal of organic pollutants and inorganic

IntechOpen

heavy metal ions from wastewater. Many traditional materials can be used for the removal of these pollutants from wastewater as an adsorbent like rice husk charcoal [17], saw dust [18], activated alumina, silica gels, activated carbon, and nut shells [19]. However, the limitation of traditional adsorbents demanded the introduction of novel materials with low cost, biocompatibility, biodegradability, easy synthesis, simple regeneration, and recycling with good efficiency. Thus, the introduction of SAHs makes possible the solution of aforementioned limitations. SAHs are the polymeric materials having 3D cross-linked polymer network (chemical or physical cross-linked) having the capability to absorb a large amount of water in its network, which is thousand times more than its dry state [20]. The dry state of SAHs is because of collapsing and dominancy of hydrophobic interactions of the polymer chains, which strongly depend on the nature and composition of the materials. But, when the dry SAH is in contact with water or any other solution (aqueous medium), it expands significantly to a considerable large size while retaining the water inside the 3D network. The better objectives to synthesize the SAH are to once absorb the water and then maintain the water inside the gels for a long time [21].

The swelling ability of these SAHs is due to the presence of physical and electrostatic interactions among polymer chains and water molecules. The presence of hydrophilic groups on polymer chains makes the network hydrophilic and sensitive toward environmental temperature [22]. The ionic groups like sulfonic ($-SO_3H$), amines ($-NH_2$), carboxyl ($-COOH$), and hydroxyl ($-OH$) [23] within the network are also responsible for the swelling of polymer networks by undergoing ionization. It generates the electrostatic repulsions and increases the osmotic pressure within the polymer network and increases the size of SAHs. Thus taking the advantage of this swelling ability, SAHs can be successfully applied for the removal of heavy metal ions from wastewater.

The two main processes involved in the removal of heavy metal ions from wastewater by SAHs are the diffusion and electrostatic attractions. The diffusion process causes the penetration of ions inside the SAH network through concentration gradient, while the ionizable groups set up an electrostatic attraction with heavy metal ions and remove them from the aqueous medium by attraction mechanism as shown in **Figure 1**.

The hydrogels are responsive toward the external stimuli including pH [25], temperature [10, 11], intensity of light [26, 27], pressure fluctuation, etc. These stimuli in turn bring a volume phase transition with reversible property in aqueous medium [28].

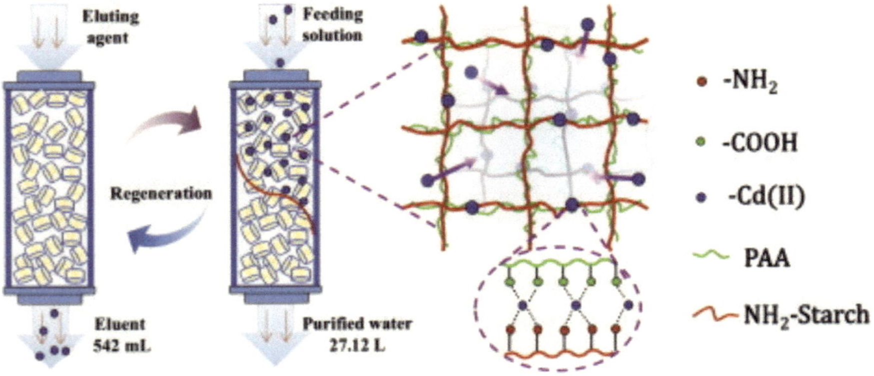

Figure 1.
Sorption mechanism, interaction, and regeneration of hydrogel network with heavy metal ions and removal process [24].

2. Synthesis of SAHs

Different polymerization methods are used for the synthesis of SAHs, i.e., grafting, free radical, suspension, etc.; among them the free radical polymerization process is widely used. It is a single-step synthesis process which requires monomers (cationic, anionic, and neutral), cross-linkers, free radical initiators, and accelerators. Accelerators are required for the generation of free radicals from initiators. This is done by heat or using accelerating agents like N,N,N′,N′-tetramethylethylenediamine (TEMED), $NaSO_3$, etc. Currently, our group synthesized SAH based on acrylic acid (AAc) and acrylamide (AAm) with different compositions having good thermal and mechanical properties by thermal free radical polymerization process [29]. Liu et al. [30] prepared a novel chitosan-g-poly(acrylic acid)/sodium humate SAH via grafting polymerization process using N,N′-methylenebisacrylamide (MBA) as a cross-linking agent and ammonium persulfate (APS) as an initiator. Similarly, Thakur and Arotiba [31] synthesized poly acrylic acid (p(AA)) grafted on sodium alginate (SA-cl-PAA) superabsorbent hydrogel. The grafted hydrogels were synthesized while grafting acrylic acid (AA) monomer on polymer sodium alginate in the presence of N,N′ methylenebisacrylamide (MBA) as cross-linking agent and potassium persulfate (KPS) as an initiator. The proposed mechanism of the reaction was reported by the author and shown in **Figure 2**. The selection of synthesis process is strongly related with the final application of materials.

A copolymer series of SAHs was synthesized and characterized by Mohana Raju and Padmanabha Raju [32] based on monomers acrylamide (AAm), sodium acrylate (SA), and calcium acrylate (CA) p(AAm-co-SA-co-CA) in the presence of MBA as a cross-linker and APS (initiator) and stirrer for 2 h at 80°C.

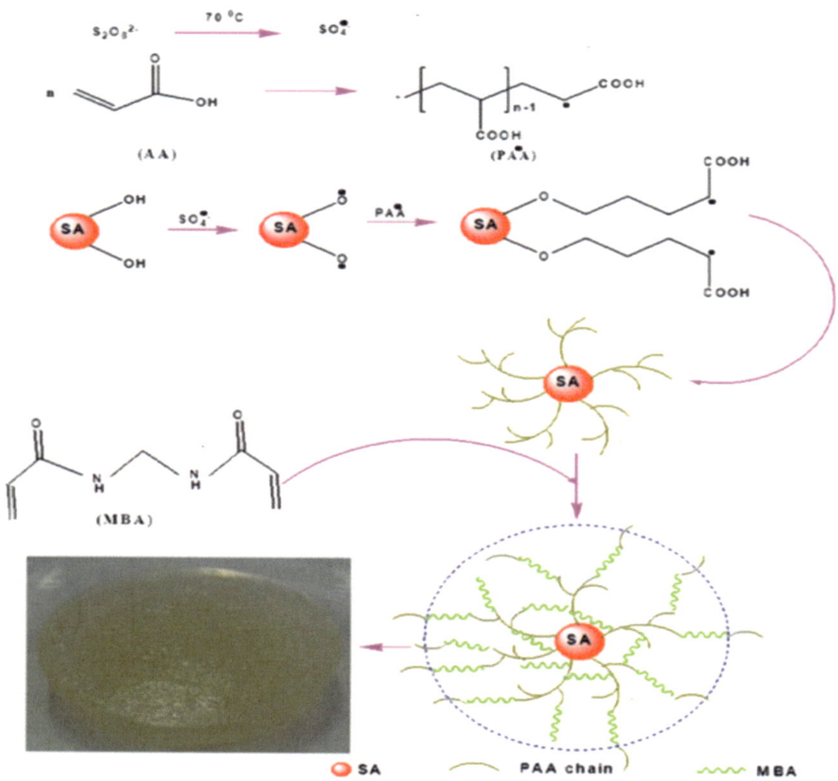

Figure 2.
Proposed mechanism for the formation of a SA-cl-PAA hydrogel network.

3. Properties of SAHs

The SAHs show unique properties which are associated with the structural changes of their network. This makes the hydrogels versatile for various applications. Reversible swelling and deswelling transition is the most important property shown by SAHs. It occurs due to the hydrophilic nature of polymer chains, which is the presence of hydrophilic groups in SAHs and enables them to interact with water. So, the penetration or uptaking of water molecules inside the polymer network causes expanding or swelling of SAHs. The absorption property of SAHs toward water can be evaluated from the swelling ratio (Eq. (1)) and depends on the temperature of the medium. The SAHs composed of thermosensitive monomers like N-isopropylacrylamide (NIPAm), acrylamide (AAm), vinylcaprolactam, etc. show reversible swelling properties with temperature of the medium. At above and below the lower critical solution temperature (LCST), these SAHs behave differently. The temperature-sensitive hydrogel system was reported for the removal of metal-extractant complex at temperature below LCST, followed by desorption from p(NIPAM) network at temperature higher than LCST. The model system for this study was Cu^{2+} as a heavy metal ion, sodium n-dodecyl benzene sulfonate (SDBS), and p(NIPA) hydrogel, while a novel proposed temperature-swing solid-phase extraction (TS-SPE) technique has been introduced effectively [33]. The mechanism can be best seen in **Figure 3**.

Thakur et al. studied the absorbency and swelling of water by the dried hydrogel samples in aqueous solution at different pH values at room temperature. The % swelling was found to be 41,298% which is quite a greater value. The % swelling of the investigated samples was calculated by Eq. (1):

$$\% \text{ Swelling} = \frac{W_s - W_d}{W_d} \times 100 \quad (1)$$

where W_s is swollen and W_d is dry weight of the SAH sample.

The SAHs consist of various different positive and negative functional groups as an active site also causes swelling and deswelling transitions by undergoing the ionization of active sites with respect to change in the pH of the medium. For example, —OH and —COOH groups in the network of the hydrogels undergo protonation and deprotonation at low and high pH of the medium, respectively. At high pH (pH > pKa of monomer used), the deprotonation causes the surface to become negatively charged, and an electrostatic repulsions occur, which in turn expands/

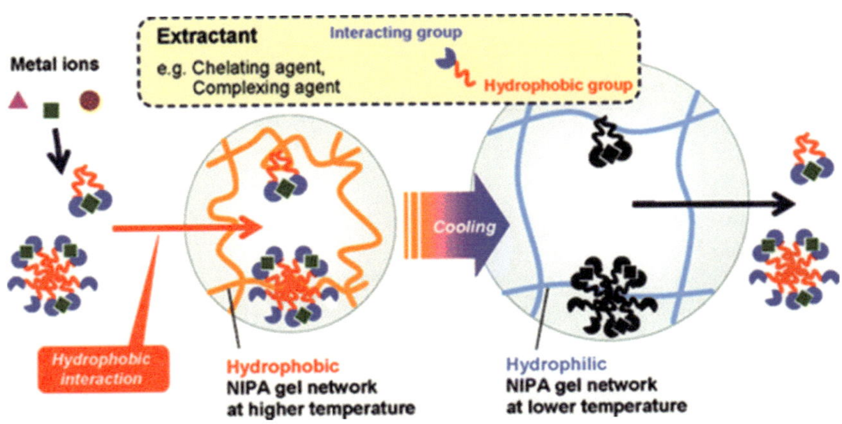

Figure 3.
Extraction of metal-extractant complex onto p(NIPAM) hydrogel by TS-SPE technique [33].

swells the SAH network. Similarly, the process becomes reversed where the low pH of the medium causes the protonation of active sites, resulting in the disappearance of surface charge and the collapse of SAH network. The same results but with opposite trends can be observed for basic groups like —NH_2 functional groups. The major application of pH-responsive hydrogels is used in drug delivery, removal of organic pollutants, dyes, and heavy metal ions from polluted and toxic water. In this chapter we are compiling the importance of SAHs in the removal of heavy metal ions from wastewater.

4. Removal of heavy metal ions by SAHs

The heavy metal ions are important and can be used up to certain permissible limit. But, when the concentration of these metal ions increases, then these become toxic and produce a huge number of disease in both terrestrial and aquatic animals. The accumulation of heavy metal ions in water bodies for a long time acts as a pollutant and undergoes in living bodies through different food chains. Therefore, the removal of these toxic heavy metal ions from wastewater is important and a hot area for researchers to work in. Various traditional adsorbents are used for the removal of these impurities from water. However, in the present the synthesis of SAHs has been increasing due to its easy synthesis route, better performance, selectivity, and recyclability with good efficiency toward absorption of heavy metal ions. A general sketch for entrapment of heavy metal ions by SAHs through swelling mechanism is presented in **Figure 4**.

SAH network consists of various functional groups (both positive and negative) as an active site for entrapping and removing of metal ions from the medium. Thus the researchers choose to test and work on the removal of these heavy metals from wastewater by incorporating opposite charge density on the network. The sorption process will be more if opposite charge exists on SAH network to that of heavy metal ions. As the charge nature on SAHs depends on the nature (cationic, anionic, and neutral) of monomers used for the synthesis process. So, for efficient removal of heavy metal ions, the hydrogels must have negative active sites for electrostatic interactions. In most of cases, the SAHs contain —OH, —COOH, and —SO_3H functional groups as an active site for electrostatic interactions. Similarly, in some cases the sorption mechanism follows the ion exchange route for the removal of metal ions, and the salt group (—COO^- Na^+) exchanges the heavy metal ions between the

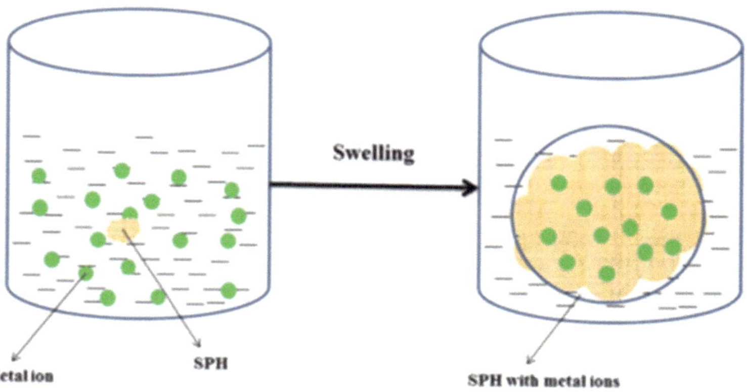

Figure 4.
Swelling and sorption process of SAHs toward heavy metal ions [29].

SAH network and that of wastewater. The heavy metal ion removal capacities of SAHs can be calculated by following Eq. (2):

$$\text{Removal capacity (mg/g)} = \frac{(C_o - C_f) \times V}{m_{SAH}} \quad (2)$$

where C_o and C_f are the initial and final concentrations (mg/L) of heavy metal ions, V is the volume of the solution (L), and m_{SAH} is the mass (g) of superabsorbent hydrogels used as an absorbent.

The effect of charge nature on hydrogels toward the removal of heavy metal ions was demonstrated previously by our group [12]. The synthesis of anionic [poly(methacrylic acid) (P(MAA))], neutral [poly(acrylamide) (P(AAm))], and cationic [poly(3-acrylamidopropyltrimethyl ammonium chloride) (P(APTMACl))] hydrogels was carried out to study the effect of charge nature on the removal of heavy metal ions. The prepared hydrogels have the potential to remove heavy metal ions selectively in the following orders such as P(MAA) > P(AAm) > P(APTMACl). Güçlü et al. [34] reported the synthesis of starch-graft-acrylic acid/montmorillonite (S-g-AA/MMT) superabsorbent composite hydrogels. These synthetic materials were used as a superabsorbent for the removal of Cu^{2+} and Pb^{2+}. The removal affinity of Cu^{2+} was found to be greater than Pb^{2+}. Similarly, Zhao and Mitomo [35] synthesized a physically cross-linked CMC-Chitosan hydrogels which were used for Cu^{2+} absorption from wastewater. The absorption of Cu^{2+} ions on hydrogels occurs due to electrostatic attraction between positive metal ions and that —COOH and —NH_2 functional groups, which are the active sites of synthesized hydrogels. The maximum amount of Cu^{2+} ions removed by Chitosan-CMC was 169.5 mg/g. Thus the hydrogels can be used for the removal of Cu^{2+} ions. Yung et al. [36] synthesized SAH of sodium poly(acrylic acid) p(AANa) through single-step cost-effective techniques. This SAH was used to remove Cu^{2+} ions from aqueous solution. It was found that the maximum uptake capacity of Cu^{2+} ions was 243.91 mg/g, which they claim the highest uptake capacity mentioned in literature. Tang et al. [37] developed an interpenetrating network hydrogels for entrapping heavy metal ions. Especially the superabsorbents were reported by the author for efficient removal of Ni^{2+}, Cr^{2+}, and Cd^{2+} from aqueous solution. The study reveals that 102.34 mg/g of Ni^{2+} metal ions were removed, followed by chromium and cadmium divalent ions. The acrylic acid-based monomer SAH was synthesized by Bulut et al. [38] by dissolving AAc monomer in distilled water and then neutralized with NaOH solution, followed by dispersing the clay (bentonite) powder. The reaction was facilitated by adding MBA (as a cross-linker) and APS (as an initiator) into the reaction vessel and leads the reaction for 2 h at 70°C. The sample was reported by the author for the removal of heavy metal ions (Pb^{2+}, Ni^{2+}, Cd^{2+}, and Cu^{2+}) from aqueous medium. The maximum adsorption capacities of the sample as per reported were 1666.67, 270.27, 416.67, and 222.22 mg g^{-1} for Pb(II), Ni(II), Cd(II), and Cu(II) at 25°C, respectively. A novel poly(pyrrole) graft-based magnetic hydrogel nanocomposite (MHN) was reported by Hosseinzadeh and Tabatabai Asl [39]. It was mentioned that the sample is an effective adsorbent for the removal of lethal Cr(VI) ion from aqueous medium. The maximum absorption capacity of this sample was found to be 208 mg/g. Chauhan et al. [40] synthesized polycarboxylated starch-based hydrogels which were applied for the sorption of copper ions. This hydrogel is stimuli responsive in nature for the sorption of Cu^{2+} ions and strongly depends on the 3D network of hydrogels. The polycarboxylated hydrogel works as a chelating agent and causes chelation with Cu^{2+} ions. The 128 mg/g absorption capacity was reported by the author toward Cu^{2+} ions. Novel SAH nanocomposites were reported by Kaşgöz et al. [41] based on acrylamide (AAm)-2-acrylamido2-methylpropane sulfonic acid (AMPS) sodium

Hydrogel	Maximum adsorption (mg/g) of copper ion	Reference
Starch-graft-acrylic acid/montmorillonite (S-g-AA/MMT)	130.1	[34]
Chitosan-CMC	169.6	[35]
Sodium poly (acrylic acid) p(AANa)	243.9	[36]
Acrylic acid/clay	222.2	[38]
Polycarboxylated starch p(CS)	128	[40]
Acrylamide (AAm)-2-acrylamido-2-methylpropane sulfonic acid (AMPS) sodium salt/clay	81.66	[41]

Table 1.
Maximum absorption capacity of different hydrogel samples toward Cu^{2+} ions.

salt along with clay and were synthesized through in situ copolymerization process. The synthesized materials were successfully used for removal of heavy metal ions including Cu(II), Cd(II), and Pb(II) from the aqueous medium. The absorption capacity of the samples was 1.07, 1.28, and 1.03 mmol/g for Cu(II), Cd(II), and Pb(II) ions, respectively. The comparative removal of copper divalent metal ions by different hydrogel samples are tabulated in **Table 1**.

5. Factor affecting the absorption of heavy metal ions by SAHs

The removal ability of SAHs toward heavy metal ions depends on the environmental conditions. The factors which directly affect the sorption of heavy metals ions are discussed in this chapter.

5.1 Effect of pH

The modified polysaccharide superabsorbent (MPSA) hydrogel was synthesized by Guilherme et al. [42] following free radical polymerization protocol by first treating KOH with AAc, followed by addition of known amount of AAm, modified gum arabic (MAG), and 84 mmol sodium persulfate (SPS) as an initiator. The synthesized SAHs were used as an ionic absorbent for the removal of two metal ions. It was found that the swelling of these SAHs is dependent on the pH of the medium. They studied the removal of Cu^{2+} and Pb^{2+} from aqueous medium, and it was found that the removal of heavy metals increases with the increase in pH value from 3 to 5. It is because of the deprotonation of carboxyl groups above the pKa value (4.0) present in hydrogel network. Thus the basic media generated anionic atmosphere which in turn established an ionic bond with positive metal ions in the medium and favor the process of removal. Another pH-sensitive hydrogels were reported by Peng et al., which were synthesized by grafting acrylic acid (AA) on hemicellulose having xylene. The samples were used for the removal of heavy metal ions like Pb^{2+}, Cd^{2+}, and Zn^{2+} from the medium. The maximum adsorption capacity of the sample for the heavy metal ions was reported by the authors as Pb^{2+} (859 mg/g), Cd^{2+} (495 mg/g), and Zn^{2+} (274 mg/g). It was also reported by the authors that the increase in the uptake capacity of the heavy metal ions from the solution by hydrogels was due to the increase in the pH value of the medium, which is due to the presence of —COOH group. The —COOH group deprotonated into —COO— at high pH value; thus more active sites are available on hydrogel network to remove the positive heavy metal ions and vice versa.

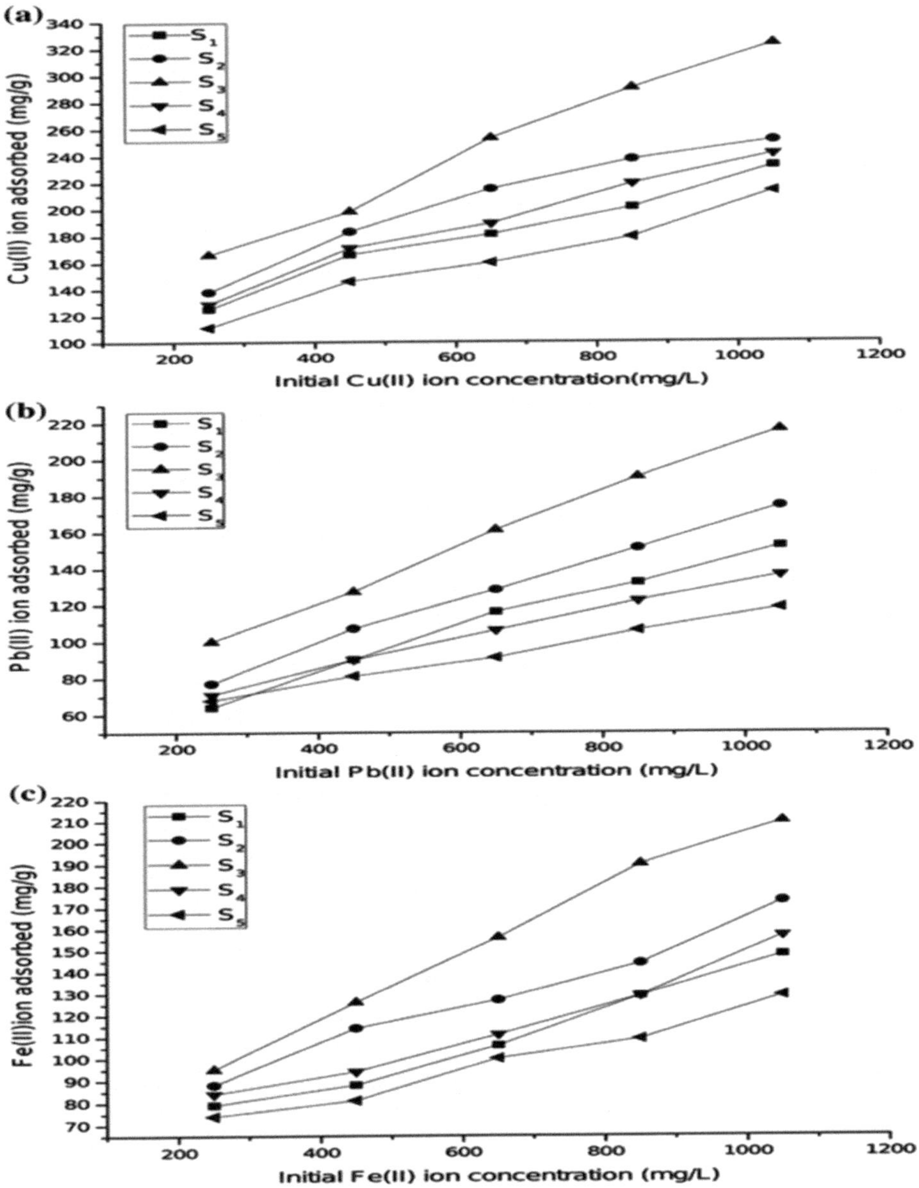

Figure 5.
(a) Variation in the adsorption capacity as function of initial Cu(II) ion concentration using AAc/NaAlg/SH SAHs with the different amount of SH. (b) Variation in the adsorption capacity as function of initial Pb(II) ion concentration using (AAc/NaAlg/SH) SAHs with the different amount of SH. (c) Variation in the adsorption capacity as function of initial Fe(II) ion concentration using (AAc/NaAlg/SH) SAHs with the different amount of SH [43].

5.2 Effect of initial concentration of heavy metal ions

A novel synthesis of SAH consisting of acrylic acid (AA), sodium alginate (SA), and sodium humate (SH) (AAc/SA/SH) to produce poly (AAc-co-NaAlg-co-SH) superabsorbent hydrogels via copolymerization process in the presence of MBA at 70°C for 6 h was reported by Agnihotri and Singhal [43]. These SAHs were used as sorbent for adsorption of heavy metal ions (Cu^{2+}, Pb^{2+}, and Fe^{2+}) and dyes (MB and CV) from aqueous solution. They observed that the adsorption capacity (mg/g) increased with the increasing initial concentration of heavy metal ions, which is due to force established by the heavy metal ions as a mass transport from liquid to adsorbent surface. The increased trends observed in the adsorption capacity are shown in **Figure 5**.

5.3 Effect of temperature on removal of heavy metal ions

Zendehdel et al. [44] reported about the synthesis of SAH of poly acrylamide-co-acrylic acid (p(AAm-co-AA/Na zeolite)) for the successive removal of heavy metal ions (Pb^{2+} and Cd^{2+}) and studied the effect of temperature (ranging from 285 to 323 K) on the removal of these heavy metal ions from the aqueous solution. They reported that raising temperature from 285 to 303 K, the adsorption of heavy metal ions increased, which indicates that the process is endothermic. However, further increase in temperature up to 323 K, the adsorption of heavy metal ions was decreased, which is due to the desorption of heavy metal ions from the hydrogel network.

5.4 Effect of dosage of adsorbent on the removal of heavy metal ions

The synthesis of a copolymer composed of itaconic acid (IA) and acryl amide (AAm) [p(IA-co-AAm)] has been reported by Sharma and Tiwari [45]. The sample was successfully tested for the removal of Mn^{2+} ions from contaminated water at various environmental conditions. They studied and reported briefly about the effect of adsorbent dose on the removal of Mn^{2+} ions. It was observed that increasing the absorbent dose ranging from 0.04 to 0.10 g increases the % removal of the Mn^{2+} ions from contaminated wastewater.

6. Conclusion

The industrial growth for manufacturing of goods not only enhances the manufacturing process but also promotes the market for economic development. However, on the other hand, many industries severely contaminated the water bodies with toxic heavy metals and other pollutants. Therefore, the treatment and removal of toxic heavy metals from the wastewater is necessary before going into stream and rivers. The adsorption of heavy metal ions by superabsorbent hydrogels is more popular due to easy operation, low cost, more effectiveness, biodegradable, reusability, and recyclability. The removal of heavy metal ions by these superabsorbent hydrogels is more recent and attracts the interest of the scientists working in this field.

Acknowledgements

L.A. Shah is highly grateful and acknowledges the financial support by the Higher Education Commission of Pakistan under research grant no: 7309.

Abbreviations

AcGGM	acetyl-galacto glucomannan
P(AA)	poly acrylic acid
PEG	poly ethylene glycol
PEDOT	poly(3,4-ethylenedioxythiophene)

P(NIPAM)	poly N-Isopropyl acryl amide
SAHs	superabsorbent hydrogels
MMT	Montmorillonite
AAc	acrylic acid
CMC	carboxyl methyl cellulose
S	starch
g	graft
SA	sodium alginate
AMPS	2-acrylamido2-methylpropane sulfonic acid
APTMACl	3-acrylamidopropyltrimethyl ammonium chloride
AAm	acryl amide
MAG	modified gum arabic
MB	methylene blue
CV	crystal violet
LCST	lower critical solution temperature
IT	itaconic acid
APS	ammonium persulfate
KPS	potassium per sulfate
TS-SPE	temperature-swing solid-phase extraction
MBA	N,N′-methylenebisacrylamide

Author details

Tanzil Ur Rehman[1], Luqman Ali Shah[1*], Noor Saeed Khattak[1], Abbas Khan[2], Noor Rehman[3] and Sultan Alam[4]

1 Polymer Laboratory, National Centre of Excellence in Physical Chemistry, University of Peshawar, KPK, Pakistan

2 Department of Chemistry, Abdulwali Khan University Mardan, KPK, Pakistan

3 Department of Chemistry, Shaheed Benazir Bhutto University, KP, Pakistan

4 Department of Chemistry, University of Malakand, KPK, Pakistan

*Address all correspondence to: luqman_alisha@yahoo.com; luqman_alisha@uop.edu.pk

IntechOpen

© 2019 The Author(s). Licensee IntechOpen. This chapter is distributed under the terms of the Creative Commons Attribution License (http://creativecommons.org/licenses/by/3.0), which permits unrestricted use, distribution, and reproduction in any medium, provided the original work is properly cited.

References

[1] Jaishankar M et al. Toxicity, mechanism and health effects of some heavy metals. Interdisciplinary Toxicology. 2014;**7**(2):60-72

[2] Herrmann J-M. Heterogeneous photocatalysis: Fundamentals and applications to the removal of various types of aqueous pollutants. Catalysis Today. 1999;**53**(1):115-129

[3] Fisher AJ et al. Spectroscopic evaluation of removal efficiency for a pharmaceutical pollutant in water using a magnetite-activated carbon nanocomposite. Journal of Chemical Education. 2019;**96**(4):751-755

[4] Krishnan U et al. MoS_2/Ag nanocomposites for electrochemical sensing and photocatalytic degradation of textile pollutant. Journal of Materials Science: Materials in Electronics. 2019;**30**(4):1-11

[5] Banister MP. Polymer and polymer loaded materials for scavenging environmental pollutants from natural water sources. Google Patents. 2018

[6] Gupta VK et al. Chemical treatment technologies for waste-water recycling—An overview. RSC Advances. 2012;**2**(16):6380-6388

[7] Khan S et al. Advanced oxidation and reduction processes. In: Advances in Water Purification Techniques. US: Elsevier; 2019. pp. 135-164

[8] Sayed M et al. Decomposition of clofibric acid in aqueous media by advance oxidation techniques: Kinetics study and degradation pathway. Journal of the Chemical Society of Pakistan. 2016;**38**(4):638-645

[9] Teh CY et al. Recent advancement of coagulation-flocculation and its application in wastewater treatment. Industrial & Engineering Chemistry Research. 2016;**55**(16):4363-4389

[10] Shah LA et al. Synthesis of sensitive hybrid polymer microgels for catalytic reduction of organic pollutants. Journal of Environmental Chemical Engineering. 2016;**4**(3):3492-3497

[11] Shah LA et al. Thermal and pH dual responsive copolymer and silver nanoparticle composite for catalytic application. Chinese Journal of Chemistry. 2015;**33**(4):467-472

[12] Javed R et al. Uptake of heavy metal ions from aqueous media by hydrogels and their conversion to nanoparticles for generation of a catalyst system: Two-fold application study. RSC Advances. 2018;**8**(27):14787-14797

[13] Shah LA. Developing Ag-tercopolymer microgels for the catalytic reduction of p-nitrophenol and EosinY throughout the entire pH range. Journal of Molecular Liquids. 2019;**288**:111045

[14] Asghar A, Raman AAA, Daud WMAW. Advanced oxidation processes for in-situ production of hydrogen peroxide/hydroxyl radical for textile wastewater treatment: A review. Journal of Cleaner Production. 2015;**87**:826-838

[15] Sayed M et al. Efficient photocatalytic degradation of norfloxacin in aqueous media by hydrothermally synthesized immobilized TiO_2/Ti films with exposed {001} facets. The Journal of Physical Chemistry A. 2016;**120**(50):9916-9931

[16] Ali I, Gupta V. Advances in water treatment by adsorption technology. Nature Protocols. 2006;**1**(6):2661

[17] Ahmaruzzaman M, Gupta VK. Rice husk and its ash as low-cost adsorbents

in water and wastewater treatment. Industrial & Engineering Chemistry Research. 2011;**50**(24):13589-13613

[18] Gupta S, Babu B. Removal of toxic metal Cr (VI) from aqueous solutions using sawdust as adsorbent: Equilibrium, kinetics and regeneration studies. Chemical Engineering Journal. 2009;**150**(2-3):352-365

[19] Salam OEA, Reiad NA, ElShafei MM. A study of the removal characteristics of heavy metals from wastewater by low-cost adsorbents. Journal of Advanced Research. 2011;**2**(4):297-303

[20] Barakat M. New trends in removing heavy metals from industrial wastewater. Arabian Journal of Chemistry. 2011;**4**(4):361-377

[21] Chi H et al. Swelling, thermal stability, antibacterial properties enhancement on composite hydrogel synthesized by chitosan-acrylic acid and ZnO nanowires. Polymer-Plastics Technology and Materials. 2019;**58**(15):1-13

[22] Shah M et al. Synthesis, fabrication and characterization of polymer microgel/photochromic dye-based sandwiched sensors. Iranian Polymer Journal. 2019;**28**(6):1-11

[23] Askari F et al. Synthesis and characterization of acrylic-based superabsorbents. Journal of Applied Polymer Science. 1993;**50**(10):1851-1855

[24] Zhou G et al. Rapid and efficient treatment of wastewater with high-concentration heavy metals using a new type of hydrogel-based adsorption process. Bioresource Technology. 2016;**219**:451-457

[25] Mahdavinia G et al. Modified chitosan 4. Superabsorbent hydrogels from poly(acrylic acid-co-acrylamide) grafted chitosan with salt-and pH-responsiveness properties. European Polymer Journal. 2004;**40**(7):1399-1407

[26] Jin C et al. Temperature and pH responsive hydrogels using methacrylated lignosulfonate cross-linker: Synthesis, characterization, and properties. ACS Sustainable Chemistry & Engineering. 2018;**6**(2):1763-1771

[27] Laftah WA, Hashim S, Ibrahim AN. Polymer hydrogels: A review. Polymer-Plastics Technology and Engineering. 2011;**50**(14):1475-1486

[28] Meng H, Hu J. A brief review of stimulus-active polymers responsive to thermal, light, magnetic, electric, and water/solvent stimuli. Journal of Intelligent Material Systems and Structures. 2010;**21**(9):859-885

[29] Shah LA et al. Superabsorbent polymer hydrogels with good thermal and mechanical properties for removal of selected heavy metal ions. Journal of Cleaner Production. 2018;**201**:78-87

[30] Liu J, Wang Q, Wang A. Synthesis and characterization of chitosan-g-poly(acrylic acid)/sodium humate superabsorbent. Carbohydrate Polymers. 2007;**70**(2):166-173

[31] Thakur S, Arotiba OA. Synthesis, swelling and adsorption studies of a pH-responsive sodium alginate-poly (acrylic acid) superabsorbent hydrogel. Polymer Bulletin. 2018;**75**(10):4587-4606

[32] Mohana Raju K, Padmanabha Raju M. Synthesis of novel superabsorbing copolymers for agricultural and horticultural applications. Polymer International. 2001;**50**(8):946-951

[33] Tokuyama H, Iwama T. Temperature-swing solid-phase extraction of heavy metals on a poly (N-isopropylacrylamide) hydrogel. Langmuir. 2007;**23**(26):13104-13108

[34] Güçlü G et al. Removal of Cu^{2+} and Pb^{2+} ions from aqueous solutions by starch-graft-acrylic acid/montmorillonite superabsorbent nanocomposite hydrogels. Polymer Bulletin. 2010;**65**(4):333-346

[35] Zhao L, Mitomo H. Adsorption of heavy metal ions from aqueous solution onto chitosan entrapped CM-cellulose hydrogels synthesized by irradiation. Journal of Applied Polymer Science. 2008;**110**(3):1388-1395

[36] Yu Y et al. Eco-friendly and cost-effective superabsorbent sodium polyacrylate composites for environmental remediation. Journal of Materials Science. 2015;**50**(17):5799-5808

[37] Tang Q et al. Synthesis of polyacrylate/polyethylene glycol interpenetrating network hydrogel and its sorption of heavy-metal ions. Science and Technology of Advanced Materials. 2009;**10**(1):015002

[38] Bulut Y et al. Synthesis of clay-based superabsorbent composite and its sorption capability. Journal of Hazardous Materials. 2009;**171**(1-3):717-723

[39] Hosseinzadeh H, Tabatabai Asl M. Enhanced removal of Cr (VI) from aqueous solutions using poly (Pyrrole)-g-poly (acrylic acid-co-acrylamide)/Fe_3O_4 magnetic hydrogel nanocomposite adsorbent. Chemical and Biochemical Engineering Quarterly. 2019;**33**(1):19-33

[40] Chauhan K, Chauhan GS, Ahn J-H. Novel polycarboxylated starch-based sorbents for Cu^{2+} ions. Industrial & Engineering Chemistry Research. 2010;**49**(6):2548-2556

[41] Kaşgöz H, Durmuş A, Kaşgöz A. Enhanced swelling and adsorption properties of AAm-AMPSNa/clay hydrogel nanocomposites for heavy metal ion removal. Polymers for Advanced Technologies. 2008;**19**(3):213-220

[42] Guilherme MR et al. Superabsorbent hydrogel based on modified polysaccharide for removal of Pb^{2+} and Cu^{2+} from water with excellent performance. Journal of Applied Polymer Science. 2007;**105**(5):2903-2909

[43] Agnihotri S, Singhal R. Synthesis and characterization of novel poly (acrylic acid/sodium alginate/sodium humate) superabsorbent hydrogels. Part II: The effect of SH variation on Cu^{2+}, Pb^{2+}, Fe^{2+} metal ions, MB, CV dye adsorption study. Journal of Polymers and the Environment. 2018;**26**(1):383-395

[44] Zendehdel M, Barati A, Alikhani H. Removal of heavy metals from aqueous solution by poly(acrylamide-co-acrylic acid) modified with porous materials. Polymer Bulletin. 2011;**67**(2):343-360

[45] Sharma N, Tiwari A. Nanomagnetite-loaded poly (acrylamide-co-itaconic acid) hydrogel as adsorbent for effective removal of Mn^{2+} from contaminated water. Desalination and Water Treatment. 2016;**57**(12):5654-5672

Chapter 11

Use of Porous no Metallic Minerals to Remove Heavy Metals, Precious Metals and Rare Earths, by Cationic Exchange

Juan Hernandez-Avila, Edgar Omar Serrano-Mejía, Eleazar Salinas-Rodríguez, Eduardo Cerecedo-Sáenz, María Isabel Reyes-Valderrama, María del Pilar Gutiérrez-Amador and Ventura Rodríguez-Lugo

Abstract

This chapter is related with the preliminary study of some non-metallic minerals to evaluate their cationic exchange capacity, to remove heavy and precious metals, as well as rare earths elements. The minerals and materials used to execute the ion metals removal were bentonite, phosphorite, and diatomite. The chapter shows the physicochemical behavior of all these minerals, which were used to remove the mentioned elements from solutions coming from ore leaching. It was found that in all cases, the removal of heavy and precious metals, as well as rare earths elements reached over 90%. Although, there were minimal differences in efficiency for all minerals used (bentonite, phosphorite, and diatomite), it could be pointed that the phosphorite has the best results going from 99.43% of removal of Gd, to 99.95–100% for the case of Ce, Nd, La, Yb, Eu, Er, Sm, Tb, Ge, Pd, Pt, and Au.

Keywords: metals removal, cationic exchange, non-metallic minerals, bentonite, phosphorite, diatomite

1. Introduction

As is known, the cation exchange capacity (CEC) is the property of a solid to adsorb cations from the liquid phase, exchanging them for an equivalent amount of other cations. In the solid-liquid system, a dynamic equilibrium is established between the cations of the solution and those adsorbed on the active points of the surface of solid [1]. Similarly, a cation exchange capacity (CEC) can be defined as the measurement of the adsorption equilibria of anions. Cations retained in clays can be replaced by other cations; therefore, they are interchangeable.

There are many kinds of materials used for ion exchange of metals; firstly, the natural and synthetic inorganic minerals such as zeolites, clay minerals, and permutites can be considered; secondly, a great variety of organic materials like synthetic

IntechOpen

resins, polysaccharides, and proteins can be found; and finally, carbonic materials can be found. Among all the above mentioned materials used for ion exchange, natural clays are of great interest because they are cheaper than the synthetic and organic materials; therefore, the evaluation of the ion exchange capability of some clays and natural materials could be of great interest to recover heavy and precious metals and also rare earths elements easily.

Particularly, the word "rare earth" comes from some oxides that are classified as "rare," not by the difficulty of finding them, but because it was difficult to separate the elements from their minerals. Although rare earths are found in hundreds of minerals, only about 20 have favorable conditions to process them, such as bastnasite, monazite, aluminous clays, xenotima, loparita, and parisita [2]. A group of rare earths belong to a series of elements that are very difficult to separate by traditional techniques, since they have very similar chemical properties. Currently, rare earths are in great demand, because in the study of their properties, and have been discovered multiple industrial and technological applications, which currently have placed them as strategic chemical elements for the development of high-tech products and in high demand [2, 3].

The increasing demand for these elements in modern industry, especially for various advanced technologies, has required obtaining them in a very pure state, impossible to achieve with conventional techniques, which is why a novel researchers and efforts have been carried out, into important development in the different fields of obtaining and recovering rare earths, from the different mineral resources that contain them [3–7]. For the recovery of these elements, there are different processes of separation of rare earths, such as photochemical separation, supercritical extraction, and extraction with solvents and ion exchange [8].

According to heavy metals, almost all of them are too toxic due to their industrial sources and the permitted levels in environment for their discharge. Toxicity of heavy metals depends on their bioaccumulation in environment, that also, depends about the chemical speciation, persistence and tendency of accumulation or bioaccumulation [9]. The classification of the techniques for the treatment of heavy metals in wastewaters depends on different factors, and these technique are conventional and nonconventional, principally. Among the conventional techniques, filtration by membrane, electrodialysis, inverse osmosis, nanofiltration, ultrafiltration, ionic exchange, chemical precipitation, and others can be found [10]. So, ion exchange using natural minerals could be of importance, due to the low cost of these minerals and the possibility of finding a better way to remove heavy metals from contaminated effluents.

According to the precious metals ions, their recovery traditionally was carried out by chemical precipitation; however, the process showed some disadvantages like the formation of toxic subproducts which avoid a complete recovery of the precious metals such as Ag and Au [11]. For the abovementioned, some processes of ion exchange have been developed, using resins and activated carbon, interchanging Ag and Au ions contained in residues or dissolved solutions from cyaniding [12].

On the other hand, various works on the ion exchange of metals with natural adsorbents (aluminosilicates such as zeolites, mineral clays, feldspars, and zeolites) show that these can be good candidates for the recovery of rare earths, as well as heavy and precious metals, taking into consideration some factors that affect the ion exchange such as pH, temperature, contact time, particle size, nature of the cations (size, ionic charge, shape, and concentration), anions associated with the cations in solution, solvent (water, organic solvents) and material selectivity, as well as the ion exchange rate [13–16].

It is for this reason and due to all the efforts made for the recovery of these metals, extraordinary growth has been generated in the last two decades due to their multiple maps in vital sectors for economic development, such as those of computer science and renewable energies, but they are also of great interest in the defense sector. However, global production of rare earths is dominated by China at an alarming 97%, as well as some derived industries such as refining, obtaining alloys, and, to a lesser extent, manufacturing new generation magnets. Currently, China has slowed down its production and very serious questions for the coming years [2].

According the above, and due to the exploration and discovery of deposits with rare earths contents in the state of Hidalgo, Mexico [17, 18], that studies are proposed for the recovery of these precious and heavy metals, as well as rare earths, through the cationic exchange with natural absorbents such as non-metallic minerals, also occurring in the state of Hidalgo.

According the abovementioned, the main goal of this work is to determine the cationic exchange capacity (CEC) of different non-metallic minerals, for the recovery of precious metals, rare earths elements, and heavy metals from leached liquors. In the same way, the determination of the CEC of these minerals could be of utility for their use in the treatment of waste water containing heavy metals. Although there are some works related to the study of the CEC of diatomite, there are not many studies related with the study of the CEC for the bentonite and phosphorite, leading to an important innovation to this work.

2. Materials and methods

In this section, the minerals and materials used to do the ion exchange for the recovery of heavy and precious metals, as well as rare earths elements, will be described.

2.1 Materials

The minerals used in this work were obtained in different regions of the State of Hidalgo, Mexico. They were collected in the grinding plants from open mines of diatomite, phosphorite, and bentonite. All the minerals were wholly characterized to describe the principal physicochemical characteristics that they have to carry out for the corresponding ion exchange. Basically, the characterization was performed preparing samples of each mineral (bentonite, phosphorite, and diatomite) and crushing and grinding them until obtain a range of particle size less than 100 μm.

2.2 Ion exchange methods

To execute the ion exchange, several experiments were executed using leaching liquors of a SEDEX-type mineral located in the State of Hidalgo, Mexico, containing precious metals and rare earths [17, 18], and previously characterized by ICP-OES. The average chemical composition having these liquors is shown in **Table 1**.

To carry out the ionic exchange with the natural minerals, a two-stage procedure was done as follows: 20 g of each mineral (bentonite, phosphorite, and diatomite), at a −400 mesh particle size, were weighted and then, were added to 500 ml volume of leaching solution containing the elements shown in **Table 1**. The experiments were executed at a stirring rate of 500 rpm, with a pH of 0.3, which was controlled adding NaOH constantly during all experiment, and the reaction time used was of 24 h. After ending the reaction time, solution was filtered and dried; then, both solids as residual liquors were characterized (ICP-OES, XRD, XRF, and SEM-EDS) to evaluate the capability of ion exchange of each mineral. The analysis done by ICP-OES was executed taking the final liquid after ion exchange, and then sample was diluted taking 1 ml of solution into 100 ml of distilled water having a 1/100

ICP analysis	
Element	**Ppm**
Au	45.93
Pd	1.92
Pt	0.005
Ce	81.79
Nd	56.97
La	51.89
Yb	33.35
Tb	25
Ge	9
Ga	2.5
Sm	1.75
Er	0.9
Eu	0.35

Table 1.
Chemical composition of the leaching liquors of the SEDEX mineral, done by ICP.

dilution factor. With this, all the elements fell in a range of concentration below 1 ppm for the case of Au, Ce, Nd, L, Tb, Ge, and Yb. For the rest of the elements, a direct sample was taken using standards between 1 and 5 ppm (without dilution).

2.3 Analytical methods

2.3.1 Characterization of minerals

The analytical methods used for the characterizations of the different mineral used to the ion exchange stage were characterized by dry granulometric analysis using a standard sieves (Tyler series); X-ray diffraction using an INEL equinox 2000 equipment was located at the Autonomous University of the State of Hidalgo (AUSH), and the spectra was obtained with a Cu $K_{\alpha 1}$ radiation of wave length of 1.50056 Å, a voltage of 30 kV, an intensity of 20 mA, and a sweep speed of 22 θ/min, which were then treated with the software MATH to identify the mineral species contained in each material.

2.3.2 Chemical analysis

Then, to determine the chemical composition of standard solutions and leached liquors (before and after ion exchange), an inductively coupled plasma spectrophotometer (ICP-OES) Perkin Elmer brand/model 2100 located at the AUSH was used. In the same way, scanning electron microscopy (SEM) was used to determine morphology and particle type of minerals, using a JEOL brand microscope model JMS6300, having an Energy Dispersive Spectrometer of X-ray (EDS) for chemical microanalysis and also located at the AUSH. And finally, X-ray fluorescence (XRF) was executed to determine the chemical composition of minerals before and after the ion exchange procedure, and for this analysis, a portable spectrophotometer of X-ray brand BRUKER, model S1 TITAN was used, also using the GeoChem calibration software (equipment located at the National Autonomous University of Mexico).

3. Results and discussion

3.1 Characterization of minerals

3.1.1 Bentonite before ion exchange experiments

Bentonite is a rock composed essentially of minerals from the group of smectites, regardless of any genetic connotation. Additionally, it has the ability to inflate and increase its volume by weight several times on contact with water, to form thixotropic gels when added in small quantities. Finally, it can be said that its main interchangeable cations are Na^+, Ca^{2+}, K^+, Mg^{2+}, Li^+, and H^+.

Table 2 shows the results obtained by ICP and FRX of the elements contained in the bentonite studied, of which silicon, aluminum, iron, sodium, and potassium are present in the majority among others.

Likewise, in **Figure 1**, the mineral species identified by X-ray diffraction are observed, in which the presence of majority of mineral phases, such as quartz,

Element (oxide)	ICP (wt.%)	FRX (wt.%)
Na_2O	2.66	0.92
MgO	2.71	2.76
Al_2O_3	14.10	14.92
SiO_2	73.10	73.41
K_2O	1.32	1.67
CaO	3.11	3.93
TiO_2	0.37	0.35
Fe_2O_3	2.63	2.60

Table 2.
Chemical composition of natural bentonite.

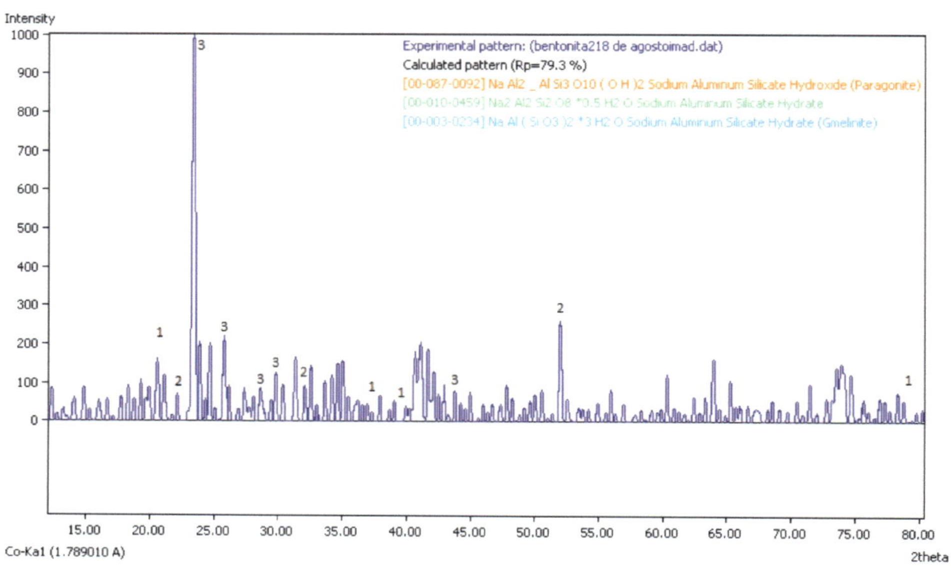

Figure 1.
X-ray diffractogram of bentonite.

anorthoclase, orthoclase, albite, and berlinite, can be seen, as well as sodium silicoaluminates (paragonite and gmelinite).

On the other hand, **Figure 2** shows an image of a bentonite particle at −400 mesh, analyzed by SEM-EDS, where the semiquantitative and point composition can be observed, showing the presence of elements such as silicon, aluminum, sodium, magnesium, potassium, and iron, which are characteristic in the bentonite (**Figure 2a**); likewise, the bentonite particle is shown in detail, having a lot of porosity, which is essentially one of the most important physical characteristics for the ion exchange that could be carried out, besides the particle size, offering a large surface and contact area, **Figure 2b**.

3.1.2 Bentonite after ion exchange experiments

The results obtained after ion exchange experiments are presented as chemical composition of liquors before and after the procedure; **Table 3** shows the chemical composition of these liquors obtained by ICP, showing also the efficiency of ion exchange for each element. It can be noticed that for all elements present in original leaching liquor, the efficiency is over 90%.

After the ion exchange, the solution was filtered and the solid residue of the bentonite was dried and analyzed by XRD. The mineral species identified by X-ray diffraction (**Figure 3**), are majority mineral phases, such as quartz, anorthoclase, orthoclase, albite, berlinite, and silico-aluminates, typical of bentonite. Likewise, signs indicating the presence of rare earths and precious metals are noted, which corroborates that the exchange of these elements was made to the natural absorbent.

Finally, **Figure 4** shows the image of a bentonite particle after performing the ion exchange with the leaching liquors of a SEDEX-type mineral, where the semi-quantitative composition and the distribution of the elements are shown by X-ray mapping, observing that, indeed, the aforementioned elements were absorbed in the bentonite ore.

3.1.3 Phosphorite before ion exchange experiments

Phosphorites are sedimentary rocks that contain at least 20% P_2O_5 in the form of cryptocrystalline fluorapatite, apatite, or some other minerals containing

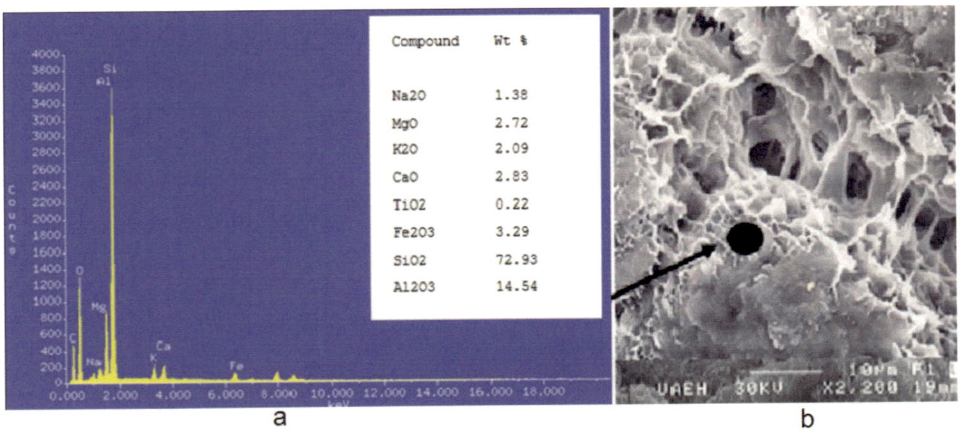

Figure 2.
SEM analysis of bentonite: (a) point microanalysis by SEM-EDS and (b) general image (SEM-SE).

Element	Content before ion exchange (ppm)	Content after ion exchange (ppm)	Efficiency of the ion exchange (%)
Au	45.93	0.072	99.84
Ce	81.79	0.000	100.00
La	51.89	0.000	100.00
Nd	56.97	0.020	99.96
Pd	1.92	0.108	94.37
Yb	33.35	0.009	99.97
Ge	9	0	100
Gd	1.4	0.111	92.07
Tb	0.25	0	100
Sm	1.75	0	100
Er	0.9	0	100
Eu	0.35	0.011	96.86
Pt	0.005	0	100

Table 3.
Results of the ion exchange done using bentonite (ICP).

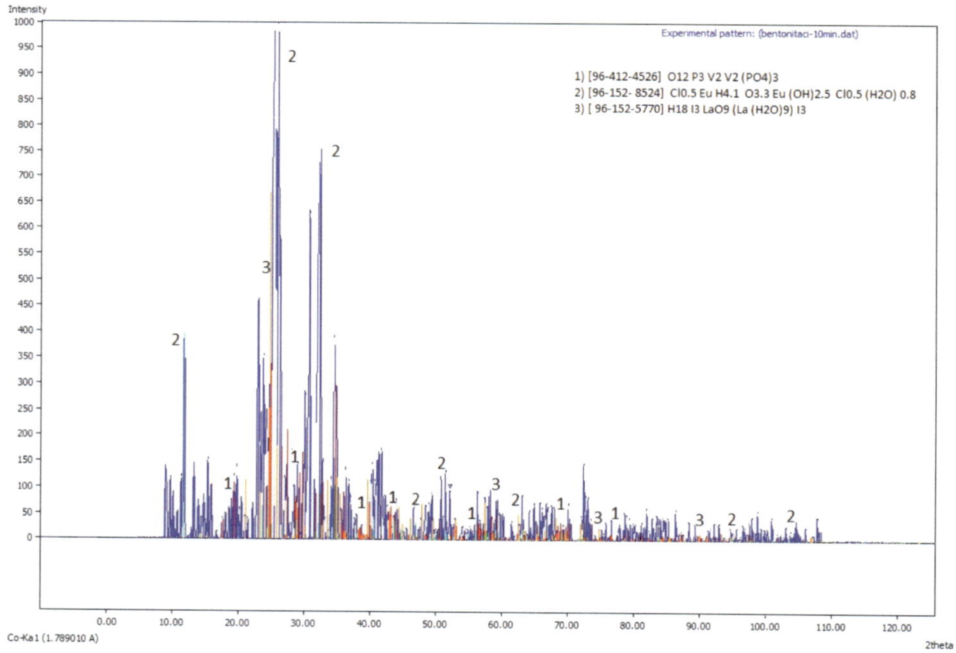

Figure 3.
X-ray diffractogram of bentonite after ion exchange.

phosphorus. They usually occur in layers, and it can also be present in the form of crusts, spherulites, and nodules in sedimentary horizons. Rocks with less than 10% phosphate receive the adjective of phosphates. Its main interchangeable cations are: Ca^{2+}, K^{+}, and Mg^{2+}.

Before carrying out the ion exchange experiments, the phosphorite was characterized to know its composition and compare it at the end of the ion exchange

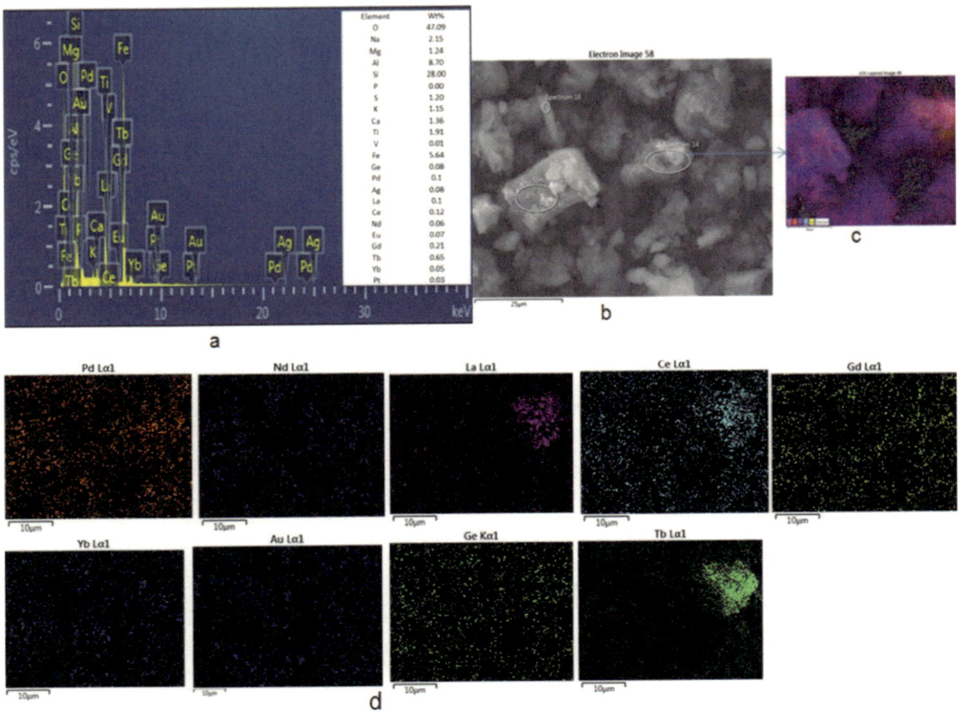

Figure 4.
Photomicrographs of bentonite after ion exchange: (a) general microanalysis SEM-EDS; (b) general image, 2000×, SEM-SE; (c) point image where the microanalysis was taken; and (d) distribution mapping of rare earth elements and precious metals in the bentonite.

Element	ICP (wt.%)	XRF (wt.%)
Na_2O	0	0.16
Al_2O_3	2.43	2.40
P_2O_5	36.03	36.08
K_2O	0.15	0.21
CaO	59.99	59.95
TiO_2	0.48	0.31
Fe_2O_3	0.92	0.89

Table 4.
Average chemical composition of phosphorite before ion exchange.

experiments, and thus to evaluate its exchange capacity for the elements present in the leaching liquors. **Table 4** shows the results obtained by ICP and XRF of the chemical composition for the phosphorite, where the average contents of 36% of P_2O_5 are shown, as well as the majority of aluminum, calcium, and minor elements such as oxides of sodium, potassium, titanium, and iron.

On the other hand, the mineral species identified by X-ray diffraction (**Figure 5**) are mainly mineral phases of calcium and phosphorus.

Finally, in **Figure 6**, an image of phosphorite particles at −400 mesh can be seen, which was analyzed by SEM-EDS, and here, also the presence of major elements of phosphorus, aluminum, and calcium oxides, as well as minor elements of sodium, titanium and iron, can be observed, as is shown in **Figure 6a**; similarly, the photomicrographs of the phosphorite particle are shown in **Figure 6b**.

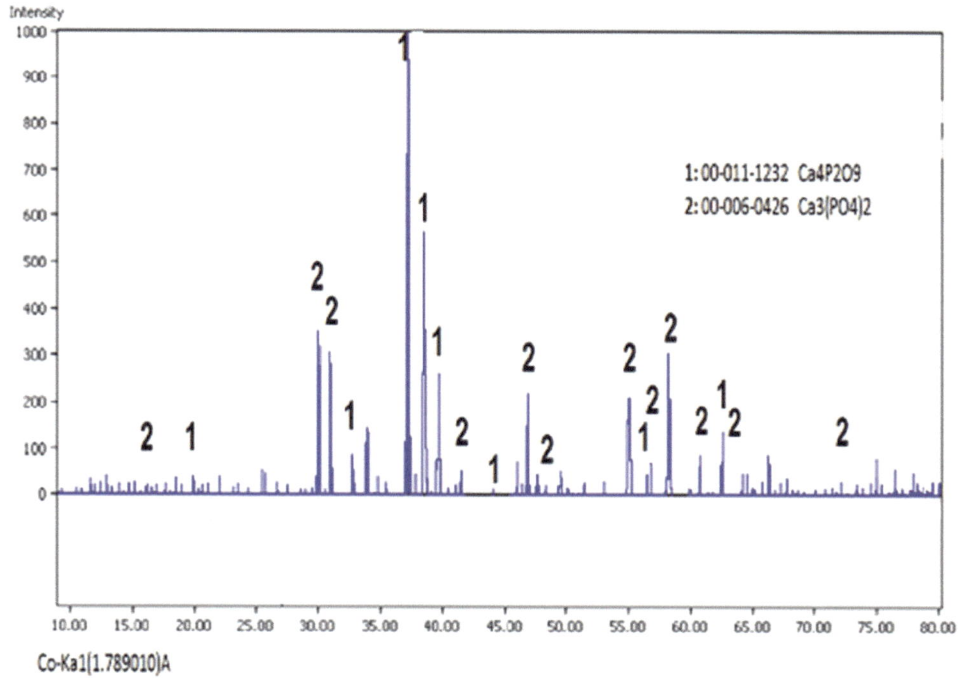

Figure 5.
X-ray diffractogram of phosphorite, before ion exchange.

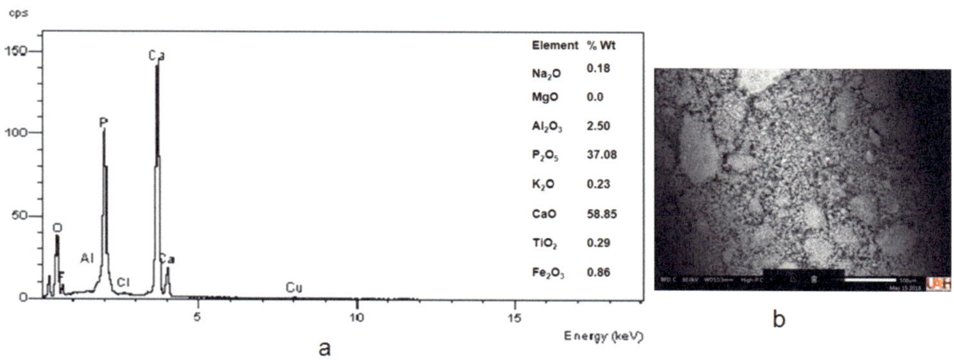

Figure 6.
Photomicrographs of the phosphorite −400 meshes, (a) SEM-EDS microanalysis and (b) general image, 2200×, SEM-SE.

3.1.4 Phosphorite after ion exchange experiments

After carrying out the ion exchange, using phosphorite as an absorbent mineral, the solution was filtered and analyzed by ICP, comparing the results with the initial composition of the leaching liquors. The results are shown in **Table 5**, where a comparison is made between the original leaching liquors before and after the exchange, as well as the efficiency of the cation exchange of this mineral; it is observed that in all cases, efficiencies are greater than 99% of the elements exchanged, which can determine that phosphorite is a good natural absorbent for rare earths and precious metals.

Likewise, the solids obtained after the ion exchange were separated and dried to be later analyzed by XRD. The mineral species identified by X-ray diffraction (**Figure 7**) are majority phases such as quartz, anorthoclase, orthoclase, albite, berlinite, and silicoaluminates. Similarly, the presence of rare earths, such as lanthanum and cerium, and precious metals that were all adsorbed by this mineral are noted.

Element	Content before ion exchange (ppm)	Content after ion exchange (ppm)	Efficiency of the ion exchange (%)
Au	45.93	0	100
Ce	81.79	0.037	99.95
La	51.89	0.017	99.97
Nd	56.97	0.02	99.96
Pd	1.92	0	100
Yb	33.35	0.001	99.99
Ge	9	0	100
Gd	1.4	0.008	99.43
Tb	0.25	0	100
Sm	1.75	0	100
Er	0.9	0	100
Eu	0.35	0	100
Pt	0.005	0	100

Table 5.
Results of the ion exchange done using phosphorite (ICP).

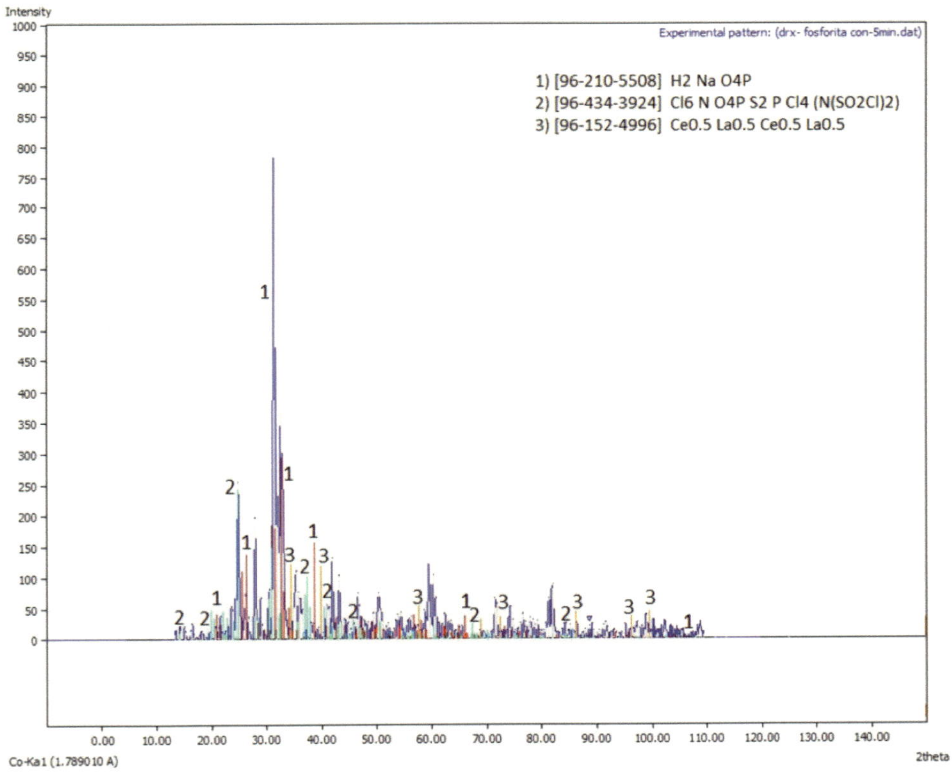

Figure 7.
X-ray diffractogram of phosphorite material after ion exchange.

Finally, **Figure 8** shows the image of phosphorite particles after performing the ion exchange with the leaching liquors of a SEDEX-type mineral, where the semiquantitative point composition and the X-ray mapping of the elements are shown. This verifies that effectively the aforementioned elements were absorbed in the mineral.

3.1.5 Diatomite before ion exchange experiments

Diatomite is a siliceous sedimentary rock of biogenic origin, composed of fossilized skeletons of diatomite frustules. It is composed by sedimentary accumulation

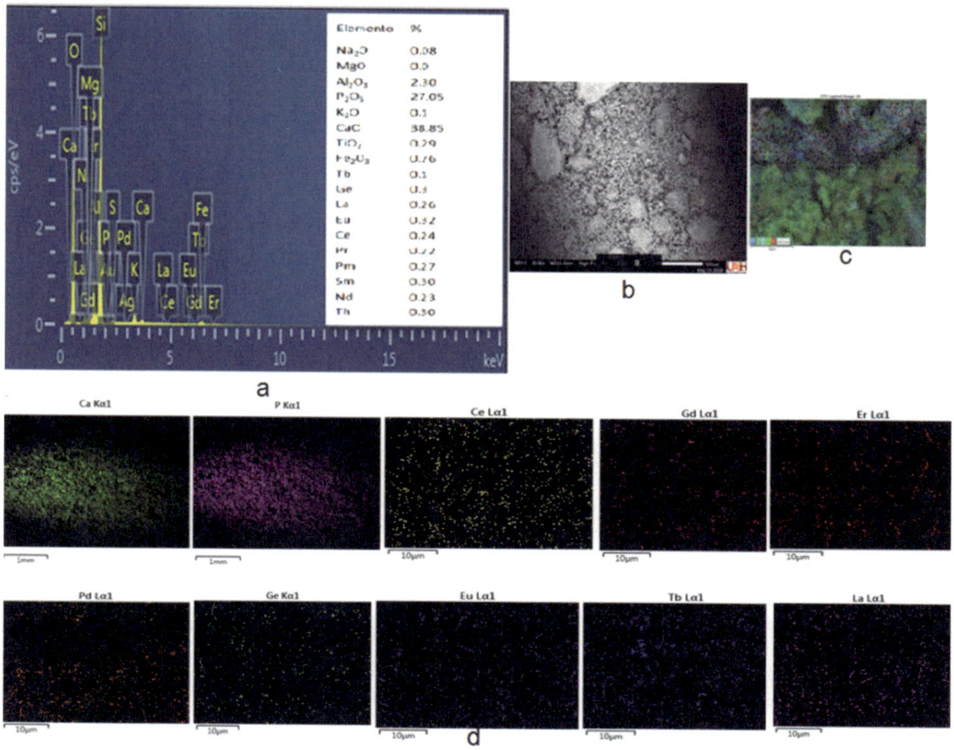

Figure 8.
Photomicrographs of the phosphorite after ion exchange: (a) SEM-EDS microanalysis; (b) general image, 2000×, SEM; (c) point image where the microanalysis was taken; and (d) distribution mapping of rare earth elements and precious metals in the phosphorite.

Element	ICP (wt.%)	XRF (wt.%)
Na_2O	0.61	0.923
MgO	1.79	1.87
Al_2O_3	11.63	7.77
SiO_2	76.00	77.36
K_2O	2.41	1.2
CaO	0.85	1.04
TiO_2	0.50	0.348
Fe_2O_3	1.95	2.35

Table 6.
Average chemical composition of diatomite before ion exchange.

to form large deposits with a sufficient thickness to have a commercial potential. Its main interchangeable cations are Na^+, Ca^{2+}, K^+, Mg^{2+}, and Si^{2+}.

Firstly, before carrying out the ion exchange experiments, the diatomite mineral was characterized to evaluate, at the end, its exchange capacity. **Table 6** summarizes the results obtained by ICP and XRF of the original elements contained in the diatomite, presenting average contents of 76.68% of silicon, as well as the majority contents of alumina, hematite, potassium oxide, magnesium oxide, calcium oxide, and minor elements such as sodium oxide and titanium.

Likewise, the mineral species present in the diatomite were identified by X-ray diffraction (**Figure 9**), observing the presence of majority mineral phases such as quartz, albite and berlinite.

Finally, in **Figure 10**, an image of a diatomite particle at −400 mesh is shown, in which an analysis was carried out by SEM-EDS. The presence of major elements such as silicon, aluminum, sodium, magnesium, potassium, and iron, are characteristic of the diatomite. Likewise, a photomicrograph of the diatomite particle can be seen in **Figure 10b** where the characteristics properties of the diatoms of the material can be observed.

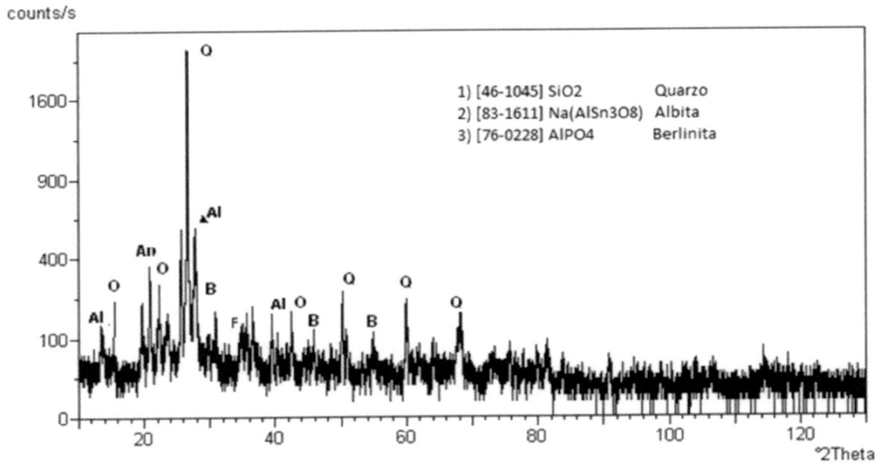

Figure 9.
X-ray diffractogram of diatomite material before ion exchange.

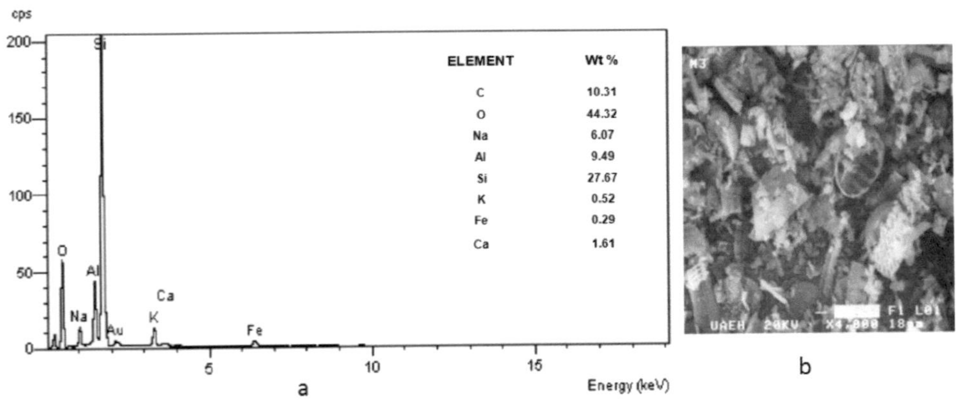

Figure 10.
Photomicrographs of the diatomite −400 meshes, (a) SEM-EDS microanalysis and (b) general image, 4000×, SEM-SE.

3.1.6 Diatomite after ion exchange experiments

After carrying out the ion exchange, the diatomite mineral was characterized to know the elements exchanged and thus calculate the ion exchange capacity of this mineral. The results are shown in **Table 7**, where a comparison is made between the original leaching liquors and after the exchange through ICP analysis. Likewise, it shows the % of efficiency of the cation exchange and as a result can be seen that

Element	Content before ion exchange (ppm)	Content after ion exchange (ppm)	Efficiency of the ion exchange (%)
Au	45.93	0	100
Ce	81.79	0.003	99.996
La	51.89	0.004	99.99
Nd	56.97	0.008	99.99
Pd	1.92	0	100
Yb	33.35	0.001	99.997
Ge	9.0	0	100
Gd	1.4	0.03	97.86
Tb	0.25	0	100
Sm	1.75	0	100
Er	0.9	0	100
Eu	0.35	0.003	99.14
Pt	0.005	0	100

Table 7.
Results of the ion exchange done using diatomite (ICP).

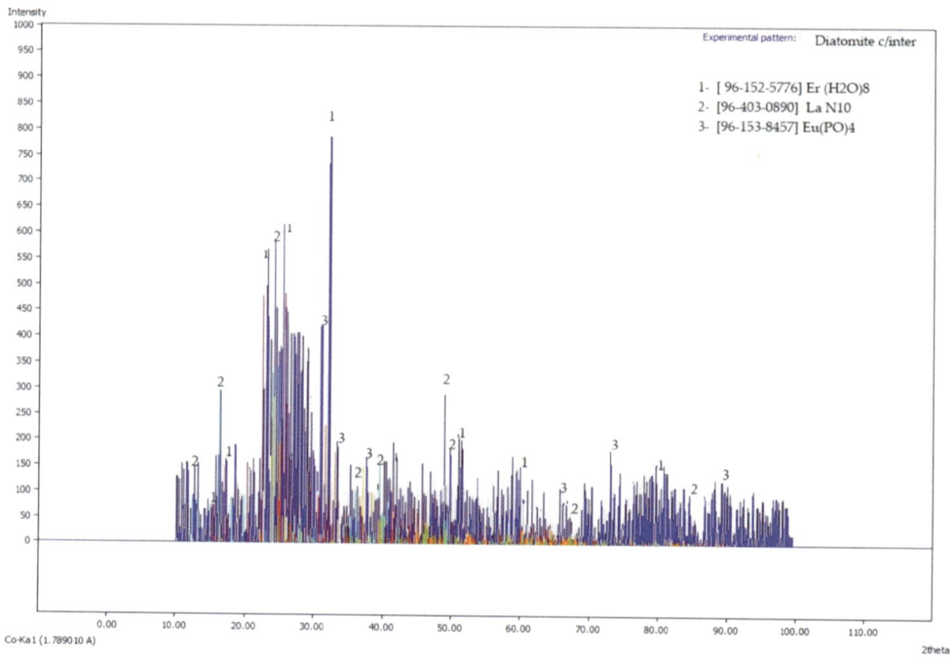

Figure 11.
X-ray diffractogram of diatomite material, after ion exchange.

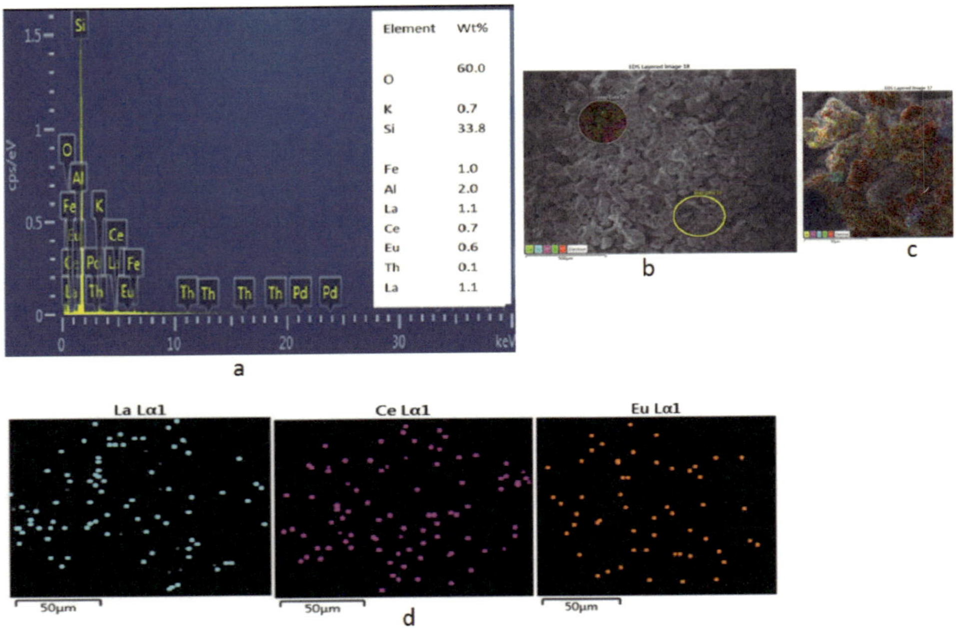

Figure 12.
Photomicrograph of the diatomite −400 meshes, (a) SEM-EDS microanalysis, (b and c) general image, SEM-SE, and (d) mappings.

diatomite shows an exchange efficiency for the precious metals of 100%, as well as the Pd, Ge, Tb, Sm, and Er. The other elements show an efficiency superior to 99%, not so for the Gd, since its exchange efficiency is 97.86, which corroborates that the diatomite has an exchange efficiency for these types of metals. On the other hand, the author also has to demonstrate the efficiency of cationic exchange of diatomite, getting 96.46% of efficiency for the removal of As^{3+}, 95.44% for Ni^{2+}, and 98.80% of Pb^{2+} [13].

Likewise, the mineral species were identified by X-ray diffraction (**Figure 11**), observing the presence of majority mineral phases such as quartz, anorthoclase, orthoclase, albite and berlinite; as well as the presence of rare earths such as Er, La, and Eu.

Finally, **Figure 12** shows the image of a diatomite particle after ion exchange with the leaching liquors of a SEDEX type mineral, where can be observed the semiquantitative point composition and the distribution of elements by X-ray mapping, can be concluded that the elements present in the leaching liquors were effectively absorbed into the diatomite particles.

4. Conclusions

The preliminary results in the cation exchange of heavy metals, precious metals, and rare earths, through the use of non-metallic porous minerals, show a good efficiency since for most of the cases, recoveries over 99% were obtained, except for some elements; in the case of bentonite, for Pd, only 94.37%, for Gd, only 92.09%, and for Eu, only 96.87% absorption were obtained, and in the case of the phosphorite, the lowest value was for Gd with a 99.43% efficiency; in the case of diatomite, Gd presented a 97.86% efficiency in the exchange (**Table 8**). Therefore, it can be concluded that these natural absorbers can be used with a high efficiency for the exchange of these metals, noting also that for most of the cases, Gd presented recoveries above 90% and for the case of the precious metals, it was above 99% for all the minerals used.

Element	Bentonite % in CEC	Phosphorite % in CEC	Diatomite % in CEC
Au	99.84	100	100
Ce	100.00	99.95	99.996
La	100.00	99.97	99.99
Nd	99.96	99.96	99.99
Pd	94.37	100	100
Yb	99.97	99.99	99.99
Ge	100	100	100
Gd	92.07	99.43	97.86
Tb	100	100	100
Sm	100	100	100
Er	100	100	100
Eu	96.86	100	99.14
Pt	100	100	100

Table 8.
Cation exchange capacity (CEC), per element, of the mineral used for the removal of precious metals, heavy metals and rare earth elements.

Acknowledgements

The authors would like to thank the PRODEP-SEP of the government of Mexico for the financial support in the publishing of this chapter. Also thy would like to thank Autonomous University of the State of Hidalgo and the CONACyT for the Master degree scholarship granted to the student Edgar Omar Mejía Serrano who worked this theme, during his thesis project (number 627910).

Conflict of interest

The authors declare no conflict of interest.

Notes

This chapter not only shows unpublished preliminary results obtained by the authors where they have evaluated the capacity of ion exchange for some natural minerals found in the state of Hidalgo Mexico, but also exhibits an exhalative sedimentary mineral found by the same authors in the north of the state of Hidalgo, where ore mineral has important contents of precious metals and some rare earths elements.

Author details

Juan Hernandez-Avila[1], Edgar Omar Serrano-Mejía[1], Eleazar Salinas-Rodríguez[1*], Eduardo Cerecedo-Sáenz[1], María Isabel Reyes-Valderrama[1], María del Pilar Gutiérrez-Amador[2] and Ventura Rodríguez-Lugo[1]

1 Academic Area of Earth Sciences and Materials, Autonomous University of the State of Hidalgo, Hidalgo, Mexico

2 Apan High School, Autonomous University of the State of Hidalgo, Hidalgo, Mexico

*Address all correspondence to: salinasr@uaeh.edu.mx

IntechOpen

© 2019 The Author(s). Licensee IntechOpen. This chapter is distributed under the terms of the Creative Commons Attribution License (http://creativecommons.org/licenses/by/3.0), which permits unrestricted use, distribution, and reproduction in any medium, provided the original work is properly cited.

References

[1] López-Acevedo M, Roquerode-Laburu C. Edafología. 3ª ed. Madrid, Spain: Mundi-Prensa; 2003. 234 p. ISBN 84-8476-148-7

[2] Martínez-Cortés JI, Del Valle-Giles AV. Las tierras raras: Un sector estratégico para el desarrollo tecnológico de China. National Autonomous University of Mexico—(In Spanish). Universidad Nacional Autónoma de México Faculty of Economics Center for Chinese Studies-Mexico. N° 6; 2014

[3] Haskin LA, Paster TP. In: Gschneindner KA Jr, Eyring YL, editors. Handbook on the Physics and Chemistry of Rare Earths. Vol. 3. Amsterdam: North-Holland; 1979. pp. 1-80. ISBN: 978-0-444-50185-1

[4] Bautista RGY, Wong MM. Rare Earths, Extraction, Preparation and Applications. TMS 1989. Las Vegas, Nevada, USA: Warrendale (EEUU); 1989

[5] Moldoveanu GA, Papangelakis VG. Recovery of rare earth elements adsorbed on clay minerals: I. Desorption mechanism. Hydrometallurgy. 2012;**117-118**:71-78. DOI: 10.1016/j.hydromet.2012.02.007

[6] Gómez-Caballero A, Elías-Herrera M. Study on Rare Earths in Mexico, Part 1 (in Spanish). Mexico DF: National Autonomous University of Mexico (In Spanish). Bulletin 108 (In Spanish); 1990

[7] Alshameri A, He H, Xin C, Zhu J, Xinghu W, Zhu R, et al. Understanding the role of natural clay minerals as effective adsorbents and alternative source of rare earth elements: Adsorption operative parameters. Hydrometallurgy. 2019;**189**:149-161. DOI: 10.1016/j.hydromet.2019.02.016

[8] Awal MR, Jyo A, Ihara T, Seko N, Tamada M, Lim KT. Enhanced trace phosphate removal from water by zirconium (IV) loaded fibrous adsorbent. Water Research. 2011;**45**:4592-4600. DOI: 10.1016/j.watres.2011.06.009

[9] Domenech X, Peral J. Environmental Chemistry of Terrestrial Systems (in Spanich). Barcelona, España: Reverté Editorial; 2008. 256 p. ISBN13: 978842919064

[10] Caviedes-Rubio DI, Muñoz-Calderón RA, Perdomo-Gualtero A, Rodríguez-Acosta D, Sandoval-Rojas IJ. Treatments for removal of heavy metals commonly founf in industrial wastewaters. A review (in Spanish). Engineering and Regio Journal (In Spanish). 2015;**13**(1):73-90. DOI: 10.25054/22161325.710

[11] Bourges J, Broudic JC, Dauby J, Leboucher I, Leclerc O, Baticle P. Assignors to Commisariat de l'Énergie Atomique, Paris, and Compagnie Générale des Matériels Nucleaires, Valizy—Villacoublary. U.S. Patent 6428679. 2002

[12] Petrova TM, Tzaneva B, Fachikov L, Hristov J. Silver recovery from spent photographic solutions by a magnetically assisted particle bed. Chemical Engineering and Processing. 2013;**71**:83-96. DOI: 10.1016/j.cep.2013.03.014

[13] Hernández-Ávila J, Salinas-Rodríguez E, Cerecedo-Sáenz E, Reyes-Valderrama M, Arenas-Flores A, Román-Gutiérrez A, et al. Diatoms and their capability for heavy metal removal by cationic exchange. Metals. 2017;7:169. DOI: 10.3390/met7050169

[14] Hernández-Ávila J, Moreno-Pérez F, Espinoza-Borgel I, Salinas-Rodríguez E. Characterization of diatoms from

Hidalgo, Mexico. Possible use as absorbent in the decrease of arsenic in drinking water (In Spanish). In: XL Mexican Congress of Chemistry (in Spanish); September 2005; Guadalajara, Jalisco, México. pp. 38-42

[15] Kononenko N, Nikonenko V, Grande D, Larchet C, Dammak L, Fomenko M, et al. Porous structure of ion exchange membranes investigated by various techniques. Advances in Colloid and Interface Science. 2017;**246**:196-216. DOI: 10.1016/j.cis.2017.05.007

[16] Niftaliev SI, Kozaderova OA, Kim KB. Electroconductance of heterogeneous ion-exchange membranes in aqueous salt solutions. Journal of Electroanalytical Chemistry. 2017;**794**:58-63. DOI: 10.1016/j.jelechem.2017.03.046

[17] Roldán-Contreras E, Hernández-Ávila J, Cerecedo-Sáenz E, Reyes-Valderrama MI, Salinas-Rodríguez E. Use of the system $S_2O_3^{(2)}$-O_2 for the leaching of precious metals contained in a mineral from Molango in the state of Hidalgo, Mexico. European Scientific Journal. 2018;**14**(3):48-54. DOI: 10.19044/esj.2018.v14n3p48

[18] Cerecedo-Sáenz E, Rodríguez-Lugo V, Hernández-Ávila J, Mendoza-Anaya D, Reyes-Valderrama MI, Moreno-Pérez E, et al. Mineralization of rare earths, platinum and gold in a sedimentary deposit, found using an indirect method of exploration. Apects in Mining & Mineral Science. 2018;**1**(2):1-9. DOI: 10.31031/AMMS.2018.01.000510

Chapter 12

Bioremediation Techniques for Polluted Environment: Concept, Advantages, Limitations, and Prospects

Indu Sharma

Abstract

Environmental pollution has been rising in the past few decades due to increased anthropogenic activities. Bioremediation is an attractive and successful cleaning technique to remove toxic waste from polluted environment. Bioremediation is highly involved in degradation, eradication, immobilization, or detoxification diverse chemical wastes and physical hazardous materials from the surrounding through the all-inclusive and action of microorganisms. The main principle is degrading and converting pollutants to less toxic forms. Bioremediation can be carried out ex-situ and in-situ, depending on several factors, which include but not limited to cost, site characteristics, type, and concentration of pollutants. Hence, appropriate bioremediation technique is selected. Additionally, the major methodologies to develop bioremediation are biostimulation, bioaugmentation, bioventing, biopiles, and bioattenuation provided the environmental factors that decide the completion of bioremediation. Bioremediation is the most effective, economical, eco-friendly management tool to manage the polluted environment. All bioremediation techniques have its own advantage and disadvantage because it has its own specific applications.

Keywords: bioremediation, environment, pollutants soil, ground water, waste-water, applications, limitations

1. Introduction

Bioremediation and natural reduction are also seen as a solution for emerging contaminant problems; microbes are very helpful to remediate the contaminated environment. Number of microbes including aerobic, anaerobic bacteria and fungi are involved in bioremediation process. Bioremediation is highly involved in degradation, eradication, immobilization, or detoxification diverse chemical wastes and physical hazardous materials from the surrounding through the all-inclusive and action of microorganisms. The main principle is degrading and converting pollutants to less toxic forms. There are two types of factors these are biotic and abiotic conditions are determine rate of degradation. Currently, different methods and strategies are applied for bioremediation process.

2. Environmental pollution

Environmental pollution has been on the rise in the past few decades due to increased human activities such as population explosion, unsafe agricultural practices, unplanned urbanization, deforestation, rapid industrialization and non-judicious use of energy reservoirs and other anthropogenic activities. Among the pollutants that are of environmental and public health concerns due to their toxicities are: chemical fertilizer, heavy metals, nuclear wastes, pesticides, herbicides, insecticides greenhouse gases, and hydrocarbons. Thousands of hazardous waste sites have been identified and estimated is that more will be identified in the coming decades. Release of pollutants into the environment comes from illegal dumping by chemical companies and industries. Many of the techniques utilized for site clean-up in the past, such as digging up the contaminated soil and hauling it away to be land filled or incinerated have been prohibitively expensive and do not provide permanent solution. More recent techniques such as vapor extraction and soil venting are cost effective but incomplete solution.

2.1 Definition of bioremediation

Bioremediation is a process where biological organisms are used to remove or neutralize an environmental pollutant by metabolic process. The "biological" organisms include microscopic organisms, such as fungi, algae and bacteria, and the "remediation"—treating the situation.

In the Earth's biosphere, microorganisms grow in the widest range of habitats. They grow in soil, water, plants, animals, deep sea, and freezing ice environment. Their absolute numbers and their appetite for a wide range of chemicals make microorganisms the perfect candidate for acting as our environmental caretakers.

> *"Bioremediation is a waste management technique that includes the use of living organisms to eradicate or neutralize pollutants from a contaminated site."*
>
> *"Bioremediation is a 'treatment techniques' that uses naturally occurring organisms to break down harmful materials into less toxic or non-toxic materials."*

2.2 Bioremediation

Bioremediation technologies came into extensive usage and continue growing today at an exponential rate. Remediation of polluted sites using microbial process (bioremediation) has proven effective and reliable due to its eco-friendly features. In the past two decades, there have been recent developments in bioremediation techniques with the decisive goal being to successfully restore polluted environments in an economic, eco-friendly approach. Researchers have developed different bioremediation techniques that restore polluted environments. The micro-organisms used in bioremediation can be either indigenous or non-indigenous added to the contaminated site. Indigenous microorganisms present in polluted environments hold the key to solving most of the challenges associated with biodegradation and bioremediation of pollutant [1]. Environmentally friendly and cost effective are among the major advantages of bioremediation compared to both chemical and physical methods of remediation.

A mechanism of bioremediation is to reduce, detoxify, degrade, mineralize or transform more toxic pollutants to a less toxic. The pollutant removal process depends mainly on the pollutant nature, which includes pesticides, agrochemicals, chlorinated compounds, heavy metals, xenobiotic compounds, organic halogens, greenhouse gases, hydrocarbons, nuclear waste, dyes plastics and sludge. Cleaning

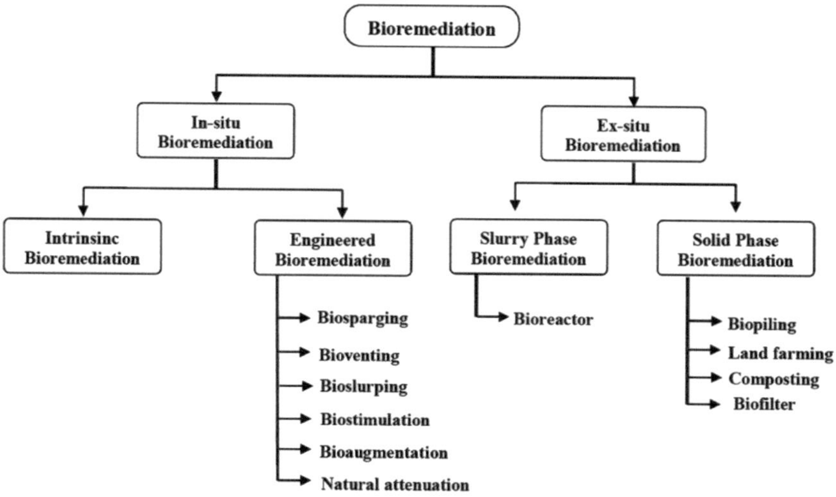

Figure 1.
Bioremediation approaches for environmental clean-up.

technique apply to remove toxic waste from polluted environment. Bioremediation is highly involved in degradation, eradication, immobilization, or detoxification diverse chemical wastes and physical hazardous materials from the surrounding through the all-inclusive and action of microorganisms (**Figure 1**).

3. Microorganisms used in bioremediation

Microorganisms play an important role on nutritional chains that are important part of the biological balance in life. Bioremediation involves the removal of the contaminated materials with the help of bacteria, fungi, algae and yeast. Microbes can grow at below zero temperature as well as extreme heat in the presence of hazardous compounds or any waste stream. The two characters of microbes are adaptability and biological system made them suitable for remediation process [2]. Carbon is the main requirement for microbial activity. Bioremediation process was carried out by microbial consortium in different environments. These microorganisms comprise *Achromobacter*, *Arthrobacter*, *Alcaligenes*, *Bacillus*, *Corynebacterium*, *Pseudomonas, Flavobacterium*, *Mycobacterium*, *Nitrosomonas*, *Xanthobacter*, etc. [3].

There are groups of microbes which are used in bioremediation such as:

Aerobic: aerobic bacteria have degradative capacities to degrade the complex compounds such as *Pseudomonas, Acinetobacter, Sphingomonas, Nocardia, Flavobacterium, Rhodococcus*, and *Mycobacterium.* These microbes have been reported to degrade pesticides, hydrocarbons, alkanes, and polyaromatic compounds. Many of these bacteria use the contaminants as carbon and energy source.

Anaerobic: anaerobic bacteria are not as regularly used as aerobic bacteria. There is an increasing interest in aerobic bacteria used for bioremediation of chlorinated aromatic compounds, polychlorinated biphenyls, and dechlorination of the solvent trichloroethylene and chloroform, degrading and converting pollutants to less toxic forms.

3.1 Factors affecting microbial bioremediation

Bioremediation process is degrading, removing, changing, immobilizing, or detoxifying various chemicals and physical pollutants from the environment through the activity of bacteria, fungi, algae and plants. Enzymatic metabolic pathways of microorganisms facilitate the progress of biochemical reactions that help in degradation of the pollutant. Microorganisms are act on the pollutants only when they have contact to the compounds which help them to generate energy and nutrients to multiply cells. The effectiveness of bioremediation depends on many factors; including, the chemical nature and concentration of pollutants, the physicochemical characteristics of the environment, and their accessibility to existing microorganisms [4].

The factors are mainly microbial population for degrading the pollutants, the accessibility of contaminants to the microbial population and environment factors like type of soils, pH, temperature, oxygen and nutrients.

3.2 Biotic or biological factors

Biotic factors are helpful for the degradation of organic compounds by microorganisms with insufficient carbon sources, antagonistic interactions among microorganisms or the protozoa and bacteriophages. The rate of contaminant degradation is frequently dependent on the concentration of the contaminant and the amount of catalyst present in biochemical reaction. The major biological factors are included enzyme activity, interaction (competition, succession, and predation), mutation, horizontal gene transfer, its growth for biomass production, population size and its composition [5, 6].

3.3 Abiotic or environmental factors

The interaction of environmental contaminants with metabolic activity, physicochemical properties of the microorganisms targeted during the process. The successful interaction between the microbes and pollutant depends on the environmental situations. Microbial growth and activity are depended on temperature, pH, moisture, soil structure, water solubility, nutrients, site conditions, oxygen content and redox potential, deficiency of resources and physico-chemical bioavailability of pollutants, concentration, chemical structure, type, solubility and toxicity. This above factors are control degradation kinetics [5, 7].

Biodegradation of pollutant can occur under range of pH (6.5–8.5) is generally optimal for biodegradation in most aquatic and terrestrial environment. Moisture affects the metabolism of contaminant because it depends on the kind and amount of soluble constituents that are accessible as well as the pH and osmotic pressure of terrestrial and aquatic systems [8].

4. Bioremediation technique

Superficially, bioremediation techniques can be carried out ex-situ and in-situ site of application (**Figure 1**). Pollutant nature, depth and amount of pollution, type of environment, location, cost, and environmental policies are the selection standards that are considered for selecting any bioremediation technique. Performance based on oxygen and nutrient concentrations, temperature, pH, and other abiotic factors that determine the success of bioremediation processes [9, 10].

Ex-situ bioremediation techniques involve digging pollutants from polluted sites and successively transporting them to another site for treatment. Ex-situ bioremediation techniques are regularly considered based on the depth of pollution, type of pollutant, degree of pollution, cost of treatment and geographical location of the polluted site. Performance standards also regulate the choice of ex-situ bioremediation techniques.

i. Solid-phase treatment

Solid-phase bioremediation is an ex-situ technology in which the contaminated soil is excavated and placed into piles. It is also includes organic waste like leaves, animal manures and agriculture wastes, domestic, industrial wastes and municipal wastes. Bacterial growth is moved through pipes that are distributed throughout the piles. Air pulling through the pipes is necessary for ventilation and microbial respiration. Solid-phase system required huge amount of space and cleanups require more time to complete as compared to slurry-phase processes. Solid-phase treatment processes include biopiles, windrows, land farming, composting, etc. [11].

ii. Slurry-phase bioremediation

Slurry-phase bioremediation is a relative more rapid process compared to the other treatment processes. Contaminated soil is combined with water, nutrient and oxygen in the bioreactor to create the optimum environment for the microorganisms to degrade the contaminants which are present in soil. This processing involves the separation of stones and rubbles from the contaminated soil. The added water concentration depends on the concentration of pollutants, the biodegradation process rate and the physicochemical properties of the soil. After completion of this process the soil is removed and dried up by using vacuum filters, pressure filters and centrifuges. The subsequent procedure is soil disposition and advance treatment of the resultant fluids.

5. Types of bioremediations

There are far more than nine types of bioremediation, but the following are the most common ways in which it is used.

5.1 Biopile

Bioremediation includes above-ground piling of dug polluted soil, followed by aeration and nutrient amendment to improve bioremediation by microbial metabolic activities. This technique comprises aeration, irrigation, nutrients, leachate collection and treatment bed systems. This specific ex-situ technique is progressively being measured due to its useful features with cost effectiveness, which allows operative biodegradation conditions includes pH, nutrient, temperature and aeration are effectively controlled. The biopile use to treat volatile low molecular weight pollutants; it can also be used effectively to remediate polluted very cold extreme environments [12–14]. The flexibility of biopile allows remediation time to be shortened as heating system can be integrated into biopile design to increase microbial activities and contaminant availability thus increasing the rate of biodegradation [15]. Additionally, heated air can be injected into biopile design to deliver air and heat in tandem, in order to facilitate enhanced bioremediation. Bulking agents such as straw saw dust, bark or wood chips and other organic materials have been added to enhance remediation process in a biopile construct. Although biopile systems connected to additional field ex-situ bioremediation techniques, such as land farming, bioventing, biosparging, robust engineering, maintenance and operation cost, lack of power supply at remote sites, which would facilitate constant air circulation in contaminated piled soil through air pump. Additional, extreme heating of air can lead to soil drying undertaking bioremediation, which will inhibit microbial activities and which stimulate volatilization than biodegradation [16].

5.2 Windrows

Windrows is bioremediation techniques depends on periodic rotating the piled polluted soil to improve bioremediation by increasing microbial degradation activities of native and transient hydrocarbonoclastic present in polluted soil. The periodic turning of polluted soil increase in aeration with addition of water, uniform distribution of nutrients, pollutants and microbial degradation activities, accordingly increase the rate of bioremediation, which can be proficient through acclimatization, biotransformation and mineralization. Windrow treatment as compared to biopile treatment, showed higher rate of hydrocarbon removal however, the effectiveness of the windrow for hydrocarbon removal from the soil [17]. However, periodic turning associated with windrow treatment not the best selection method to implement in bioremediation of soil polluted with toxic volatiles compounds. The use of windrow treatment has been associated in greenhouse gas (CH_4) release due to formation of anaerobic zone inside piled polluted soil, which frequently reduced aeration [18].

5.3 Land farming

Land farming is the simplest, outstanding bioremediation techniques due to its low cost and less equipment requirement for operation. It is mostly observed in ex-situ bioremediation, while in some cases of in-situ bioremediation technique. This consideration is due to the site of treatment. Pollutant depth is important in land farming which can be carried out ex-situ or in-situ. In land farming, polluted soils are regularly excavated and tilled and site of treatment speciously regulates the type of bioremediation. When excavated polluted soil is treated on-site, it is ex-situ as it has more in common than other ex-situ bioremediation techniques. Generally, excavated polluted soils are carefully applied on a fixed layer support above the ground surface to allow aerobic biodegradation of pollutant by autochthonous microorganisms [19]. Over all, land farming bioremediation technique is very simple to design and implement, requires low capital input and can be used to treat large volume of polluted soil with minimal environmental impact and energy requirement [20].

5.4 Bioreactor

Bioreactor is a vessel in which raw materials are converted to specific product(s) following series of biological reactions. There are different operational modes of bioreactors, which include: batch, fed-batch, sequencing batch, continuous and multistage. Bioreactor provides optimal growth conditions for bioremediation. Bioreactor filled with polluted samples for remediation process. The bioreactor based treatment of polluted soil has several advantages as compared to ex-situ bioremediation procedures. Bioreactor-based bioremediation process having excellent control of pH, temperature, agitation and aeration, substrate and inoculum concentrations efficiently reduces bioremediation time. The ability to control and manipulate process parameters in a bioreactor implies that biological reactions. The flexible nature of bioreactor designs allows maximum biological degradation while minimizing abiotic losses [21].

Advantages of ex-situ bioremediation

- Suitable for a wide range of contaminants
- Suitability relatively simple to assess from site investigation data

- Biodegradation are greater in a bioreactor system than or in solid-phase systems because the contaminated environment is more manageable and more controllable and predictable.

Disadvantages

- Not applicable to heavy metal contamination or chlorinated hydrocarbons such as trichloroethylene.
- Non-permeable soil requires additional processing.
- The contaminant can be stripped from soil via soil washing or physical extraction before being placed in bioreactor.

5.4.1 In-situ bioremediation techniques

These techniques comprise treating polluted substances at the pollution site. It does not need any excavation and by little or no disturbance in soil construction. Perfectly, these techniques should to be cost effective compared to ex-situ bioremediation techniques. Some in-situ bioremediation techniques like bioventing, biosparging and phytoremediation may be enhanced, while others may be progress without any form of improvement such as intrinsic bioremediation or natural attenuation. In-situ bioremediation techniques have been effectively used to treat chlorinated solvents, heavy metals, dyes, and hydrocarbons polluted sites [22–24].

5.4.2 Types of in-situ bioremediation

In-situ bioremediation is two types; these are intrinsic and engineered bioremediation.

i. Intrinsic bioremediation

Intrinsic bioremediation also known as natural reduction is an in-situ bioremediation technique, which involves passive remediation of polluted sites, without any external force (human intervention). This process deals with stimulation of indigenous or naturally occurring microbial population. The process based on both microbial aerobic and anaerobic processes to biodegrade polluting constituents containing those that are recalcitrant. The absence of external force implies that the technique is less expensive compared to other in-situ techniques.

ii. Engineered in-situ bioremediation

The second approach involves the introduction of certain microorganism to the site of contamination. Genetically Engineered microorganisms used in the in-situ bioremediation accelerate the degradation process by enhancing the physicochemical conditions to encourage the growth of microorganisms.

5.5 Bioventing

Bioventing techniques involve controlled stimulation of airflow by delivering oxygen to unsaturated (vadose) zone in order to increase activities of indigenous microbes for bioremediation. In bioventing, amendments are made by adding nutrients and moisture to increase bioremediation. That will achieve microbial transformation of pollutants to a harmless state. This technique has gained popularity among other in-situ bioremediation techniques [25].

5.6 Bioslurping

This technique combines vacuum-enhanced pumping, soil vapor extraction and bioventing to achieve soil and ground water remediation by indirect providing of oxygen and stimulation of contaminant biodegradation [26]. This technique is planned for products recovery from remediating capillary, light non-aqueous phase liquids (LNAPLs), unsaturated and saturated zones. This technique used to remediate soils which are contaminated with volatile and semi-volatile organic compounds. The method uses a "slurp" that spreads into the free product layer, which pulls up liquids from this layer. The pumping machine transports LNAPLs to the surface by upward movement, where it becomes separated from air and water. In this technique, soil moisture bounds air permeability and declines oxygen transfer rate, which reducing microbial activities. Although this technique is not suitable for low permeable soil remediation, it is cost effective operation procedure due to less amount of ground water, minimizes storage, treatment and disposal costs.

5.7 Biosparging

This technique is similar to bioventing in this air is injected into soil subsurface to improve microbial activities which stimulate pollutant removal from polluted sites. However, in bioventing, air is injected in saturated zone, which can help in upward movement of volatile organic compounds to the unsaturated zone to stimulate biodegradation process. The efficiency of biosparging depends on two major factors specifically soil permeability and pollutant biodegradability. In bioventing and soil vapor extraction (SVE), biosparing operation is closely correlated technique known as in-situ air sparging (IAS), which depend on high air-flow rates for volatilization of pollutant, whereas biosparging stimulates biodegradation. Biosparging has been generally used in treating aquifers contaminated with diesel and kerosene.

5.8 Phytoremediation

Phytoremediation is depolluting the contaminated soils. This technique based on plant interactions like physical, chemical, biological, microbiological and biochemical in contaminated sites to diminish the toxic properties of pollutants. Which is depending on pollutant amount and nature, there are several mechanisms such as extraction, degradation, filtration, accumulation, stabilization and volatilization involved in phytoremediation. Pollutants like heavy metals and radionuclides are commonly removed by extraction, transformation and sequestration. Organic pollutants hydrocarbons and chlorinated compounds are mostly removed by degradation, rhizoremediation, stabilization and volatilization, with mineralization being possible when some plants such as willow and alfalfa are used [27, 28].

Some important factors of plant as a phytoremediator include: root system, which may be fibrous or tap depending on the depth of pollutant, above ground biomass, toxicity of pollutant to plant, plant existence and its adaptability to predominant environmental conditions, plant growth rate, site monitoring and above all, time mandatory to achieve the preferred level of cleanliness. In addition, the plant must be resistant to diseases and pests [29]. In phytoremediation removal of pollutant includes uptake, translocation from roots to shoots. Further, translocation and accumulation depends on transpiration and partitioning [30]. However, the process is possible to change, depending on other factors such as nature of contaminant and plant. The mostly plants growing in any polluted site are good phytoremediators. Therefore, the success of any phytoremediation method mainly depends on improving the remediation potentials of native plants growing in polluted sites either by bioaugmentation with endogenous or exogenous plant. One of the major advantages of using plants to remediate polluted site is that some precious metals can bioaccumulate in some plants and recovered after remediation, a process known as phytomining.

5.9 Permeable reactive barrier (PRB)

This technique is commonly observed as a physical method for remediating contaminated groundwater. However, biological mechanisms are precipitation degradation and sorption of pollutant removal used in PRB method. The substitute terms such as biological PRB, bio-enhanced PRB, passive bioreactive barrier, have been suggested to accommodate the biotechnology and bioremediation aspect of the technique. In general, PRB is an in-situ technique used for remediating heavy metals and chlorinated compounds in groundwater pollution [31, 32].

5.10 Advantages of in-situ bioremediation

- In-situ bioremediation methods do not required excavation of the contaminated soil.

- This method provides volumetric treatment, treating both dissolved and solid contaminants.

- The time required to treat sub-surface pollution using accelerated in-situ bioremediation can often be faster than pump and treat processes.

- It may be possible to completely transform organic contaminants to innocuous substances like carbon dioxide, water and ethane.

- It is a cost effective method because there is minimal site disruption.

5.11 Limitation of in-situ bioremediation

Depending on specific site, some contaminants may not be absolutely transformed to harmless products.

If transformation stops at an intermediate compound, the intermediate may be more toxic and/or mobile than parent compound some are recalcitrant contaminants cannot be biodegradable.

When incorrectly applied, injection wells may become blocked by profuse microbial growth due to addition of nutrients, electron donor and electron acceptor.

Heavy metals and organic compounds concentration inhibit activity of indigenous microorganisms.

In-situ bioremediation usually required microorganism's acclimatization, which may not develop for spills and recalcitrant compounds.

6. Bioremediation prospects

Bioremediation techniques are varied and have demonstrated effective in restoring polluted sites. Microorganisms play fundamental role in bioremediation;

consequently, their diversity, abundance and community structure in polluted environments offer insight into the chance of any bioremediation technique providing other environmental factors, which can inhibit microbial activities. Advanced Molecular techniques such as 'Omics' includes genomics, proteomics, metabolomics and transcriptomics have contributed towards microbial identification, functions, metabolic and catabolic pathways, with microbial based methods. Nutrient availability, low population or absence of microbes with degradative capabilities, and pollutant bioavailability may delay the achievement of bioremediation. Since bioremediation depends on microbial process, biostimulation and bioaugmentation approaches speed up microbial activities in polluted sites. Biostimulation increase microbial activities by the addition of nutrients to a polluted sample. Microorganisms are abundantly present in different type of environmental condition, it is noticeable that pollutant degrading microbes are naturally present in polluted contaminated sites, their growth and metabolic activities may depends on pollutant type and concentration; later, we can use of agro-industrial wastes, which contains nitrogen, phosphorus and potassium as a nutrient source most polluted sites. Microbial consortium has been reported to degrade pollutants more efficiently than pure isolates [33].

This activity due to metabolic diversities of individual isolates, which potency create from their isolation source, adaptation process, pollutant composition, and synergistic effects, which may lead to complete and rapid degradation of pollutants when such isolates are mixed together [34]. Additional so, both bioaugmentation and biostimulation were effective in removing pollutant such as polyaromatic hydrocarbons (PAHs) from heavily polluted sample compared to non-amended setup (control) [35].

Although bioaugmentation has recognized effective method, it has been shown to increase the degradation of many compounds. If proper biodegrading microorganisms are not present in soil or if microbial populations decreased because of contaminant toxicity, specific microorganisms can be added as "introduced organisms" to improve the current populations and the possibility that the inoculated microorganisms may not survive in the new environment make bioaugmentation a very uncertain method. This process is known as bioaugmentation. Bioremediation technique in which natural or genetically engineered bacteria with unique metabolic profiles are used to treat sewage or contaminated water or soil. The use of alginate, agar, agarose, gelatin, gellan gum and polyurethane as carrier materials solve some of the challenges associated with bioaugmentation [36].

Biosurfactants are chemical equivalents having ecofriendly and biodegradable properties. However, high construction cost and low scalability application of biosurfactants to polluted site are uneconomical. Agro-industrial wastes combination are nutrient sources for development of biosurfactant producers during fermentation process. Application of several bioremediation techniques will help increase remediation efficiency [37].

Enhancing bioremediation ability with organized use of genetically engineered microorganisms (GEM) is a favorable approach. This is due to possibility of engineering a designer biocatalyst target pollutant including recalcitrant compounds by combining a novel and efficient metabolic pathways, widening the substrate range of existing pathways and increasing stability of catabolic activity [38].

However, parallel gene transfer and multiplication of GEM in an environmental application are encouraging approach. Bacterial containment systems, in which any GEM escaping an environment to reconstruct polluted environment.

Further, derivative pathway of genetically engineering microorganisms with a target polluted compound using biological approach could increase bioremediation efficiency. Nanomaterials decline the toxicity of pollutant to microorganisms because nanomaterials having increase surface area and lower activation energy, which reduce time and cost of bioremediation [39].

7. Bioremediation applications

Bioremediation must be considered as appropriate methods that can applied to all states of matter in the environment

i. Solids (soils, sediment and sludge)

ii. Liquids (ground water, surface water and industrial waste water

iii. Gases (industrial air emissions)

iv. Sub-surface environments (saturated and vadose zones).

The general approaches to bioremediation are the (i) intrinsic (natural) bioremediation, (ii) biosimulation (environmental modifications, through nutrient application and aeration, and (iii) bioaugmentation (addition of microbes).

The biological community exploited for bioremediation generally consists of the native soil microflora. However, higher plants can also be manipulated to enhance toxicant removal (phytoremediation), especially for remediation of metal contaminated soils.

7.1 Advantage and disadvantage

All bioremediation techniques have its own advantage and disadvantage because it has its own specific applications.

7.1.1 The advantage of bioremediation

- It is a natural process; it takes a little time, as an adequate waste treatment process for contaminated material such as soil. Microbes able to degrade the contaminant, the biodegradative populations become reduced. The treatment products are commonly harmless including cell biomass, water and carbon dioxide.

- It needs a very less effort and can commonly carry out on site, regularly without disturbing normal microbial activities. This also eradicates the transport amount of waste off site and the possible threats to human health and the environment.

- It is functional in a cost effective process as comparison to other conventional methods that are used for clean-up of toxic hazardous waste regularly for the treatment of oil contaminated sites. It also supports in complete degradation of the pollutants; many of the toxic hazardous compounds can be transformed to less harmful products and disposal of contaminated material.

- It does not use any dangerous chemicals. Nutrients especially fertilizers added to make active and fast microbial growth. Because of bioremediation change harmful chemicals into water and harmless gases, the harmful chemicals are completely destroyed.

- Simple, less labor intensive and cheap due to their natural role in the environment.

- Contaminants are destroyed, not simply transferred to different environmental.
- Nonintrusive, possibly allowing for continued site use.
- Current way of remediating environment from large contaminates and acts as ecofriendly sustainable opportunities.

7.1.2 The disadvantage of bioremediation

- It is restricted for biodegradable compounds. Not all compounds are disposed to quick and complete degradation process.
- There are particular new products of biodegradation may be more toxic than the initial compounds and persist in environment.
- Biological processes are highly specific, ecofriendly which includes the presence of metabolically active microbial populations, suitable environmental growth conditions and availability of nutrients and contaminants.
- It is demanding to encourage the process from bench and pilot-scale to large-scale field operations. Contaminants may be present as solids, liquids and gases. It often takes longer than other treatment preferences, such as excavation and removal of soil or incineration.
- Research is needed to develop and engineer bioremediation technologies that are appropriate for sites with complex mixtures of contaminants that are not evenly dispersed in the environment.

7.2 Limitations of bioremediation

Bioremediation is limited to those compounds that are biodegradable. This method is susceptible to rapid and complete degradation. Products of biodegradation may be more persistent or toxic than the parent compound in the environment.

1. Specificity

Biological processes are highly specific. Important site factors mandatory for success include the presence of metabolically capable microbial populations, suitable environmental growth conditions, and appropriate levels of nutrients and contaminants.

2. Scale up limitation

It is difficult to scale up bioremediation process from batch and pilot scale studies applicable to large scale field operations.

3. Technological advancement

More research is required to develop modern engineer bioremediation technologies that are suitable for sites with composite combinations of contaminants that are not equally distributed in the environment. It may be present as solids, liquids and gases forms.

4. Time taking process

Bioremediation takes longer time compare to other treatment options, such as excavation and removal of soil from contaminated site.

5. Regulatory uncertainty

We are not certain to say that remediation is 100% completed, as there is no accepted definition of clean. Due to that performance evaluation of bioremediation is difficult, and there is no acceptable endpoint for bioremediation treatments.

8. Conclusion

Biodegradation is very fruitful and attractive option to remediating, cleaning, managing and recovering technique for solving polluted environment through microbial activity. The speed of undesirable waste substances degradation is determined in competition with in biological agents like fungi, bacterial, algae inadequate supply with essential nutrient, uncomfortable external abiotic conditions (aeration, moisture, pH, temperature), and low bioavailability. Bioremediation depending on several factors, which include but not limited to cost, site characteristics, type and concentration of pollutants. The leading step to a successful bioremediation is site description, which helps create the most suitable and promising bioremediation technique (ex-situ or in-situ). Ex-situ bioremediation techniques tend to be more costly due to excavation and transportation from archeological site. However, they can be used to treat wider range of pollutants. In contrast, in-situ techniques have no extra cost for excavation; however, on-site installation cost of equipment, attached with effectively and control the subsurface of polluted site can reduce some ineffective in-situ bioremediation methods. Geological characteristics of polluted sites comprising soil, pollutant type and depth, human habitation site and performance of every bioremediation technique should be integrated in determining the most appropriate and operative bioremediation technique to successfully treatment of polluted sites.

Author details

Indu Sharma
Department of Biotechnology, Maharishi Markandeshwar (Deemed to be University), Mullana, Ambala, Haryana, India

*Address all correspondence to: endusharma@gmail.com

IntechOpen

© 2020 The Author(s). Licensee IntechOpen. This chapter is distributed under the terms of the Creative Commons Attribution License (http://creativecommons.org/licenses/by/3.0), which permits unrestricted use, distribution, and reproduction in any medium, provided the original work is properly cited.

References

[1] Verma JP, Jaiswal DK. Book review: Advances in biodegradation and bioremediation of industrial waste. Frontiers in Microbiology. 2016;**6**:1-2. DOI: 10.3389/fmicb.2015.01555

[2] Prescott LM, Harley JP, Klein DA. Microbiology. 5th ed. New York: McGraw Hill; 2002. pp. 10-14

[3] Singh R, Singh P, Sharma R. Microorganism as a tool of bioremediation technology for cleaning environment: A review. International Academy of Ecology and Environmental Science. 2014;**4**(1):1-6

[4] El Fantroussi S, Agathos SN. Is bioaugmentation a feasible strategy for pollutant removal and site remediation? Current Opinion in Microbiology. 2005;**8**:268-275. Available at: https://goo.gl/y6kLsc

[5] Madhavi GN, Mohini DD. Review paper on parameters affecting bioremediation. International Journal of Life Science and Pharma Research. 2012;**2**:77-80. Available at: https://goo.gl/tBP2C6

[6] Boopathy R. Factors limiting bioremediation technologies. Bioresource Technology. 2000;**74**:63-67. Available at: https://goo.gl/eQhPh7

[7] Adams GO, Fufeyin PT, Okoro SE, Ehinomen I. Bioremediation, biostimulation and bioaugmention: A review. International Journal of Environmental Bioremediation & Biodegradation. 2015;**3**:28-39. Available at: https://goo.gl/9XY7ni

[8] Cases I, de Lorenzo V. Genetically modified organisms for the environment: Stories of success and failure and what we have learned from them. International Microbiology. 2005;**8**:213-222. Available at: https://goo.gl/3oaxJT

[9] Frutos FJG, Pe'rez R, Escolano O, Rubio A, Gimeno A, Fernandez MD, et al. Remediation trials for hydrocarbon-contaminated sludge from a soil washingprocess: Evaluation of bioremediation technologies. Journal of Hazardous Materials. 2012;**199**:262-271. DOI: 10.1016/j.jhazmat.2011.11.017

[10] Smith E, Thavamani P, Ramadass K, Naidu R, Srivastava P, Megharaj M. Remediation trials for hydrocarbon-contaminated soils in arid environments: Evaluation of bioslurry and biopiling techniques. Integrated Journal of Biodeterioration Biodegradation. 2015;**101**:56-65. DOI: 10.1016/j.ibiod.2015.03.029

[11] Kulshreshtha A, Agrawal R, Barar M, Saxena S. A review on bioremediation of heavy metals in contaminated water. IOSR Journal of Environmental Science Toxicology and food Technology (IOSR-JESTFT). 2014;**8**(7):44-50

[12] Gomez F, Sartaj M. Optimization of field scale biopiles for bioremediation of petroleum hydrocarbon contaminated soil at low temperature conditions by response surface methodology (RSM). Integrated Journal of Biodeterioration Biodegradation. 2014;**89**:103-109. DOI: 10.1016/j.ibiod.2014.01.010

[13] Dias RL, Ruberto L, Calabró A, Balbo AL, Del Panno MT, Mac Cormack WP. Hydrocarbon removal and bacterial community structure in on-site biostimulated biopile systems designed for bioremediation of diesel-contaminated Antarctic soil. Polar Biology. 2015;**38**:677-687. DOI: 10.1007/s00300-014-1630-7

[14] Whelan MJ, Coulon F, Hince G, Rayner J, McWatters R, Spedding T, et al. Fate and transport of petroleum hydrocarbons in engineered biopiles

in polar regions. Chemosphere. 2015;**131**:232-240. DOI: 10.1016/j. Chemosphere.2014.10.088

[15] Aislabie J, Saul DJ, Foght JM. Bioremediation of hydrocarbon contaminated polar soils. Extremophiles. 2006;**10**:171-179. DOI: 10.1007/s00792-2005-0498-4

[16] Sanscartier D, Zeeb B, Koch I, Reimer K. Bioremediation of diesel-contaminated soil by heated and humidified biopile system in cold climates. Cold Regions Science and Technology. 2009;**55**:167-173. DOI: 10.1016/j.coldregions.2008.07.004

[17] Coulon F, Al Awadi M, Cowie W, Mardlin D, Pollard S, Cunningham C, et al. When is a soil remediated? Comparison of biopiled and windrowed soils contaminated with bunker-fuel in a full-scale trial. Environmental Pollution. 2009;**158**:3032-3040. DOI: 10.1016/j.envpol.2010.06.001

[18] Hobson AM, Frederickson J, Dise NB. CH_4 and N_2O from mechanically turned windrow and vermincomposting systems following in-vessel pre-treatment. Waste Management. 2005;**25**:345-352. DOI: 10.1016/j.wasman.2005.02.015

[19] Silva-Castro GA, Uad I, Gȯnzalez-Lȯpez J, Fandiño CG, Toledo FL, Calvo C. Application of selected microbial consortia combined with inorganic and oleophilic fertilizers to recuperate oil-polluted soil using land farming technology. Clean Technology of Environment Policy. 2012;**14**:719-726. DOI: 10.1007/s10098-011-0439-0

[20] Maila MP, Colete TE. Bioremediation of petroleum hydrocarbons through land farming: Are simplicity and cost-effectiveness the only advantages? Review. Environmental Science Biotechnology. 2004;**3**:349-360. DOI: 10.1007/s111157-004-6653

[21] Mohan SV, Sirisha K, Rao NC, Sarma PN, Reddy SJ. Degradation of chlorpyrifos contaminated soil by bioslurry reactor operated in sequencing batch mode: Bioprocess monitoring. Journal of Hazardous Materials. 2004;**116**:39-48. DOI: 10.1016/j. jhazmat.2004.05.037

[22] Folch A, Vilaplana M, Amado L, Vicent R, Caminal G. Fungal permeable reactive barrier to remediate groundwater in anartificial aquifer. Journal of Hazardous Materials. 2013;**262**:554-560. DOI: 10.1016/j. jhazmat.2013.09.004

[23] Frascari D, Zanaroli G, Danko AS. In situ aerobic cometabolism of chlorinated solvents: A review. Journal of Hazardous Materials. 2015;**283**:382-399. DOI: 10.1016/j. jhazmat.2014.09.041

[24] Roy M, Giri AK, Dutta S, Mukherjee P. Integrated phytobial remediation for sustainable management of arsenic in soil and water. Environment International. 2005;**75**:180-198. DOI: 10.1016/j. envint.2014.11.010

[25] Höhener P, Ponsin V. In situ vadose zone bioremediation. Current Opinion in Biotechnology. 2014;**27**:1-7. DOI: 10.1016/j.copbio.2013.08.018

[26] Gidarakos E, Aivalioti M. Large scale and long term application of bioslurping: The case of a Greek petroleum refinery site. Journal of Hazardous Materials. 2007;**149**:574-581. DOI: 10.1016/j. jhazmat.2007.06.110

[27] Meagher RB. Phytoremediation of toxic elemental organic pollutants. Currant Opinion Plant Biology. 2000;**3**:153-162. DOI: 10.1016/ S1369-5266(99)00054-0

[28] Kuiper I, Lagendijk EL, Bloemberg GV, Lugtenberg BJJ. Rhizoremediation: A beneficial

plant-microbe interaction. Molecular Plant Microbe Interactions. 2004;7:6-15. DOI: 10.1094/MPMI.2004.17.1.6

[29] Lee JH. An overview of phytoremediation as a potentially promising technology for environmental pollution control. Biotechnology and Bioprocess Engineering. 2013;**18**:431-439. DOI: 10.1007/s12257-013-0193-8

[30] Miguel AS, Ravanel P, Raveton M, A comparative study on the uptake and translocation of organochlorines by *Phragmites australis*. Journal of Hazardous Materials. 2013;**244**:60-69. DOI: 10.1016/j.jhazmat.2012.11.025

[31] Thiruvenkatachari R, Vigneswaran S, Naidu R. Permeable reactive barrier for groundwater remediation. Journal of Industrial and Engineering Chemistry. 2007;**14**:145-156. DOI: 10.1016/j.jiec.2007.10.001

[32] Obiri-Nyarko F, Grajales-Mesa SJ, Malina G. An overview of permeable reactive barriers for in situ sustainable groundwater remediation. Chemosphere. 2014;**111**:243-259. DOI: 10.1016/j.chemosphere.2014.03.112

[33] Silva-Castro GA, Uad I, Gonzalez-Lopez J, Fandino CG, Toledo FL, Calvo C. Application of selected microbial consortia combined with inorganic and oleophilic fertilizers to recuperate oil-polluted soil using land farming technology. Clean Technology of Environmental Policy. 2012;**14**:719-726. DOI: 10.1007/s10098-011-0439-0

[34] Bhattacharya M, Guchhait S, Biswas D, Datta S. Waste lubricating oil removal in a batch reactor by mixed bacterial consortium: A kinetic study. Bioprocess and Biosystems Engineering. 2015;**38**:2095-2106. DOI: 10.1007/s00449-015-1449-9

[35] Sun G-D, Xu Y, Jin J-H, Zhong Z-P, Liu Y, Luo M, et al. Pilot scale ex situ bioremediation of heavily PAHs-contaminated soil by indigenous microorganisms and bioaugmentation by a PAHs-degrading and bioemulsifier-producing strain. Journal of Hazardous Materials. 2012;**233**:72-78. DOI: 10.1016/j.jhazmat.2012.06.060

[36] Tyagi M, da Fonseca MMR, de Carvalho CCCR. Bioaugmentation and biostimulation strategies to improve the effectiveness of bioremediation processes. Biodegradation. 2011;**22**:231-241. DOI: 10.1007/s10532-010-9394-4

[37] Cassidy DP, Srivastava VJ, Dombrowski FJ, Lingle JW. Combining in situ chemical oxidation, stabilization, and anaerobic bioremediation in a single application to reduce contaminant mass and leachability in soil. Journal of Hazardous Materials. 2015;**297**:347-355. DOI: 10.1016/j.jhazmat.2015.05.030

[38] Paul D, Pandey G, Pandey J, Jain RK. Accessing microbial diversity for bioremediation and environmental restoration. Trends in Biotechnology. 2005;**23**:135-142

[39] Rizwan M, Singh M, Mitra CK, Morve RK. Ecofriendly application of nanomaterials: Nanobioremediation. Journal of Nanoparticles. 2014:1-7. DOI: 10.1155/2014/431787

Chapter 13

Sustainable Treatment of Heavy Metals by Adsorption on Raw Chitin/Chitosan

Boukhlifi Fatima

Abstract

The objective of this chapter was to treat metal pollution of wastewater rich in Pb^{2+}, Cd^{2+}, Cu^{2+}, and Zn^{2+} ions by adsorption tests on the raw chitin/chitosan. Different origin namely shrimp (Ccre), crab (Ccra) and lobster (Clan). Raw shrimp chitin had a strong affinity for Pb^{2+} and Cd^{2+}. The adsorption capacity of zinc on the crabs chitin is twice as great as that on the shrimp chitin. The kinetic study showed that more than 50% of these ions are adsorbed before equilibrium is reached (20 minutes). The adsorption kinetics also showed that the hardness of the shells has a negative effect on the kinetics of the adsorption process. Indeed, the adsorption of Pb^{2+} on the raw chitin shrimp requires only 30 minutes, while on the raw chitin lobster; the equilibrium time is 60 minutes. To ensure a sustainable treatment, sludge generated by adsorption of heavy metals was incinerated at high temperature. Incineration has led to calcite phases, which do not represent any toxicity on the environment and it can be recycled in the industry of solid materials (ceramics, cement, etc.). However, the regeneration of sludge by the acid changes the structure of the material and gives new adsorbent supports.

Keywords: raw chitin, adsorption, heavy metals, sustainable treatment

1. Introduction

The adsorption of heavy metals on the differently charged surfaces has been widely studied; on the one hand, mineral surfaces such as oxides, clays, soils, and activated carbons; and on the other hand, organic surfaces and biological surfaces.

Mineral surfaces: the most commonly used adsorbent is activated carbon, which has become a potential absorbent for wastewater [1–5]. Huang [1] studied the adsorption of Cd on activated carbon, which showed that the elimination efficiency increases when the pH increases; Ku [2] studied the coupling between two techniques adsorption on the CA and precipitation by hydroxides, which showed that the percentage of adsorption of Zn (II) increases with pH; and the same study was carried out on Cd, which found that the maximum adsorption of Cd is at pH = 9.5. Saito [6] used CA and sulphonated cottons for the removal of Cu and Cd from wastewater. The percentage of elimination reaches 98%. Most studies done on the CA have been undertaken for water treatment. However, most of the studies carried out on oxide surfaces have been undertaken to assimilate the behavior of heavy metals in the soil or the mechanism of transport and fixation of metals by these types of surfaces. Swallow [7] studied the adsorption of Cu and Pb on the hydrous ferric oxide, he noted that this support shows a great affinity toward Pb and Cd. Anjana [8] showed that the adsorption of Zn on Fe (III) hydroxide increases with increasing pH. Spark [9] studied the adsorption of heavy metals on oxides and oxyhydroxides, which has shown that this adsorption is very dependent on pH. The study of Cd adsorption on Fe $(OH)_2$, δ Al_2O_3, and Fe_2O_3 showed that the percentage of removal exceeds 80%. The hydroxide aluminosilicates were used for the adsorption of Zn (II) in alkaline medium [10]. Benjamin [11] studied the adsorption of Cd^{2+} on four oxide surfaces in the presence of certain complexing agents such as SO_4^{2-}, Cl^-, and $S_2O_3^{2-}$, and concluded that the interaction between metal ions and the complexing ligands in the presence of adsorbent surfaces can be divided into three categories based on the origin and the interaction force.

- Metal-ligand complex can form in the solution and weakly adsorb.
- Species can interact indirectly on the surface by altering the electrical properties of the surface.
- Metal-ligand complex can be strongly adsorbed by improving the removal of metals.

With regard to the clays, among the studies carried out, the adsorption of Cd on illite by Comans [12]. Schindler [13] showed that the importance of Cu, Cd, and Pb adsorption on kaolinite increases with increasing pH. Cadena [14] studied the adsorption of Pb on bentonite, which showed that the percentage of adsorption is 35%. There are also comparative adsorption studies between clay and sand, such as the study by El Khatib [15], which found that the amount of adsorbed Pb is very important in the case of clay as in the case of sand. Petersen [16] showed that the percentage of Cu adsorption on kaolinite exceeds 90%. Spark [17] found that the concentration of kaolinite has a small effect on the pH at which the adsorption process of Cd, Zn, Cu, and Co begins. Mrad [18] showed that the adsorption of Cd on montmorillonite is accompanied by a proton departure.

Biological surfaces: the most used adsorbents are active sludge and seaweed; indeed, many studies have been carried out for the removal of heavy metals by biological surfaces. Filip [19] studied the adsorption of Cu and Cd synthetic solutions and the elimination reached was 98% Cu and 100% Cd. Sloan [20] exploited different kinds of algae to eliminate Pb, Cd, Cu, and Zn at different concentrations. Nicolas [21] studied adsorption of heavy metals on active sludge. Roy [22] carried out the adsorption of the heavy metals, Cd, Co, Cr, Pb, Ni, and Zn, on two types of biomass, and found that the green alga Chlorella minutissima adsorbs more than 90% of the initial quantity of Pb, while Rice Hulls has an adsorption efficiency of more than 99%. However, Brown [23, 24] showed that the main constituents of active sludge surfaces are polysaccharides and proteins, while the dewaxed sludge mainly consists of lipopolysaccharides and proteins.

Organic surfaces: several studies on the fate of heavy metals in nature have shown that organic surfaces are primarily responsible for the transport, fixation, and distribution of heavy metals in nature. Thus, Lester [25, 26] found that adsorption of metal ions from wastewater is mainly due to organic matter consisting of polysaccharides, nucleic acids, and polypeptides. Among the most important adsorbents are cellulose and polysaccharides. Thus, Irwin [27] showed that wood cedar flours mixed with a base reduce the Cd concentration from 8.7 to 0.4 mg/l of a synthetic solution with a 95.4% removal rate. Recently, Deans and Dixon [28] carried out a comparative study of the adsorption of heavy metals on several organic surfaces, cellulose, alginic acid, chitin, chitosan, carrageenan, carboxymethyl, hydroxamic acid, terephthalic acid, and its hydroxamic, which showed that the hydroxamic acid is the most effective adsorbent for the removal of metals under different conditions. Among the polysaccharides is chitin, which plays an important role in the transport of heavy metals to sediments [29].

The applications of chitin are diverse and numerous [30], which is used in the treatment of water, for its flocculating properties or for its retention properties of heavy metals. The importance of chitin as an unused resource has increased in recent years, due to its unique structure and numerous properties. Among them is the ability of this chelating polysaccharide, which seems to be the most promising property of chitin. Chitin is prepared in the form of a flocculant for the purification of municipal water [31, 32]. It was exploited in the Cu assay by adsorption-elution on a column [33]. It has also been exploited for the collection of heavy metals in seawater and these metals are then recovered by an appropriate washing of chitin [34]. Muzzarelli [35, 36] and Yang [37] also used chitin for the removal of heavy metals from wastewater and natural waters. Finally, Later Melchor [38, 39] studied the adsorption of Cu, Pb, and Cd on pure chitin extracted from crab shells as a function of pH, temperature, and salinity of the medium, which showed that this adsorption is very efficient and very fast and is not affected by the presence of other metals.

In parallel, it is found that chitin derivatives are used for selective removal of metal ions [35]. Thus chitosan, which is only deacetylated chitin, has been known especially as the most powerful adsorbent of natural origin [40]. Chitosan is very effective for the elimination of uranium [41]. Muzzarelli [42] also studied the elimination of Co and Cu from water using three chitin derivatives: serine glucan, aspartate glucan, and glycine glucan. Kurita [43] showed that chitin derivatives are very effective for the elimination of Cu.

Studies on the raw chitin adsorption are few and concern in general the crude chitin extracted from the mushrooms. Huang [44, 45] tried to eliminate Cd by several species of chitin-rich fungi such as *Aspergillus oryzae*, *Aspergillus. clavatus*, and *Candida utilis*, showing an adsorption capacity according to the model of Langmuir, it has also studied the adsorption of Cu, Cd, Zn, and Ni on 12 species of fungi. Boukhlifi et al. [46, 47] studied the elimination of metals ions Zn, Pb, Cu, and Cd by raw chitin from different sources. The affinity of each metal for the adsorbent depends on the source used. A study on the competitive adsorption of metal ions on crude chitin showed that the retention of an ion is strongly influenced by the number and nature of the other ions present in the wastewater [48]. Boukhlifi et al. [49–51] applied the raw chitin shrimp for the treatment of real industrial discharges. Wales [52] also recovered Zn, Pb, Cu, and Cd by chitin/chitosan micro fungi. Until now, the work on the raw chitin from shellfish, which represents 75% of fisheries waste are rare. From then, we were interested in this study to the valuation of gross chitin some crustaceans (shrimp, crabs, and lobsters).

This chapter studied the behavior and adsorbent properties of raw chitin against Cd, Cu, Pb, and Zn metal ions. First of all, we will study the kinetic and isothermal study of the adsorption of metal ions on the raw chitin of different origins namely shrimp (Ccre), lobster (Clan), and crab (Ccra), then the study of series adsorption of metal ions, and finally, the study of sludge recovery generated from treatment by trying to study the regeneration of the adsorbent support by several cations or by incineration.

2. Isothermal study of heavy metals adsorption

2.1 Preparations

2.1.1 Preparation of adsorbent carriers

Our study focused on an economic and abundant adsorbent, the raw chitin. This polymer exists at concentrations ranging between 20% and 50% in the skeletons of arthropods and more than 60% in the skeletons of crabs and shrimp [53, 54]. The studied crustaceans [crabs, lobsters, and shrimps (**Figure 1**)] belong to the phylum of the arthropods, it is the most important branch of the animal kingdom (75%) about 1 million species, are invertebrate animals metazoans, triploblastics, acoelomates, protostomials with metamerized body, bilateral symmetry and heteronome segmentation, covered by a protective epicuticle serving as external skeleton/exoskeleton, rigid but flexible in some places, and very rich in chitin. This richness is relative to each species. The systematics of these crustaceans is summarized **Table 1**.

The selected species, Squinado, Panulirus, and Kerathurus (**Figure 1**) are the most abundant species in Moroccan markets. Their carapaces consist of an external epicuticle followed inward by a pigmented zone, a calcified layer, another noncalcified, and an epidermis.

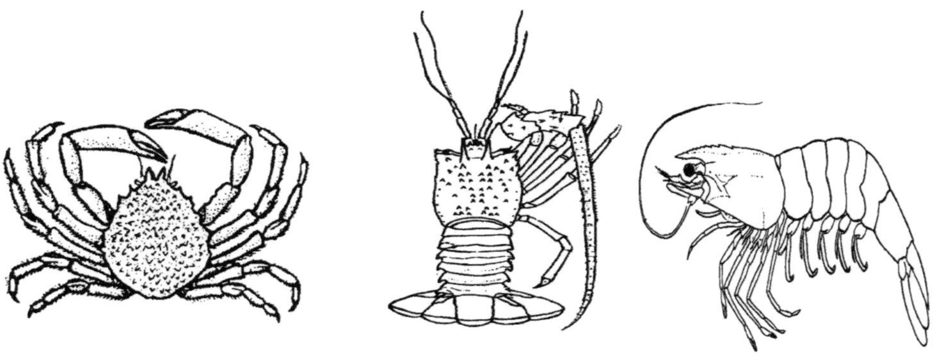

crab (Squinado) - lobster (Panulirus) - shrimp (Penaeus Kerathurus)

Figure 1.
Crustaceans used in the preparation of raw chitin.

	Crabs	Lobsters	Shrimps
Branch	Arthropods	Arthropods	Arthropods
Class	Shellfish	Shellfish	Shellfish
Superorder	Fucarides	Fucarides	Fucarides
Order	Decapoda	Decapoda	Decapoda
Under order	Reptantia	Reptantia	Natantia
Section	Brachyoures	Macroura	—
Kind	Maia	Palinutus	Pehaens
Cash	Maia (Squinado)	Panulirus	*Penaeus kerathurus*

Table 1.
Systematic crustaceans used for the treatment.

The raw chitin of shrimp origin (Ccre), crab origin (Ccra), and lobster origin (Clan) is extracted from the shells of these crustaceans according to the following mode: the shells are isolated from their soft parts containing the proteins, washed with bidistilled water, dried in an oven at 100°C for 48 hours, and finally they are crushed and sieved at particle size well defined.

2.1.2 Preparation of synthetic wastewater

For each metal, a stock solution of 1 g/l [expressed in (g) metal cations per liter of test solution] was prepared using the metallic substances used in the most soluble mineral form (M^{2+}, $2NO_3^-$) × H_2O with [M = Zn (x = 6), Pb (x = 0), Cu (x = 3), and Cd (x = 4)]. For each adsorption test, the solution containing the metals to be studied is obtained from the daughter solutions.

2.2 Experimental study of the adsorption phenomenon

Adsorption is a surface phenomenon and to evaluate the adsorption of a compound on an adsorbent, it is necessary to start the following studies:

2.2.1 Adsorption kinetics

The adsorption kinetics was established by stirring for varying periods of time, the M^{2+} solutions introduced into 250 ml Erlenmeyer flasks and added to constant amounts of the adsorbent. After a determined contact time of each suspension, they are filtered; the amount of M^{2+} not removed by the support is subsequently assayed.

2.2.2 Adsorption isotherm

The adsorption mechanism can be described using an adsorption isotherm. An adsorption isotherm is the curve, which represents the variation of the adsorbed quantity "Qe" as a function of the equilibrium concentration "Ce" with Eq. (1)

$$Qe = ma / m \quad et \quad ma = (Co - Ce).V \tag{1}$$

where, C_0: initial concentration of adsorbate (mg/l); m: mass of adsorbent used (mg); v: volume of the solution (ml); and ma: mass of retained adsorbate (mg).

The adsorption can be studied, either by examining the decrease of the concentration of the adsorbate in the solution, or by a direct determination of the amount fixed on the solid; the second case requires a washing of the solid in order to remove all the unadsorbed molecules, which can cause destruction in the case of the reversible process, and thus leads to erroneous results, and for these reasons, we opted for the first possibility.

To obtain an adsorption isotherm for a solute on an adsorbent, a constant temperature was used, by mixing a determined quantity of the adsorbent with a solution whose solute concentration is known. After vigorous stirring (500 rpm) and prolonged, the equilibrium time between the solid and liquid phases is determined. The mixture is filtered, and by appropriate dosage, the residual concentration of the solute in the solution is measured. This equilibrium concentration will be noted Ce (mg/l). The amount fixed on the adsorbent is deduced, by difference with the initial concentration. This quantity will be symbolized by Qe (mg/g): amount of solute adsorbed per unit mass of the adsorbent. The measurement is repeated several times using solutions of different concentrations, which allow us to draw the isothermal curve. The operating conditions are summarized in **Table 2**. The supernatant is filtered and the equilibrium

Concentration of the material	**1 g/l**
Granulometry	$100\ \mu \leq \Phi \leq \mu\ 125$ m
Concentration range	10–100 mg/l
Stirring speed	500 rpm
Temperature	20 ± 2°C
Stirring time	Equilibrium time of each metal

Table 2.
Operating conditions of adsorption isotherms.

Metals	Pb^{2+}	Cd^{2+}	Cu^{2+}	Zn^{2+}
Co (mg/l)	100.02	100.05	99.99	99.98
This (mg/l)	99.98	99.97	99.97	99.92

Table 3.
Results of the blank tests.

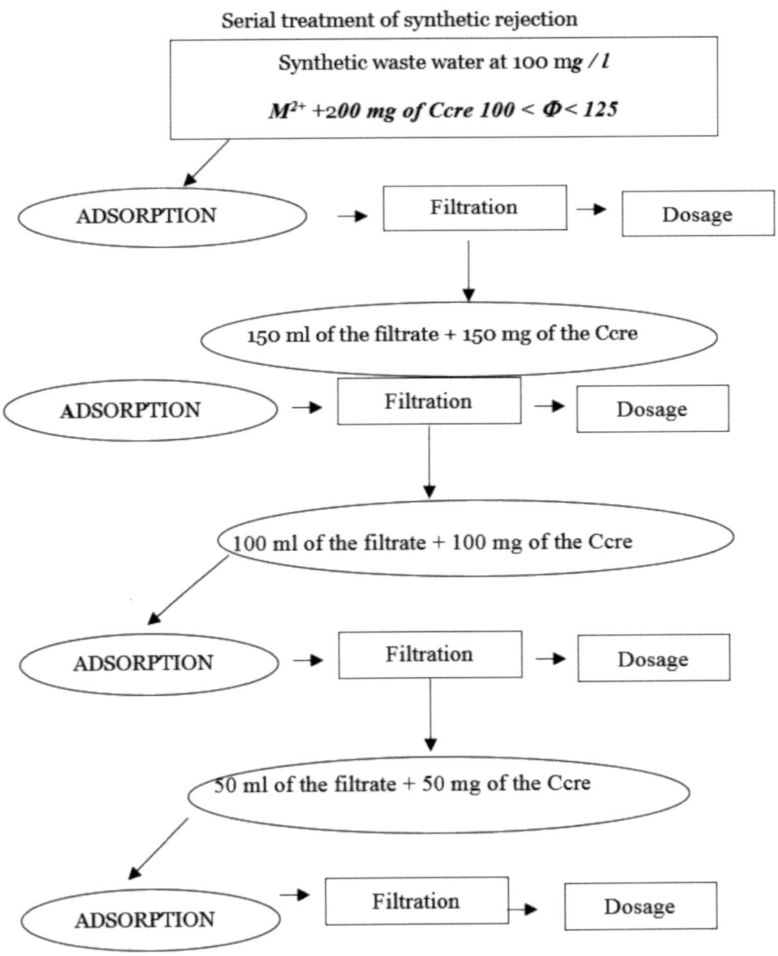

Figure 2.
Serial adsorption process.

concentration Ce is determined by a suitable assay. The precipitates [M^{2+}/adsorbent] are dried in an oven for 24 hours to be mineralized. The percentage of elimination is calculate according to Eq. (2):

$$\text{Percentage of reduction} = \text{percentage of abatement} = \text{percentage of elimination} = (Co - Ce/Co).100 \quad (2)$$

Blank tests are carried out by stirring the same adsorbent solution (Cd^{2+}, Pb^{2+}, Cu^{2+}, and Zn^{2+}) at a concentration of 100 mg/l during equilibrium time, and this is filtered. Then dosed to determine the amount of adsorbate retained by the Erlenmeyer flasks and the filter, the results of these tests are grouped in **Table 3**.

According to this table, we can neglect the amount retained by the filter accessories.

2.3 Serial metal ion adsorption test

To demonstrate the effectiveness of the Ccre in retaining the metal ions, we proceeded to a simple treatment using 500 mg of the chitin and a series treatment using amounts of adsorbent whose sum is equal to 500 mg. Starting with an initial concentration of 100 mg/l M^{2+}, after each adsorption test, the measured filtrate is again adsorbed on the material. This operation is repeated four times for four metals according to the following flowchart **Figure 2**.

2.4 Dosage of heavy metals

Depending on the results sought and also to cross-check the information obtained, we used different techniques and equipment:

- Atomic absorption spectrophotometry with flame (cadmium, copper, lead, and zinc) by spectrophotometers VARIAN AA-475, PHLIPS PU-900, PERKIN ELMER 305, and SHIMADZU AA-680.
- Graphite furnace atomic absorption spectrophotometry (lead and cadmium) type SHIMADZU GFA-4B.

Prior to the adsorption experiments of heavy metals, it appeared useful to follow their adsorption kinetics to determine the times required to reach equilibrium.

3. Study of cadmium adsorption

3.1 Kinetic study of the Cd^{2+} adsorption

To determine the time required to reach the Cd^{2+} adsorption equilibrium, we monitored the change in Cd^{2+} concentration as a function of time. The results are shown in **Figure 3**.

This study shows that equilibrium is established rapidly for all adsorbents (**Figure 3**). It reaches its maximum for 30 and 20 minutes, respectively for Ccra and Ccre, while the adsorption kinetics of Cd^{2+} on Clan is relatively slow.

From these curves, it is also found that 50% of Cd^{2+} is adsorbed on the Ccre before the equilibrium reached 20 minutes. The equilibrium times (teq) are

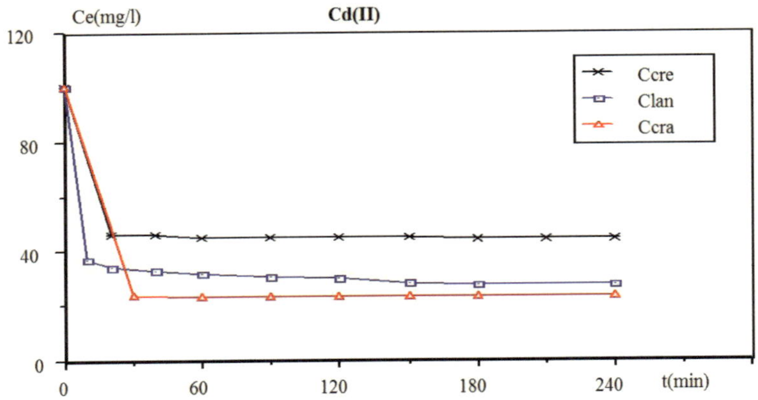

Figure 3.
Kinetic study of Cd (II) adsorption on raw chitin.

classified in the following order: teq Ccre < teq Ccra <teq Clan. This kinetics is very fast in comparison with other works cited in the literature concerning the adsorption of Cd^{2+} on mineral surfaces, which require a very large equilibrium time. For example, for the adsorption of Cd^{2+} on the illite, the equilibrium is reached after 54 days [1] and in the case of a natural clay [55, 56], the equilibrium is reached after 50 days. Melchor [39] having studied the adsorption of Cd^{2+} on chitin, mentions a time of equilibrium (40 minutes) comparable to our result.

3.2 Study of Cd^{2+} adsorption isothermals on raw chitin

The curves in **Figure 4a** show the evolution of the adsorption capacity (Qe) as a function of the equilibrium concentration (Ce) of Cd^{2+} and **Figure 4b** shows the removal efficiency by source of chitin.

When the residual concentration increases, the isotherm of Cd^{2+} on the Ccre deviates from the other two isotherms, which presents two different curvatures. For low concentrations, the slope is steep up to 40 mg/g, which shows that adsorption is important for low concentrations as for high concentrations. Indeed, the increase of the ionic strength of the system reduces the adsorption through the effect of the coefficient of activity. More than 40 mg/g, the curve no longer follows Henry's law. The Cd^{2+} adsorption isotherm on the Ccra is almost straight; Henry's hypothesis (low surface coverage) is true for all the concentrations studied, whereas for the Clan,

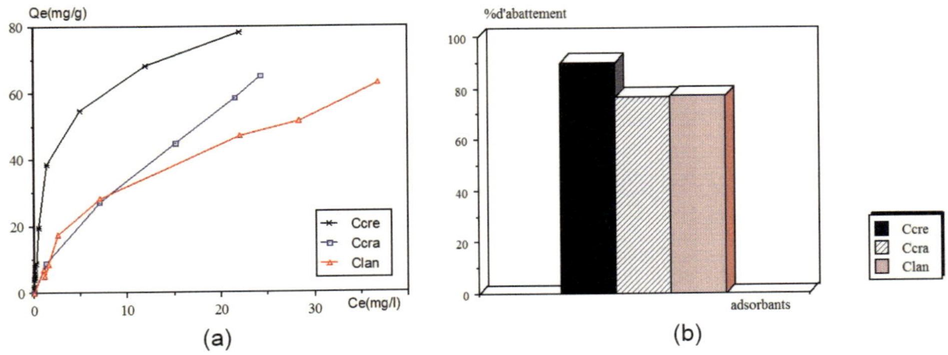

Figure 4.
(a) Experimental isotherms of Cd (II) adsorption on raw chitin, (b) Effect of the origin of raw chitin on the removal efficiency of Cd (II).

	Kf mg/g	1/n	R
Ccre	34.81	0.27	0.98
Ccra	7.38	0.69	0.99
Clan	18.92	0.34	0.97

Table 4.
Values of the Freundlich parameters deduced from the adsorption isotherms of Cd^{2+} *on raw chitin.*

the isothermal loss loses its linearity above 18 mg/g. For a residual concentration of 22 mg/l, the quantity retained depends on the origin of the raw chitin. Thus, it decreases successively from 78 mg/g for the Ccre to 57 mg/g for the Ccra and 47 mg/g for the Clan. From an initial concentration of Cd^{2+} equal to 100 mg/l, one can reach 22 mg/l after adsorption on the Ccre and 37 mg/l after adsorption on the Clan.

From **Figure 4b**, the percentage of removal of Cd on the Ccre reaches 90%. The selectivity of Cd for three supports is as follows: Ccre > Clan > Ccra.

The modeling of the isotherms in **Table 4** showed that the adsorption of Cd on these different supports perfectly satisfies the Freundlich model **Table 1**. The adsorption capacity of Cd on Ccre is greater than that relative to the Ccra and Clan (**Table 4**). This capacity is comparable to that found by Wales [52] by studying the adsorption of Cd^{2+} on chitin-rich fungi.

In this study **Figure 4b**, we found that the percentage of abatement decreases with increasing concentration. Thus, it varies from 78% for 100 mg/l to 98% for a concentration of 10 mg/l after adsorption on the Ccre.

4. Study of the Cu (II) adsorption

4.1 Kinetic study of Cu (II) adsorption

From **Figure 5**, we observe that the adsorption of Cu^{2+} is very fast initially and quickly reaches equilibrium time, 60 minutes for Ccra and 90 minutes for Clan. We also found that the hardness (mineral part) of the carapaces plays a role in the kinetics of the adsorption process by comparing these results with the equilibrium times found by Melchor [38].

According to Davis [57], the shape of these kinetic curves shows that the adsorption of Cu^{2+} is done according to the following model: Cu^{2+} binds rapidly in one time on the surface and groups co-ordinates very slowly to groups of chitin after scattering inside wall surfaces of the shells. Such adsorption kinetics is also

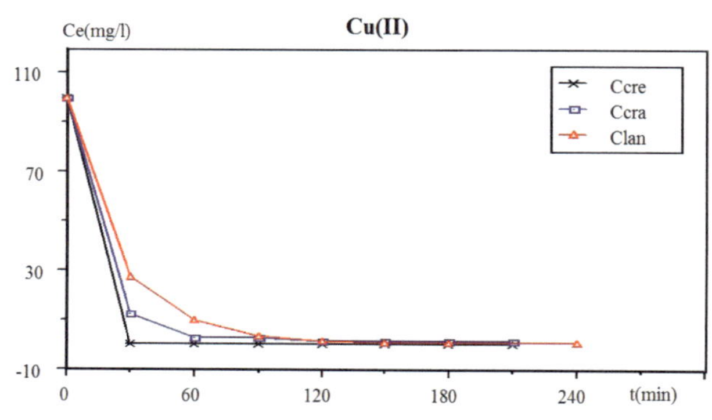

Figure 5.
Kinetic study of Cu (II) adsorption on raw chitin.

observed in the toxicological and physiological effect of copper. A similar behavior of copper retention was observed according to Balistrieri [58] on metal oxide surfaces, according to Fisher [29] on phytoplankton, and according to Xue et al. [59] on the algae.

4.2 Study of the Cu^{2+} isotherms adsorption on the raw chitin

Figure 6a shows the evolution of the adsorption capacity (Qe) expressed in milligrams of metal per gram of material as a function of the residual content (Ce) of Cu^{2+} at equilibrium. The results of calculating the average removal percentages for Cu^{2+} adsorption assays are shown in **Figure 6b**.

Examination of the different isotherms shows that Cu has a high affinity for Clan. For a residual concentration of 1 mg/l, the adsorbed amount reaches 95 mg/g for Clan and 40 mg/g for Ccre. This example proves that the adsorption of Cu^{2+} depends on the origin of the raw chitin.

From **Figure 4b**, it is clear that Cu^{2+} shows a particularly great preference for Clan. The percentage of Cu^{2+} adsorption on Clan varies between 99.07% to 5 mg/l and 93.2% to 100 mg/l, whereas on the Ccre, it varies between 81% to 5 mg/l and 68% at 100 mg/l; this shows that the adsorption of Cu^{2+} on these supports increases when the concentration decreases, and this is in agreement with the study carried out by Okieimen [60], which found that the percentage of $CuCl_2$ adsorption on the original Arachis Hypogea cellulose varies between 70% to 200 mg/l and 30% to 500 m/l. Likewise, Elliott [61] having studied the adsorption of Cu^{2+} on Al_2O_3 has argued that an increase in the ionic strength of the adsorbent-adsorbent system reduces the adsorption of Cu^{2+} through the effect of the activity coefficients and the double electric surface layer.

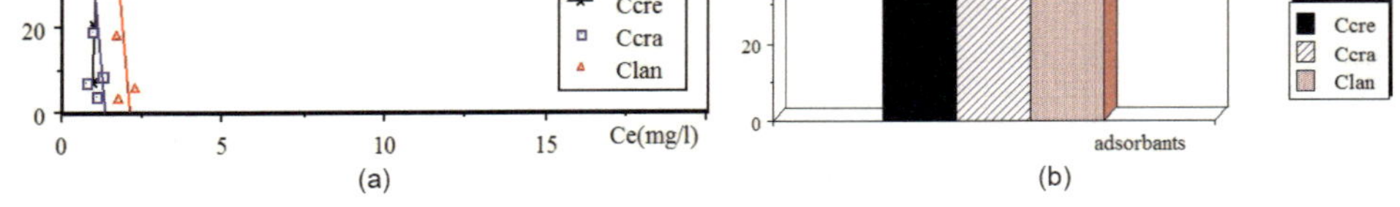

Figure 6.
(a) Experimental Cu (II) adsorption isotherms on the raw chitin, (b) Effect of the origin of the raw chitin on the Cu (II) elimination efficiency.

	Kf mg/g	1/n	R
Ccre	42.25	0.14	1.00
Ccra	17.19	−3.45	0.99
Clan	204.67	−4.58	1.00

Table 5.
Values of Freundlich parameters deduced from Cu^{2+} adsorption isotherms on the raw chitin.

The application of the linearized formula of the Freundlich equation (**Table 5**) in the case of Cu^{2+} adsorption allows us to say that this model is perfectly applicable. Indeed, the correlation coefficients R (**Table 2**) obtained are very satisfactory. Note that the adsorption capacity of Cu^{2+} on Clan is greater than that on Ccre and Ccra (**Table 5**).

Adsorption capacities deduced from Freundlich prove that copper is more adsorbed by Clan.

5. Study of the zinc adsorption

5.1 Kinetic study of the Zn^{2+} adsorption

As with other metals, the kinetic study shows that equilibrium is established rapidly. According to the curves of **Figure 7**, the contact times Ccre-Zn^{2+}, Ccra-Zn^{2+}, and Clan-Zn^{2+} required to reach equilibrium are around 40, 60, and 50 minutes, respectively. The times are very low in comparison with those necessary to reach equilibrium on mineral surfaces such as on the hydroxide of Fe (III), which requires 48 hours [8] and the goethite, which requires 42 days. Indeed, Balistrieri [58] having studied the adsorption of Zn^{2+} and Cd^{2+} on goethite, found that initially, there is a fast adsorption kinetics (a few hours) followed by a very slow diffusion controlling the penetration of the metal.

The tendency of surfaces to fix Zn^{2+} ions decreases with an increase in ion surface charge, since Zn^{2+} binds rapidly early with high affinity to the surface of the materials, and then slowly to with less activity. In addition, the increase in surface charge may induce some electrostatic repulsion. This effect on surface coverage is less pronounced with Cd^{2+} and Cu^{2+}. The same remarks were recorded for the kinetics of Cu^{2+} adsorption on algae [60].

5.2 Study of Zn^{2+} adsorption isothermals on raw chitin

Monitoring the amount of Zn^{2+} adsorbed as a function of the residual concentration (**Figure 8a**) shows that the adsorption phenomenon obeys Henry's law. For Ccra and Clan, isotherms are straight with high slopes. This time, Zn^{2+} is strongly retained by the Ccra than by the Ccre, the concentration decreases from 100 to 1.85 mg/l by adsorption on the Ccra, which proves a particular affinity of the Zn (II) to the Ccra. This affinity is stronger than that exhibited by mineral surfaces such as illite, montmorillonite, and kaolinite [2]. **Figure 8b** shows the calculation of the average removal percentages for six adsorption tests.

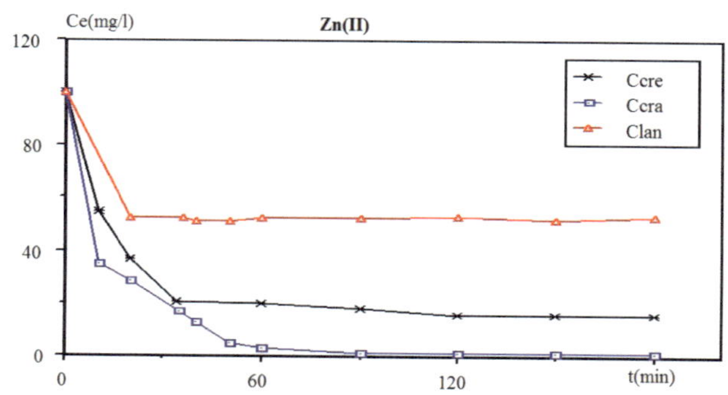

Figure 7.
Kinetic study of the Zn (II) adsorption on the raw chitin.

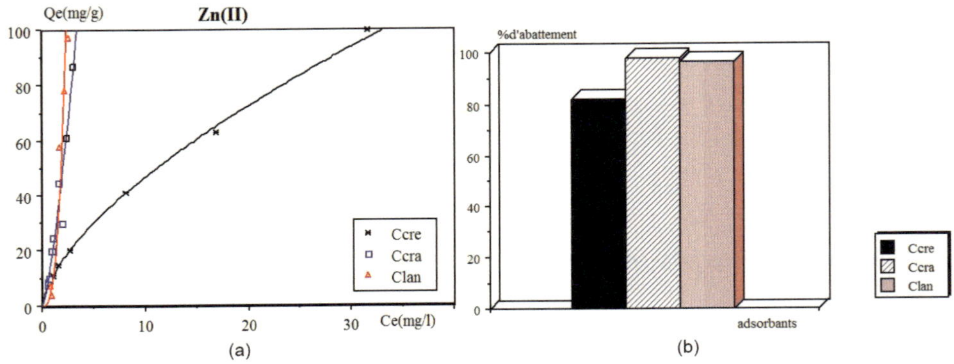

Figure 8.
(a) Experimental isotherms of Zn (II) adsorption on raw chitin (b) Effect of the origin of raw chitin on the elimination efficiency of Zn.

	Kf mg/g	1/n	R
Ccre	10.82	0.63	0.99
Ccra	23.74	1.72	0.99
Clan	11.49	1.99	0.99

Table 6.
Values of Freundlich parameters deduced from Zn^{2+} adsorption isotherms on the raw chitin.

Elimination percentages reach 98% for Ccra and 97% for Clan. These results are compared to those found for mineral surfaces such as Al_2O_3, zeolite, and albeit whose percentages of elimination do not go beyond 30% [10].

Linear adsorption transforms according to the Freundlich model—**Table 6** show that Zn has a great affinity for hard shells. The Freundlich model makes it possible to describe the adsorption of Zn: $R^2 > 0.95$. In this case, the adsorption is hydrophobic in nature through Van Der Waals interaction and hydrogen bonds.

The perfectly applicable Freundlich model shows that the adsorption capacity of Zn^{2+} on the Ccra is twice as large as that on the Ccre (**Table 6**).

According to the isotherms of **Figure 8a**, the percentage of adsorption decreases when the concentration of Zn^{2+} increases, the same remark was recorded by Spark and Bourg [17, 62] during the study of Zn adsorption on kaolinite and clay mineral.

6. Study of the lead adsorption

6.1 Kinetic study of the Pb^{2+} adsorption

The evolution of equilibrium concentration as a function of time is shown in **Figure 9**. The kinetic study of the adsorption has shown that more than 50% of the initial amount of Pb^{2+} is absorbed on the Ccre before the time is reached 5 minutes. The concentration of Pb^{2+} decreases rapidly in a few minutes, from 100 mg/l to a concentration of less than 1 mg/l for the Ccre. A pseudo-equilibrium with the surface of the supports appears to be reached, the equilibrium times are estimated at about 18 minutes for the Ccre, 30 minutes for the Clan, and 60 minutes for the Ccra.

The study conducted by Melchor [39] on the adsorption of Pb^{2+} on chitin extracted from crabs has shown that this is a reaction of the first order and gives an equilibrium time t = 44 minutes relatively low compared to the time found for the raw chitin 60 minutes; this implies that the mineral part plays a role in adsorption kinetics.

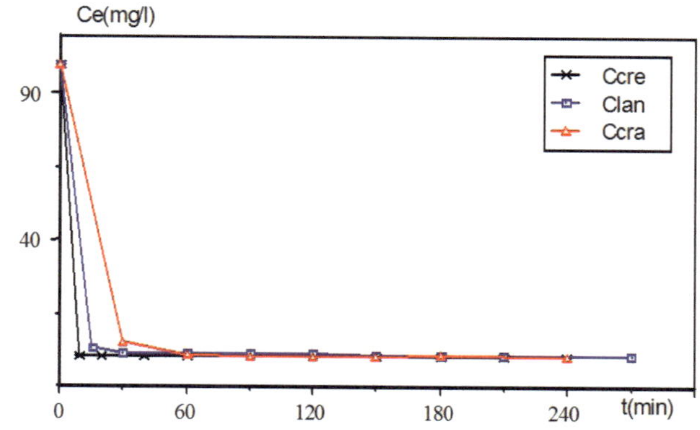

Figure 9.
Kinetic study of Pb (II) adsorption on raw chitin.

After equilibrium, more than 80% of the initial amount of Pb^{2+} (100 mg/l) is adsorbed on the Ccra, whereas only 41% of the initial amount of Pb^{2+} is adsorbed on pure chitin [39]; this shows that the mineral part of the crab shells plays a role in the retention of Pb^{2+}. The percentages of adsorption as a function of time are in good agreement with the results obtained by Yang [37].

6.2 Study of Pb^{2+} adsorption isothermals on raw chitin

According to **Figure 10a**, the curve representing the variation of the quantity of Pb^{2+} adsorbed on the Ccra as a function of residual concentration is linear up to 80 mg/g, whereas for Ccre and Clan, the adsorption obeys Henry's law in the range (10–100 mg/l).

Pb adsorption on the Ccre is peculiar, which is a very important slope that shows Ccre exhibits a strong affinity toward this metal. This feature was also noticed during the adsorption of Pb^{2+} on the zeolites [63, 64] and during adsorption on clinoptilolite [65]. Pb adsorption on all three supports perfectly the Freundlich model in comparison with the adsorption of Pb on activated carbon [66], which gives a capacity of 1.046 mg/g.

The calculation of the percentage of average abatement of Pb^{2+} (**Figure 10b**), for several adsorption tests confirms these results. So Pb is more retained by the Ccre. The concentration of Pb from 100 mg/l to only 0.37 mg/l for 10 minutes is a remarkable retention with a yield of 99.63%. From **Table 7**, the maximum

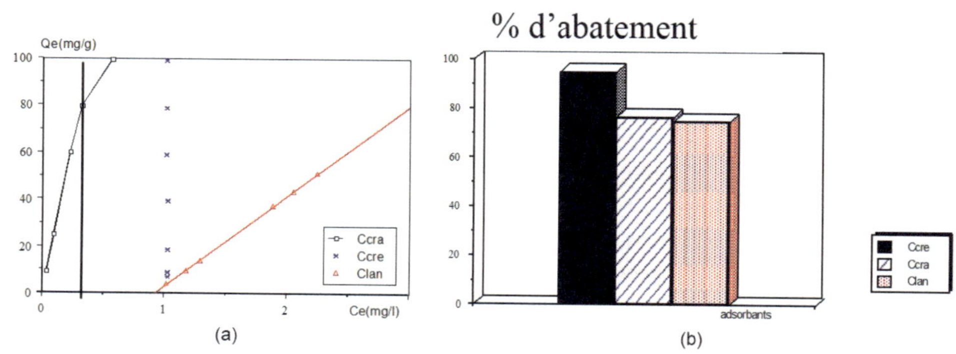

Figure 10.
(a) Experimental isotherms of Pb (II) adsorption on raw chitin, (b) effect of the origin of crude chitin on the removal efficiency of Pb (II).

	Kf mg/g	1/n	R
Ccre	9266.9	263.98	1.00
Ccra	265.07	1.04	1.00
Clan	3.59	3.69	1.00

Table 7.
Values of Freundlich parameters deduced from Pb adsorption isotherms on the raw chitin.

adsorption capacity according to the Freundlich model is 9266.86 mg/g on the Ccre, it is 35 times greater than that corresponding to the Ccra. There is also a better adsorbability for Ccra 265.07 mg/g.

We compared the rate of fixation of Pb by Ccre with adsorption results on biological surfaces such as 90% algae and rice biomass 99% Hulls [22], adsorption on mineral surfaces such as 35% bentonite and montmorillonite, which retain only 25% of Pb and to the cellulose for which the percentage of adsorption does not exceed 62.8%.

The average reduction percentage of Pb is 95% for the Ccre, and 75% for the Clan. According to the adsorption tests, the percentage of adsorption decreases when the concentration increases, the same observation was recorded for the adsorption of Pb on Kaolinite [13], a significant reduction in Pb is obtained at low initial concentrations.

7. Comparison with active charcoal powder (CAP)

The adsorbent supports are numerous and diverse, but in water treatment, activated carbon is used almost exclusively as adsorbent. To value chitin in the treatment of metallic water pollution, we have tried to compare its purification performance with that of CAP. We will then study the kinetics and adsorption isotherms of metals on CAP.

7.1 Kinetic study of adsorption of heavy metals on active coal powder (CAP)

During the treatment of liquid discharges, processes are always sought which give the maximum of treatment efficiency in a minimum of time. Since then, we have tried to compare the kinetics of adsorption on the studied materials and on the CAP. We have tried to situate ourselves under the same operating conditions as those of the raw chitin. The evolution of the adsorbed quantity as a function of time is reported in **Figure 11**.

From figure above, active carbon rapidly sets a maximum amount of metal. So for the Pb, it takes 60 minutes to reach adsorption equilibrium. Cd requires a time of about 10 minutes, while Cu and Zn only require a few minutes for their concentrations in the solution to stabilize.

Comparing the kinetic curves of the CAP with those obtained for different types of chitin, we find that the raw chitin exhibits identical kinetic behavior to that of the CAP even if the times of the adsorption kinetics on the raw chitin depend on its origin; these times are always close to those observed with the CAP, and sometimes, the equilibrium times are smaller than those of the CAP (case of the Ccre-Zn).

7.2 Comparing isothermals of metals adsorption on CAP and on raw chitin

To compare the reduction of metallic pollution by crude chitin to that by CAP, we carried out adsorption tests of heavy metals on the CAP under same conditions

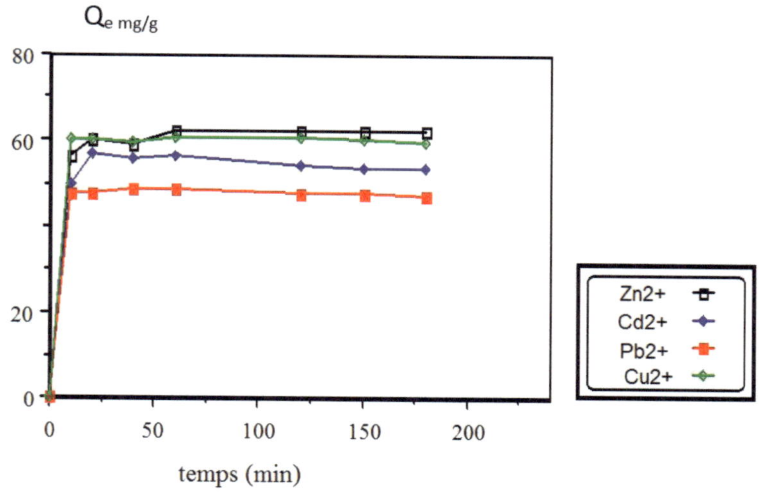

Figure 11.
Kinetic study of heavy metals adsorption on the CAP.

as those of crude chitin. **Table 8** represents the values of the Freundlich parameters for this study.

According to this table, Pb is more adsorbed by CAP than other metals, it is about 10 times more adsorbed than Cu. The retention of Zn and Cu is low, 1 g of CAP retains only 4.35 mg of Cu^{2+} and 3.38 mg of Zn. The selectivity of CAP toward these metals is as follows Pb > Cd > Cu > Zn. In order to compare these results with those of raw chitin, we have drawn up **Table 9** grouping the maximum adsorption capacities of metals on different media.

According to this table, the adsorption capacities of the metals on the CAP are low compared to the raw chitin. There is a difference between the values observed, and this difference is important in the case of Pb. Indeed, the Pb is 58 times retained by the Ccre than by the CAP, it is almost five times more retained by the Ccra than by the CAP. In the case of Cd^{2+}, the gap exists particularly between the Ccre and the CAP, Cd is more than three times retained by the Ccre than by the CAP. For Cu, 1 g of the Ccre retains a quantity of Cu 10 times higher than that retained by the CAP Similarly, Zn is more adsorbed by crude chitin than by CAP The same remarks were made by calculating the elimination percentages for six heavy metal adsorption tests; we have used the example of Cu^{2+} and Zn, their percentages compared to those observed with the raw chitin are grouped in the following **Figure 12**.

This figure clearly shows that the adsorption of the metals on the raw chitin is much stronger than the adsorption of the metals on the CAP. From this study,

	Kf (mg/g)	1/n	R
Pb^{2+}	58.69	1.03	0.98
Cd^{2+}	9.01	0.56	1.00
Cu^{2+}	4.35	0.57	1.00
Zn^{2+}	3.38	0.59	1.00

Table 8.
Values of the Freundlich parameters deduced from the adsorption isotherms of metal ions on activated carbon.

	Pb^{2+}	Cd^{2+}	Cu^{2+}	Zn^{2+}
CAP	58.69	9.01	4.35	3.38
Ccre	9266.86	34.81	42.25	10.82
Ccra	265.07	7.38	17.02	23,74
Clan	3.59	18.92	204.67	11.49

Table 9.
Values of the maximum adsorption capacity K (mg/g) of heavy metals.

it follows that the CAP can be substituted by the raw chitin for the treatment of the effluents rich in heavy metals.

7.3 Interpretation and comparison of results

According to this study, we can draw strong conclusions about the existence an affinity between biosorbent materials based on shells of crustaceans and some heavy metals.

Raw shrimp chitin exhibits a strong affinity for Pb. The adsorption capacity of zinc on the raw chitin of crabs is twice as great as that on the raw chitin of shrimp. The selectivity of the metal for each of these materials was defined by the plot of the adsorption isotherms. The metal ions are retained by these materials in the following order:

Lead: Ccre > Ccra > Clan
Cadmium: Ccre > Clan > Ccra
Copper: Clan > Ccre > Ccra
Zinc: Ccra > Clan > Ccre

The kinetic study has shown that the adsorption process is relatively fast compared to the adsorption on the supports described in the bibliography (of mineral origin). More than 50% of these ions are adsorbed before equilibrium is reached (20 minutes). The adsorption kinetics has also shown that the mineral part of the raw chitin is partly responsible for the retention of heavy metals. The hardness of the shells has a negative effect on the kinetics of the adsorption process, given the time required for equilibrium. Indeed, the adsorption of Pb on the raw chitin shrimp requires only 30 minutes; while the raw chitin lobsters the equilibrium time is 60 minutes.

Studies cited in the literature have shown that to obtain a given abatement, it is necessary to introduce a sufficient quantity of the material. Thus, our treatment tests have shown that to achieve a removal rate of 99% copper contained in

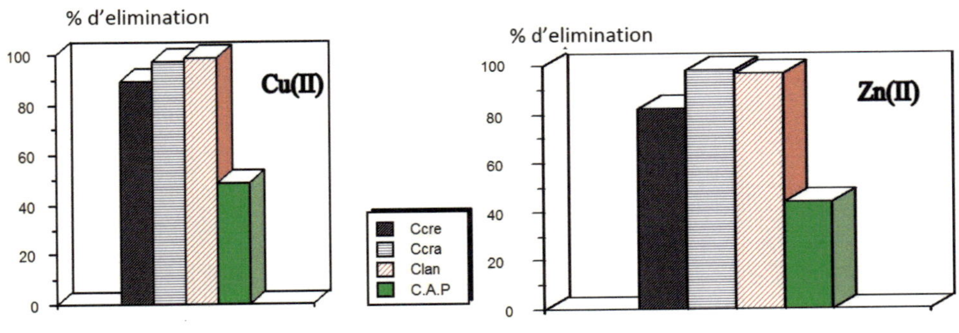

Figure 12.
Evolution of the average percentages of elimination according to the supports adsorbents.

a synthetic release with an initial concentration of 100 mg/l, it is necessary to add 1 g of raw chitin. The adsorption percentages of the metal ions on the raw chitin increase when the concentration of these ions decreases.

The comparative study with activated carbon powder (CAP) indicates that the substitution of the latter by chitin may be possible. However, chitin can be coupled to CAP during a chain of treatment of industrial discharges of various loads. As a result, the quantity of heavy metals not retained by the CAP will be retained by the raw chitin.

7.4 Comparative study of the selectivity of heavy metals adsorption on raw chitin in relation to other adsorbent cited in the literatures

From the previous results, we can classify the retention capacity of heavy metals by raw chitin according to the following sequences:

Ccre Pb > Cu > Cd > Zn
Ccra Pb > Zn > Cu > Cd
Clan Cu > Cd > Zn > Pb

These results are in agreement with some studies, indeed Haug et al. [67] showed that the addition of metal ions to a solution of Na alginate prepared from *Laminaria digitata* leads to the order of the next affinity: Pb > Cu > Cd > Zn. Nicolas [21] found the affinities of the following heavy metals:

Zn > Cu = Ni > Cd (pyrophosphates +61 mg/g polysaccharides)
Cu = Ni > Cd > Zn (pyrophosphates +222 mg/g of proteins)

Melchor [38, 39] showed that pure chitin crabs exhibits affinity vis-à-vis heavy metals in the following order: Cu > Cd > Pb.

In general, the chitin selectivity sequence depends, on the one hand, on the origin of this chitin (algae, fungi, crustaceans, etc.) and on the other hand, on the nature of the ions, in particular the 3d orbitals. It does not depend in any case on their Cu 0.72 Å, Pb 1.2 Å, Cd 0.97 Å, and Zn 0.74 Å sizes. However, the adsorption selectivity of metal ions on mineral surfaces varies greatly from one medium to another. We give the following sequences as an example: Pb > Cu > Zn = Cd and Pb > Cd > Zn > Cu respectively for montmorillonite and Kaolinite [68].

8. Test of metallic ion adsorption in series

In order to study the effect of the number of treatments on the percentages of elimination of metal ions and to solve the problem of competitive adsorption, we proceeded to a series treatment starting with an initial concentration of 100 mg/l from M^{2+}. After each adsorption test, the measured filtrate is again adsorbed on the material. This operation is repeated four times for the four metals. In parallel, we do a simple treatment using the same quantity used previously. The results found will be compared to FAO standards.

The results of the monitoring of the evolution of the metal ion contents during the series treatment and after the simple treatment of the synthetic rejection are shown in **Table 10**.

This table shows that the equilibrium concentration decreases after each adsorption test. Thus, we go from 100 to 0.032 mg/l after the first test and from 0.032 to only 0.002 mg/l after second test for Pb. This value is 25 times lower than the FAO standard of 0.050 mg/l; it is a remarkable reduction of the Pb ion.

	Pb^{2+}	Cd^{2+}	Cu^{2+}	Zn^{2+}
Co (mg/l)	100	100	100	100
Simple treatment	0.300	1.780	1.260	3.860
First trial	0.032	22.340	23.050	31.480
Second trial	0.002	11.400	0.824	2.489
Third trial	0.001	0.465	0.780	0.915
Fourth trial	—	0.019	0.151	0.728
FAO (mg/l)	0.05	0.02	0.2	5
PVL (mg/l)	0.5	0.2	0.5	5

Table 10.
Values of equilibrium concentration in mg/l of metal ions after single treatment and after each test during serial treatment.

	Pb^{2+}	Cd^{2+}	Cu^{2+}	Zn^{2+}
First treatment	99.97	77.66	76.95	68.52
Fourth treatment	99.99	99.98	99.84	99.27
Simple treatment	99.54	98.22	98.74	96.14

Table 11.
Percentages of reduction of metals according to the number of tests.

In the case of Cd after the fourth test, the Cd ion content is below the FAO standard. In the case of Cu, the concentration ranges from 100 to 0.780 mg/l after third test and from 0.780 to 0.151 mg/l, after the fourth test, the equilibrium concentration is therefore lower than the FAO standard for Cu^{2+}.

Finally, in the case of Zn^{2+}, only two adsorption tests are required to reach the FAO standard of 5 mg/l, which is relatively higher than the standards for other metals given the biological role played by Zn^{2+} in living organisms.

After a simple treatment, using the same quantity of the adsorbent used in the series treatment, it is noted that the quantities of metal ions remaining at equilibrium are greater than those remaining after the fourth serial treatment test; this shows that serial processing is more effective than simple treatment.

In **Table 11**, we calculated the percentage of abatement after the first test, the fourth trial, and simple treatment.

According to this table, the series treatment is very effective to reduce the amount of metal ions and thus increase the reduction efficiency. For example, for Zn, we go from a percentage of 68.52% after the first test and 99.27% after fourth adsorption test.

9. Sludge management test generated from the adsorption of heavy metals

One of the problems of the environment is the accumulation of solid waste. In order not to generate a simple transfer of pollution and in order to valorize the proposed treatment process, the solid sludge component generated by this treatment process "raw chitin saturated with heavy metals" should not weigh down the pollutant mass of household waste. From then we proposed at the end of this chapter two solutions to get rid of the generated sludge:

- The first is the elimination of this sludge by calcination at high temperature.
- The second is the recovery of this sludge by trying to study the regeneration of the adsorbent support by an acid.

9.1 Examination test for sludge generated by incineration

Incineration can greatly reduce the volume and weight of household waste by transforming it into gas, heat, and sterile and inert materials, ashes and slag. Volume and weight reductions can 90 and 60%, respectively of the initial volume and weight of the garbage. The gases formed contain mainly excess air, water vapor, carbon dioxide (CO_2), nitrogen oxides (NO_x) fly ash, and in small quantities, various products from combustion: carbon monoxide (CO), organic matter, sulfur dioxide (SO_2), hydrochloric acid, etc. The incineration takes place in three phases to know: drying with evaporation of water, spraying materials organics from 200°C and gasification of combustion of the carbonaceous residue: the volatile substances emitted burn at 500°C, this combustion is considered complete at 1000°C, provided that the air/fuel contact is satisfactory and the residence time at this high temperature is sufficient.

After the adsorption tests, the precipitates obtained loaded with the heavy metals are dried at 200°C in an oven and then calcined at high temperature (800°C) using a tubular oven for 2 hours, at the entry of the sample, the release of C, CO_2, and H_2O. Generally, when burning a conventional hydrocarbon fuel, one obtains at the end of the combustion, two harmless products CO_2 and H_2O, which result from the complete oxidation of the carbon and the hydrogen contained in the fuel, by the oxygen of the air. After few seconds, this release stops. The calcined final product is white for three samples relating to three metals studied, although for the case of copper, the sample was blue. In order to identify the phases obtained, the products resulting from the calcination were characterized by X-ray diffraction.

At high temperature, the predominant phase is the calcite phase, this remark is recorded for three metals. The calcination gave rise to a new phase, which is only the metal oxide phase. The appearance of the PbO, CdO, and CuO oxide phases relating to the cure loaded, respectively by Pb, Cd, and Cu. The metal ions are finally recovered in the form of an oxide that can be used as a raw material in the field of solid materials.

9.2 Test of the chitin regeneration

An adsorbent is of real interest only if it can be easily regenerated. The raw chitin used in the retention of metals must recover its initial structure and its starting chemical composition for a possible reuse with an appreciable re-adsorption efficiency.

In most of the regeneration studies carried out on organic materials, the essential element used as regenerant is the H^+. So, Muzzarelli [34] has exploited chitin to collect heavy metals from seawater, these metals are then recovered by appropriate washing with an acid. Huang [44, 45] studied the regeneration of fungi biomass loaded by Cd^{2+}, and found that these materials can be regenerated by strong acids. Hoshi [33] exploited chitin in the determination of copper by column adsorption-elution, Cu^{2+} retained was eluted with a mixture of acetic acid and acetone. Roy [22] showed that most heavy metals can be recovered by lowering the pH of the medium, using an acid wash. In general, the efficiency of the regeneration depends on the rate of passage on the adsorbent, the direction of this passage, and the amount of the regenerating solution which is passed over the adsorbent. According to our study, it turns out that

Metal	$C_{M2+\ fixed}$ (mg/l)	$C_{(metal)}$ (mg/l)	Percentage of release
Pb^{2+}	98.91	14, 820	149.8
Cd^{2+}	86.77	23.17	26.70
Cu^{2+}	94.09	23.35	24.82

Table 12.
Variation of the release percentage according to the nature of the metal.

the raw chitin has undeniable qualities of fixing heavy metals. For the regeneration tests, we chose the raw chitin of shrimp origin. Finally, the regeneration is carried out by HCl and focused on four metals, the regeneration conditions are optimized.

Let the percentage of release defined as following Eq. (3):

$$\% \text{ of release} = C_{(metal)}/C_{M2+fixed} \tag{3}$$

with

$C_{M2+\ fixed}$: concentration of M^{2+} fixed on the adsorbent before regeneration,

$C_{(metal)}$: concentration of the released amount of M^{2+} after regeneration.

Concentration $C_{M2+\ fixed}$ is calculated according to the following Eq. (4):

$$C_{M2+fixed} = [(CiVi - CeVr)\ Ms]/MoVr \tag{4}$$

with

Mo: initial mass before treatment; Ms: mass saturated with heavy metals; Vi: volume used for saturation; Vr: volume used for regeneration; Ci: initial concentration of metals; Ce: concentration of ions at equilibrium.

The amount released by HCl remains relatively stable from a concentration of 0.08 mol/l HCl, if we increase the concentration of HCl. It should be noted that all the metals studied are easily displaced by a low concentration of H, this shows that the interactions established between the metals and the raw chitin are not strong to resist the H^+ ions. The released quantity is maximum in the case of Pb. In the case of Cu and Cd, it is almost constant and neighbor. These results are grouped in **Table 12**, indeed the percentages of release are equal.

9.3 Comparison between the two processes and interpretation of results

The incineration of raw chitin has led to metal oxides that can be recycled in other industries; in addition, it is a fast and perfectly hygienic process. It allows a significant reduction of waste in weight and volume. Therefore, it is the best solution in the case of effluents less loaded with heavy metals and whose composition is complex. However, the regeneration of raw chitin is a cumbersome process that requires the addition of reagents (acid, salt), and the use of material, in addition, it modifies the structure of the material, but in spite of this, the regeneration is the best solution in the case of mineral effluents heavily loaded with heavy metals and whose composition is simple.

10. Conclusion

The isothermal study and kinetics of adsorption of metal ions on crude chitin has shown that the adsorption capacity of a metal depends on the origin of the latter and that the adsorption process is relatively fast. The Freundlich model is perfectly applicable; in fact, the correlation coefficients obtained for the four metal ions are very satisfactory. The removal efficiency of the metal ions depends on the initial concentration. This yield is high for a low initial concentration; therefore, the treatment of metal pollution after dilution in industrial effluents is interesting.

The comparative study with activated carbon (CAP) showed us that the raw chitin has a greater power of elimination than the (CAP), which suggests a probable substitution of CAP by chitin. According to this study, it can also be seen that the high temperature incineration of sludge generated by the adsorption of heavy metals on crude chitin results in calcite phases, which do not represent any toxicity in the environment and at low levels of metal oxide phases that can be recycled in the industry of solid materials (ceramics, cement, etc.). However, the regeneration of crude chitin changes the structure of the material relatively well and gives new adsorbent supports.

In summary, it can be argued that all the results obtained show the good ability of crude chitin to remove heavy metals when working under adequate conditions. Thus, fisheries waste, rich in chitin, would therefore constitute materials that could, on the one hand, eliminate metal pollutants, especially Pb^{2+}, Cd^{2+}, Cu^{2+}, and Zn^{2+} ions and on the other hand to substitute activated carbon in metal pollution treatment facilities.

Author details

Boukhlifi Fatima
Faculty of Science, MY Ismail University, Meknes, Morocco

*Address all correspondence to: boukhlifi1@yahoo.fr

IntechOpen

© 2020 The Author(s). Licensee IntechOpen. This chapter is distributed under the terms of the Creative Commons Attribution License (http://creativecommons.org/licenses/by/3.0), which permits unrestricted use, distribution, and reproduction in any medium, provided the original work is properly cited.

References

[1] Huang CP, Wirth PK. Activated carbon for treatment of cadmium wastewater. Journal of the Environmental Engineering Division. 1982;**108**:1280-1299

[2] Ku Y, Peters RW. The use of activated carbon as a polishing step for treatment of industrial plating wastewater. In: Proc. Nat. Conf. Envron. Eng; 1982

[3] Argaman Y, Weddle CL. Fate of heavy metals in physical-chemical treatment processes. AIChE Symposium Series. 1973;**70**(136):400-414

[4] Netzer PA, Wilkinson P. Removal of trace metals from wastewater by treatment with line. Water Pollution Research. 1974;**8**(10):813-817

[5] Huang CP, Wu M. The removal of chromium VI from dilute aqueous solutions by activated carbon. Water Research. 1977;**11**(8):673-679

[6] Saito I. Removal of metals from aqueous solutions using sulfonated coal and activated carbon. Kogai Shigen Kenkyushie Iho. 1976;**5**(2):57-64

[7] Swallow KC, Hume DN, Morel FMM. Sorption of copper and lead by hydrous ferric oxide. Environmental Science & Technology. 1980;**14**(11):1326-1331

[8] Anjana S, Srivastava PC. Adsorption-desorption behavior of zinc (II) at iron (III). Hydroxyde-aqueous solution interface as influenced by pH and temperature. Environmental Pollution. 1989;**68**:171-180

[9] Spark KM, Johnson BB, Wells JD. Characterization heavy-metal adsorption on oxides. European Journal of Soil Science. 1995a;**46**:621-631

[10] Huang CP, Rhoads EP, Hao OJ. Adsorption of Zn on to hydrous aluminosilicates in the presence of EDTA. Water Research. 1988;**22**(8):1001-1009

[11] Benjamin MM, Leckie JO. Effects of complexation by Cl, SO_4 and S_2O_3 on adsorption behavior of Cd on oxide surfaces. Environmental Science and Technology. 1982;**16**:162-170

[12] Comans RNJ. Adsorption, desorption and isotopic exchange of cadmium on illite. Water Research. 1987;**21**(12):1573-1576

[13] Schindler PW, Liechti P, Westall JC. Adsorption of copper, cadmium, and lead from aqueous solution to the kaolinite/water interface. Netherlands Journal of Agricultural Science. 1987;**35**(3):219-230

[14] Cadena F, Rizvi R, Peters RW. Feasibility studies for the removal of heavy metals from solution using Tailored bentonite in hazardous and industrial wastes. In: Proceedings of the 22nd Mid-Atlantic industrial waste conference; 1990. pp. 77-94

[15] El Khatib EA, Elshebiny GM, Balba AM. Lead sorption in calcareous soils. Environmental Pollution. 1991;**69**(4):269-276

[16] Petersen W, Wallmann K, Schroer S, Schroer F. Studies on the adsorption of cadmium on hydrous iron (III) oxides in toxic sediments. Analytica Chimica Acta. 1993;**273**(1-2):323-327

[17] Spark KM, Johnson BB, Wells JD. Characterizating trace metal adsorption on kaolinite. European Journal of Soil Science. 1995;**46**:633-640

[18] Mrad I, Ghorbel A, Lambert J, Che M. Cadmium adsorption on al-pillared montmorillonite, acte du colloque Franco-Maghrebine sur la catalyse, Lille; 1996

[19] Filip DS, Peters T, Adams VD, Middlebrooks EJ. Residual heavy metal removal by an algae-intermittent sand filtration system. Water Research. 1979;**13**(3):305-313

[20] Sloan FJ, Abernathy AR, Jennett JC, Goodman GV. Removal of heavy metal ions from wastewater by algae. In: Proc. 38th Purdue Industrial waste Conf., 38; 1984. pp. 423-429

[21] Nicolas KK, Sterritt RM, Lester JN. Heavy metal binding by the polymeric organic fractions of sewage sludges. Environmental Pollution. 1990;**67**:259-278

[22] Roy D, Greenlaw PN, Shane B. Adsorption of heavy metals by green algae. Journal of Environ–mental Science and Health. 1993;**28**(1): 37-50

[23] Brown MJ, Lester JN. Metal removal in sludg. Water Research. 1979;**13**:817-837

[24] Brown MJ, Lester JN. Comparison of bacterial extracellular polymer extraction methods. Applied and Environmental Microbiology. 1980;**40**:179-185

[25] Lester JN, Sterritt RM, Rudd T, Brown MJ. Assessment of the role of bacterial extracellular polymers in controlling metal removal in biological wastewater treatment, Microbiol. In: Grainger JM, Lyunch JM, editors. Methods in Environmental Biotechnology. New York: Academic Press; 1984. pp. 197-217

[26] Lester JN, Sterrit RM. Microbiol-accumulation of heavy metals in wastewater treatment processes. The Journal of Applied Bacteriology. 1985;**59**(14)

[27] Irwin PL, Sevilla MD, Shieh JJ. ESR evidence for sequential divalent cation binding in higher plant cell walls. Biochimica et Biophysica Acta. 1984;**805**:186-190

[28] Deans JR, Dixon BG. Uptake of Pb and Cu by novel biopolymers. Water Research. 1992;**26**(4):469-472

[29] Fisher NS. Accumulation of metals by picoplankton. Marine Biology. 1985;**87**:137-142

[30] Boukhlifi F, Mamouni FZ, Razouk R. Chitin/Chitosan's bio-fertilizer: Usage in vegetative growth of wheat and potato crops. In: Dongre RS, editor. Chitin-Chitosan—Myriad Functionalities in Science and Technology, Ch 16. IntechOpen; 2018. p. 75208

[31] Boukhlifi F, Bencheikh A. Characterization of natural biosorbents used for the depollution of waste water. Annales de Chimie Science des Materiaux. 2000;**256**:153-160

[32] Muzzarelli RAA. Industrial production and uses of chitin and its derivatives, the XII International Carbohydrate Sym. 1984

[33] Hoshi S, Tanaka Y, Inoue S, Matsubara M. Spectrophotometric determination of trace copper after adsorption and elution of neocuproine complexe on chitin. Analytical Sciences. 1989;**5**:9-12

[34] Muzzarelli RAA. Chitin and chitosan as chromatographic supports and adsorbents for collection of metal ions from organic and aqueous solutions and sea water. Official Gazette of the United States Pattern Office. 1972;**894**(3):1093

[35] Muzzarelli RAA. 1977 - Carbohydrate Polymers. 1985;**5**(6):461-472

[36] Muzzarelli RAA, Tanfani F. The chelation of chitinous material from Aspergillus siger, streptomyces, mucorrouxi, phycomyces blakesleeanus

and Choanephora curbitarum, Inc. In: Proc. 2nd Intl. Conf. On Chitin and Chitosan. Sapporo, Japan; 1982. pp. 183-186

[37] Yang IC, Zall RR. Adsorption of metals by natural polymers generated from seafood, processing wastes. Industrial and Engineering Chemistry Product Research and Development. 1984;**23**:168-172

[38] Melchor GD, Frank JM. The adsorption of copper to chitin in seawater. Geochimica et Cosmochimica Acta. 1989;**54**:761-768

[39] Melchor GD, Magdalena SCJ, Frank JM. The adsorption of Cd and Pb to chitin in seawater. Journal of Colloid and Interface Science. 1990;**137**(1):102-110

[40] Muzzarelli RAA. Natural Chelation Polymers. Oxford: Perg Press; 1973

[41] Tsezos M, Volesky B. The mechanism of uranium biosorption by Arrhizuo. Biotechnology and Bioengineering. 1982b;**24**(2):385-401

[42] Muzzarelli RAA, Tanfani F, Emanuelli M. Aspartate glucan, glycine glucan and serine glucan for the removal of cobalt and copper from solutions and brines. Biotechnology and Bioengineering. 1985;**XXVII**:1115-1121

[43] Kurita K, Chicaoka S, Koyama Y. Improvement of adsorption for copper II ion by N-nonanoylation of chitosan. Chemistry Letters. 1988:9-12

[44] Huang CP, Westman D, Quirk K, Huang JP. Removal of Cd(II) from dilute aqueous solutions by fungal adsorbent. Water Science and Technology. 1988;**20**(11/12):369-376

[45] Huang CP, Morehart AL, Westman DC. Removal of toxic heavy metals from contaminated ground water by a fungal adsorption process. Nat. Tech. Infor. Service, Spring Field; VA 22161 as pb 90-138637. 1989

[46] Boukhlifi F. Etude de la rétention des micro-polluants métalliques(Pb, Cd, Cu et Zn) sur de nouveaux matériaux biosorbants: essais d'épuration d'effluents industriels liquides, thèse de doctorat. Maroc: université chouaib doukkali El jadida; 2000

[47] Boukhlifi F, Bencheikh A, Ahlafi H. Characterisation and adsorption propriety of chitin toward copper cu2+. Physical and Chemical News. 2011;**58**:67-72

[48] Boukhlifi F. et al. Etude de l'adsorption compétitive des métaux lourds sur la chitine brute: application aux eaux usées d'une industrie chimique. Vol. 55. LA Tribune de l'eau, N° 611/3; 2001. pp. 37-43

[49] Boukhlifi F, el Akili C, Moussout H, Benzakour A, Ahlafi H. Treatment of global rejection of electroplating industry by raw chitin. International Journal of Applied Environmental Sciences. 2013;**8**:13-23

[50] Boukhlifi F et al. Rhazes: Purification of the waste water of the galvanoplasty industry by natural adsorbents: Current rinsing. Green and Applied Chemistry. 2018;**2**(8):80-89

[51] Boukhlifi F, Allali M, Bencheikh A. Etude de la pollution métallique des eaux usées d'une industrie chimique et essai de traitement par la chitine brute. Marine Life. 2001;**11**(1-2):49-56

[52] Wales DS, Sagar BF. Recovery of metal ions by microfungal filters. Journal of Chemical Technology and Biotechnology. 1990;**49**(4):345-355

[53] Ornum JV. Shrimp waste must it be wasted. INFOFISH International. 1992;**6**(92):48-52

[54] Ryuji Y, Ayuji KT, Tatsuro I. Utilization of partially N-succinylated

derivatives of chitosan and glucolchitosan as supports for the immobilization of enzymes. Biotechnology and Bioengineering. 1982;**XXIV**:1081-1091

[55] Sadiq M, Zaidi TH. The adsorption characteristics of soil and removal of cadmium and nickel from wastewaters. Water, Air, and Soil Pollution. 1981; **16**(3):293-299

[56] Avnimelech Y, Ravech A. Decomposition of chelates teached from waste disposal sites. Journal of Environmental Quality. 1982;**11**(1):69-72

[57] Davis JA. Use of sewage sludge on land in the United Kingdom. Water Science and Technology. 1987;**19**:1-8

[58] Balistrieri LS, Murray JW. The adsorption of Cu, Pb, Zn and Cd on geothite from mayor seawater. Geochimica et Cosmochimica Acta. 1982;**46**:1253-1265

[59] Xue HB, Stumm W, Sigg L. The binding of heavy metals to algal surfaces. Water Research. 1988;**22**(7): 917-926

[60] Okieimen FF, Ogbeifun DE, Nwala GN, Kumsah CA. Binding of cadmium, copper and lead ions by modified cellulosic materials. Bulletin of Environmental Contamination and Toxicology. 1985;**34**(6):866-870

[61] Elliott HA, Huang CP. Factors affecting the adsorption of complexed heavy metals on hydrous Al_2O_3. Water Sience and Technology. 1985;**17**(6/7):1017-1028

[62] Bourg ACM, Filby RH. Adsorption isotherms for the uptake of zinc by clay minerals in a fresh water medium. In: Proceeding of the international conference on transport of persistent chemicals in aquatic ecosystems held in ottawa. Canada; 1974. pp. 1-3

[63] Tessa N, Tyburce B, Joly G. Journal de Chimie Physique. 1991;**88**(5):603-613

[64] Karhat Y. Etude de l'élimination de polluants NH4+, Pb2+, Cd2+, Hg2+ par échange d'ions au moyen de la zéolithe Na-P. Thèse de 3ème cycle. Rennes: Univ de Rennes I; 1994. p. 151

[65] Blanchard G, Maunaye M, Martin G. The adsorption of lead on clinoptilolite. Water Research. 1984;**18**(12):1501-1507

[66] Tant TC, Teo WK. Combined effect of carbon dosage and initial adsorbate concentration on the adsorption isotherm of heavy metals on activated carbon. Water Research. 1987;**21**(10):1183-1188

[67] Haug A, Smidsrod O. The effect of divalent metals on the properties of alginate solutions. II Comparison of different metal ions. Acta Chemica Scandinavica. 1965;**19**:341-351

[68] Yong RN, Phadung Chewit Y. Influence on selectivity and retention of heavy metals in some clay soils. Canadian Geotechnical Journal. 1993;**30**(5):821-833

Chapter 14

Removal of Heavy Metals from Water and Wastewater Using *Moringa oleifera*

Eman Noori Ali

Abstract

One of the contaminants in wastewater is the heavy metals. Treatment of heavy metals is of great importance because they can be harmful and dangerous for human being health. Conventional removal methods used include: ultra-filtration, reverse osmosis, ion exchange, solvent extraction, sedimentation, and chemical precipitation, and each method has some disadvantages besides high costs. In this chapter, Moringa oleifera cake residue, Moringa oleifera press cake, and Moringa oleifera leaves are introduced as a proposed alternative to replace conventional methods for heavy metal ions' removal. The results of using Moringa oleifera cake residue showed that iron (Fe) was fully removed; copper (Cu) and cadmium (Cd) were successfully removed up to 98% and reduction of lead (Pb) of 82.17%. The heavy metals were successfully reduced using Moringa oleifera press cake. The removal percentage of iron, copper, and chromium reached 69.99%, 88.86%, and 93.73%, respectively. Moringa oleifera leaves were used to remove Cd (II) from synthetic water; the optimization was performed and each parameter was affecting the Cd (II) removal with different percentages, but pH was insignificant. As a conclusion, the Moringa oleifera seeds and leaves can be considered as a promising alternative in water treatment for heavy metal ions removal.

Keywords: water treatment, *Moringa oleifera*, heavy metals, cadmium, copper

1. Introduction

Water is a source that is essential for life and is required by almost every living creature. Water source for human being consumption needs to be treated first, due to the contamination by different industrial advancements made over the years. In addition, there is natural contamination by soil erosion and organisms that live in water. Water pollution has contributed to negative environmental and human health impacts [1], and many pollution problems are introduced [2]. Every day, there are thousands of chemicals discharged directly and indirectly into water bodies without further treatment for elimination of the included harmful compounds. Water pollution is a serious problem in the world and needs to be solved. Aluminium sulphate and iron are the chemical coagulant salts used in the conventional method to treat water and wastewater [3, 4], but they have

IntechOpen

many drawbacks [4–6]. Therefore, researchers are working hard to find natural alternatives. Heavy metals in water can be hazardous and harmful as they can accumulate in living organism tissue. One of the most toxic and hazardous heavy metals is Cd(II). Industries of pigments, electroplating, plastic, and metal finishing are the main sources for Cd(II) presence in water which might cause kidney damage, high blood pressure, bone fraction, renal disorder, and destruction of red blood cells [7]. There are many techniques to remove Cd(II) from aqueous solutions, such as; ion exchange, chemical precipitation, and membrane separation and adsorption, but some of these methods got several restrictions [8], such as low heavy metal concentrations, not effective and not economical [9]. As a result, the search for appropriate alternative solutions is of great importance. Recently, researchers are paying attention to Cd(II) removal from aqueous solution using adsorbents derived from low-cost tree leaves such as *Moringa oleifera* leaves [10].

Heavy metals present in water are harmful and poisonous and need to be removed from water; using natural biosorbent is one of the solutions. Few studies reported the removal of some heavy metals from water by *Moringa oleifera*.

Copper can be produced from electronics plating, paint manufacturing, wire drawing, copper polishing, and printing operations. The high presence of copper can cause acute toxicity, dizziness, and diarrhoea [11, 12]. It is an important element required by humans in trace amounts, used by the human body for enzyme synthesis and tissue and bone development [13]. However, the presence of copper in excessive amount can be toxic and carcinogenic. The copper can be deposited in the body and can cause some health problems, such as liver and kidney failure, Wilson disease, and gastrointestinal bleeding [11]. The osmoregulatory mechanism of the freshwater animals can be damaged due to the presence of Cu(II) in freshwater resources and aquatic ecosystem. Usually, chemical coagulants with high cost such as aluminium sulphate and activated carbon are used in water treatment [14].

Moringa oleifera is a member of the Moringaceae family which is a single genus family of shrubs [1, 2]. It can be found in Malaysia and other tropical countries where it was imported originally from India. It is easy to plant this tree, and it grows in soil or sand, can stand high temperature, and needs less water [15]. *Moringa oleifera* is a multipurpose tree with most of its parts being useful for a number of applications (**Figure 1**). *Moringa oleifera* seeds have been found to be a natural coagulant, flocculant, softener, disinfectant, sludge conditioner in water treatment [16, 17], and heavy metal remover in water and wastewater treatment [11, 16].

Moringa oleifera leaves are unique as a good protein supplement; it contains high amounts of minerals. In *Moringa oleifera* leaves, tannins and phytates are present with 12 and 21 g/kg on a dry basis, respectively [18]. It contains 38.6% carbohydrate, 27.2% protein, and 17.1% fat on a dry basis. It contains 2098 mg calcium, 1922 mg potassium, 406 mg magnesium, 351.1 mg phosphor, 28.3 mg iron, and 5.4 mg zinc in each 100 gm of dry extracted leaves [19] and around 0.58–0.73 g of different proteins/g leaves dry weight [20]. Flavonoids and phenolic acids are found in *Moringa oleifera* leaves by HPLC analysis [21]. *Moringa oleifera* leaves are a rich source of vitamins [22].

Moringa oleifera seeds, leaves, husks, or press cake can remove heavy metals from water such as Cu, Cd, Pb, Fe, and Cr which will be covered in this chapter.

Microbial biomass is one of the effective natural materials as biosorbent [23], bagasse fly ash (from sugar cane), and peat are agricultural waste materials that can be an effective heavy metal biosorbent [24]. In addition, rice husks and straws [25],

Figure 1.
Moringa oleifera tree.

corn cobs, soya bean, walnut and cotton seed hull, sawdust, and banana peels [26] are good examples for alternative biosorbents.

2. Materials and methods

The *Moringa oleifera* tree parts such as seeds or leaves are mainly used to remove heavy metals from river water or wastewater in this chapter.

2.1 Seeds

Moringa oleifera seeds which are not harmful to human and do not have significant drawbacks have been applied for wastewater treatment [4–6]. The seeds showed natural coagulation activity which is similar and even better than aluminium sulphate (alum) [27, 28].

The seed oil is extracted by different methods either by solvent (cake residue) or by mechanical press (press cake). Both seed extracts were used in heavy metal removal.

Cake residue preparation: method (a): the seed husks are removed manually, the kernel is ground, and the kernel powder is then mixed with ethanol and mixed using a magnetic stirrer for 30 min. The mixture is then centrifuged at 4000 rpm for 10 min. The residual solid (seed cake) was dried at room temperature for 24 h [3]. The stock preparation was prepared with different concentrations: 10,000,

20,000, and 30,000 mg/L. The different concentrations were applied to wastewater, and jar test was performed with an initial speed of 150 rpm for 2 min, reduced to 50 rpm for 25 min, and then left to settle for 1 h [29].

Method (b): oil was extracted by Soxhlet extraction method [29]. Extraction thimble is filled with 10 g of *Moringa oleifera* seed powder. Hexane solvent (170 mL) was poured into a round bottom flask. The heater is turned on until the solvent is boiling; the oil was extracted within around 45 min. *Moringa oleifera* cake residue was collected from the thimble and dried in an oven at 50°C overnight. Once the oil was fully removed from the *Moringa oleifera* seed, the seed cake residue could be used in water treatment [30]. The oil was extracted from the seeds because the presence of oil in the *Moringa oleifera* seed would affect the coagulation activity and heavy metal removal [19]. The higher the oil content in *Moringa oleifera* seed, the lower the performance of the *Moringa oleifera* cake residue in water treatment process.

Press cake preparation: the seeds were pressed mechanically by Mitomasa Sdn. Bhd., Kuala Lumpur, Malaysia, and the press cake was bought from the company and applied in water treatment [31]. Press cake was soaked with water overnight to get the remaining oil. And the clean press cake was used in this experimental work. The moisture content was calculated to get the right mass for the biosorbent added to water [31]. The synthetic turbid water was used for biosorption test on Cu(II) removal from water by *Moringa oleifera* press cake.

2.2 Leaves

Moringa oleifera leaves were collected from a nearby area at Universiti Malaysia Pahang, Kuantan, Pahang, Malaysia, dried and ground, and sieved using a sieve shaker to 2, 1 mm, 500, 250, and < 250 μm.

Moringa oleifera leaf extract was used in aqueous solutions as a good sorbent for Pb(II) [32]. It was used for optimization of Cd(II), Cu(II), and Ni(II) biosorption [33] and removing of Cd(II) from wastewater [34].

2.3 Synthetic water preparation

To prepare synthetic water, kaolin with laboratory grade (k7375-500G Sigma-Aldrich) was used, and 5 gm was dissolved in 500 ml of distilled water. Sodium bicarbonate solution was prepared by dissolving 100 gm of sodium carbonate (Hamburg Chemicals) in 1000 ml of distilled water to get a concentration of 100 mg/L. 500 ml of the sodium bicarbonate solution was added to kaolin solution and mixed at 200 rpm for 1 h to get uniform dispersion of kaolin particles which was left for 24 h for complete hydration of the kaolin [29].

2.4 Jar test

Each beaker in jar test was filled with 500 mL of wastewater sample. 10 mL of *Moringa oleifera* press cake with different concentrations prepared were added into each beaker. The stirring speed was set at 200 rpm for 4 min followed by 40 rpm for 30 min [35].

2.5 Atomic absorption spectrometer

Atomic absorption spectrometer (AAnalyst 400, Perkin Elmer) was used to measure initial and residual heavy metal in treated water. Stock solution prepared from each standard at different concentrations was used to get calibration curve for each metal [30]. Wavelength of 324.80 nm was used to measure the Cu(II) concentration and 228.8 nm for Cd(II) measurement using an acetylene air flame [10, 31].

2.6 *Moringa oleifera* solution preparation

The stock solution was prepared by adding the distilled water to *Moringa oleifera* cake residue and forming a paste at different concentrations of 10,000, 20,000, and 30,000 mg/L by dissolving 50, 100, and 150 mg of the seed powder in 5 ml distilled water to obtain 1, 2, and 3%, respectively [36].

2.7 Water samples

The water samples of the Cu(II), Cd(II), Fe(II), Pb(II), and Cr(II) study were collected from "Sungai Baluk" river (Gebeng River {GR}), Gebeng Industrial Estate, Kuantan, Pahang State, Malaysia [30].

The synthetic water was prepared for measuring Cd(II) removal [10].

3. Results

The heavy metals of Cu, Cd, Fe, and Pb were high in the GR water samples. After a tenfold dilution, the concentration of these metals is still more than 1 mg/L. *Moringa oleifera* cake residue showed that iron (Fe) was fully removed, while copper (Cu) and cadmium (Cd) were successfully removed up to 98%. The reduction of lead (Pb) by 82.31% was achieved (**Table 1**). The initial amount of Fe was 1.306 mg/L and fully removed by *Moringa oleifera* cake residue. This result was totally agreed with [4] study, while [37] had reported that a reduction of Fe by using *Moringa oleifera* cake residue was up to 92% if compared to other heavy metals. Fe was fully removed, and the level of Fe achieved the water standards since concentration of Fe in water should be less than 0.30 mg/L (National Water Quality Standard from Malaysia). In this study, the removal of Pb was within the range of the previous research [37, 38] which indicated that Pb can be removed within 80–89% by using *Moringa oleifera* seed if compared to other natural sources such as beans and peanuts. Copper removal was better than [38] results of Cu which was 90%. The results of Cd removal were in contrast with [37] since the latter results showed 48% of Cd was removed which might be due to the absence of oil extraction from seeds.

In a different study, it was found in Balok River wastewater, Gebeng, Kuantan, that the concentration of Fe and Cr metal were 1 mg/L, while Cu concentration was 0.4 mg/L. *Moringa oleifera* solution was added into the wastewater with different doses to determine the optimum dose. The heavy metals were successfully reduced after the treatment with the *Moringa oleifera* press cake. The removal percentage of iron, copper, and chromium reached 69.99, 88.86, and 93.73%, respectively

Metal	Initial mg/L	Final mg/L	Removal percentage
Fe	1.306	0	100%
Cu	1.054	0.019	98%
Cd	1.718	0.035	98%
Pb	1.64	0.29	82.31%

Table 1.
Heavy metal removal from Gebeng River, Kuantan, Malaysia.

Metal	Initial mg/L	Final mg/L	Removal percentage
Fe	1	0.3001	69.99%
Cr	1	0.0627	93.73%
Cu	0.4	0.04456	88.86%

Table 2.
Heavy metal removal from Gebeng River, Kuantan, Malaysia.

Parameter	Initial mg/L	Final mg/L	Removal percentage
Particle size ≤250μm	1	0.19	81%
Contact time (min)	1	0.28	72%
Moringa oleifera dose (6gm/L)	1	0.19	81%
Turbidity (50 NTU)	1	0.166	83.40%

Table 3.
The optimum results for Cd (II) removal with different parameters.

(**Table 2**). The removal of iron in this experiment was lower than the research done by [37], which had a removal of 92.14%. The removal percentage of chromium in this research was higher than the research done by [39], which was only 60% and almost similar to [40], which was 73%. The copper removal efficiency in this research is approximately similar to the literature statement of [39], which was approximately 90%. The trend of the removal percentage of copper and chromium was found to be alike with the trend in the research done by [31, 41].

Moringa oleifera leaves were used to remove Cd(II) from water, and the results showed that there are many parameters that affect the results, such as dose effect, contact time, particle size, pH, and water turbidity effect [10]. In this study, the pH value was insignificant, and other results showed that *Moringa oleifera* leaves are a potential alternative for heavy metal removal at a different percentage as shown in **Table 3**.

4. Conclusions

This chapter focused on the potential use of natural material for heavy metal removal from water and wastewater. The *Moringa oleifera* seeds which can be processed to get the press cake (mechanical process) or cake residue (chemical process) using solvents can be applied to polluted water and remove heavy metals Fe, Cu, Cd, Pb, and Cr.

Moringa oleifera leaves can be applied as a natural adsorbent to remove heavy metals from water (without any modification or chemical treatment); it is an environmentally friendly biosorbent. It is available in Malaysia and other tropical countries which make it a low-cost adsorbent with high biosorption capacity.

Acknowledgements

The author would like to thank Research & Innovation Department/Universiti Malaysia Pahang/Malaysia, for financial support to perform the research work for a few projects which is summarised in this chapter.

DOI: http://dx.doi.org/10.5772/intechopen.89769

Author details

Eman Noori Ali
Faculty of Chemical and Natural Resources Engineering, Universiti Malaysia Pahang, Lebuhraya Tun Razak, Gambang, Kuantan, Pahang Darul Makmur, Malaysia

*Address all correspondence to: iman129@gmail.com

IntechOpen

© 2020 The Author(s). Licensee IntechOpen. This chapter is distributed under the terms of the Creative Commons Attribution License (http://creativecommons.org/licenses/by/3.0), which permits unrestricted use, distribution, and reproduction in any medium, provided the original work is properly cited.

References

[1] Jahn S. Effectiveness of traditional flocculants as primary coagulants and coagulant aids for the treatment of tropical waters with more than a thousand fold flocculation in turbidity. Water Supply. 1984;**2**(3/4) Special Subject 6:8-10

[2] Ndabigengesere A, Narasiah K, Talbot B. Active agents and mechanism of coagulant of turbid waters using Moringa oleifera. Water Research. 1995;**29**(2):703-710

[3] Gidde M, Bhalerao A, Malusare C. Comparative study of different forms of Moringa oleifera extracts for turbidity removal. International Journal of Engineering Research and Development. 2012;**2**:14-21

[4] Egbuikwem P, Sangodoyin A. Coagulation efficacy of Moringa oleifera seed extract compared to alum for removal of turbidity and *E. coli* in three different water sources. European International Journal of Science and Technology. 2013;**2**(7):13-20

[5] Joshua R, Vasu V. Characteristics of stored rain water and its treatment technology using Moringa seeds. International Journal of Life Science and Pharma Reviews. 2013;**2**:155-174

[6] Madrona G, Branco I, Seolin V, Filho B, Fagundes-Klen M, Bergamasco R. Evaluation of extracts of Moringa oleifera lam seed obtained with NACl and their effects on water treatment. Acta Scientiarum. Technology. 2012;**34**:289-293. DOI: 10.4025/actascitechnol.v34i3.9605

[7] Drasch G. An increase of cadmium body burden for this century "an investigation on human tissues". The Science of the Total Environment. 1983;**26**:111-119

[8] Abaliwano K, Ghebremichael A, Amy L. Application of the purified Moringa oleifera coagulant for surface water treatment. Water Mill Working Paper Series. 2008;**5**:1-22

[9] Kelly-Vargas K, Cerro-Lopez M, Reyna-Tellez S, Bandala E, Sanchez-Salas J. Biosorption of heavy metals in polluted water, using different waste fruit cortex. Physics and Chemistry of the Earth Parts A/B/C. 2012;**37-39**:26-29

[10] Eman N, Sabreen R, Mashita M, Md Lutfor R. Environmentally friendly biosorbent from *Moringa oleifera* leaves for water treatment. International Journal of Environmental Science and Development. 2015;**6**(3):165-169

[11] Bilal M, Shah J, Ashfaq T, Gardazi A, Tahir A, Prevez A, et al. Waste biomass adsorbents for copper removal from industrial wastewater- a review. Journal of Hazardous Materials. 2013;**263**:322-333

[12] Moh M. Drinking Water Quality Standard [Online]. Available from: http://kmam.moh.gov.my/public-user/drinking-water-quality-standard.html [Accessed: 15 June 2015]

[13] Nand N, Maata M, Koshy K, Sotheeswaran S. Water purification using Moringa oleifera and other locally available seeds in Fiji for heavy metal removal. International Journal of Applied Science and Technology. 2012;**2**:125-129

[14] Farooq U, Kozinski J, Khan M, Athar M. Biosorption of heavy metal ions using wheat based biosorbents – A review of the recent literature. Bioresource Technology. 2010;**101**: 5043-5053

[15] Karmakar A, Karmakar S, Mukherjee S. Properties of various plants and animals feedstocks for biodiesel production. Journal

of Bioresources Technology. 2010;**101**:7201-7210

[16] Suarez M, Entenza J, Doerries C, Meyer E, Bourquin L, Sutherland J, et al. Expression of a plant-derived peptide harboring water-cleaning and antimicrobial activities. Biotechnology and Bioengineering. 2003;**81**:12-20. DOI: 10.1002/bit.10550

[17] Al Bsoul A, Zeatoun L, Abdelhay A, Chiha M. Adsorption of copper ions from water by different types of natural seed materials. Desalination and Water Treatment. 2014;**52**:5876-5882. DOI: 10.1080/19443994.2013.808593

[18] Wasif N, Shahzad M, Muhammad T, Azra Y, Tehseen G, Maria A. Potential of Moringa oleifera L. as livestock fodder crop: A review. Turkish Journal of Agriculture and Forestry. 2014;**38**:1-14. DOI: 10.3906/tar-1211-66

[19] Charles W, Marcel D, Aly S, Philippe A, Sabadenedyo A. Determination of chemical composition and nutritional values of Moringa oleifera leaves. Pakistan Journal of Nutrition. 2011;**10**(3):264-268

[20] Mona S. Improvement of growth and nutritional quality of *Moringa oleifera* using different biofertilizers. Annals of Agricultural Science. 2012;**57**(1):53-62

[21] Verma A, Vijayakumar M, Mathela C, Chandana V. In vitro and in vivo antioxidant properties of different fractions of Moringa oleifera leaves. Food and Chemical Toxicology. 2009;**47**:2196-2201

[22] Makkar H, Becker K. Nutritional value and antinutritional components of whole and ethanol extracted Moringa oleifera leaves. Animal Feed Science and Technology. 1996;**63**:211-228

[23] Beveridge T, Murray R. Sites of metal deposition in the cell wall of Bacillus subtilis. Biotechnology and Applied Biochemistry. 1980;**141**:876-887

[24] Gupta V, Ali I. Utilization of bagasse fly ash (a sugar industry waste) for the removal of copper and zinc from wastewater. Separation and Purification Technology. 2000;**18**:131-140

[25] Han X, Liang C, Li T, Wang K, Huang H, Yang X. Simultaneous removal of cadmium and sulfamethoxazole from aqueous solution by rice straw biochar. Journal of Zhejiang University. Science. B. 2013;**14**:640-649

[26] Memon J, Memon S, Bhanger M, Memon G, El-Turki A, Allen G. Characterization of banana peel by scanning electron microscopy and FT-IR spectroscopy and its use for cadmium removal. Colloids and Surfaces. B, Biointerfaces. 2008;**66**:260-265

[27] Arnoldsson E, Bergman M, Matsinhe N, Persson M. Assessment of drinking water treatment by using Moringa oleifera natural coagulant. Vattern. 2008;**64**:137-150

[28] Folkard G, Sutherland J, Shaw R. Water clarification using Moringa oleifera seed coagulant. Waterlines. 1999;**17**:15-17

[29] Ali E, Muyibi S, Salleh H, Md Alam Z, Salleh M. Production of natural coagulation from *Moringa oleifera* seed for application in treatment of low turbidity water. Journal of Water Resource and Protection. 2010;**2**:259-266. DOI: 10.4236/jwarp.2010.23030

[30] Eman A, Tan C, Makky E. Impact of Moringa oleifera cake residue application on waste water treatment: A case study. Journal of Water Resource and Protection. 2014;**6**:677-687. DOI: 10.4236/jwarp.2014.67065. http://www.scrip.org/journal/jwarp

[31] Eman A. Removal of copper from water system using *Moringa*

oleifera press cake. In: Proceeding of International Conference on Agricultural, Civil and Environmental Engineering (ACEE-16). Istanbul, Turkey; 18-19 April 2016

[32] Reddy D, Harinath Y, Seshaiah K, Reddy A. Biosorption of Pb(II) from aqueous solutions using chemically modified Moringa oleifera tree leaves. Chemical Engineering Journal. 2010;**162**:626-634. DOI: 10.1016/j.cej.2010.06.010

[33] Reddy D, Seshaiah K, Reddy A, Lee S. Optimization of Cd(II), Cu(II) and Ni(II) biosorption by chemically modified Moringa oleifera leaves powder. Carbohydrate Polymers. 2012;**88**:1077-1086

[34] Eman A. Biosorption of Cd (II) from water by Moringa oleifera leaves. Advanced Materials Research. Switzerland: Trans Tech Publications. 2014;**925**:223-227. DOI: 10.4028/www.scientific.net/AMR.925.223

[35] Lea M. Bioremediation of turbid surface water using seed extract from *Moringa oleifera* lam. (drumstick) tree. Journal of. Current Protocols in Microbiology. 2010;**16**(1):1-14

[36] Alo M, Anyim C, Elom M. Coagulation and antimicrobial activities of Moringa oleifera seed storage at 3°C temperature in turbid water. Advanced Applied Science Research. 2012;**3**:887-894

[37] Sajidu S, Henry E, Kwamdera G, Mataka L. Removal of Lead, iron and cadmium ions by means of polyelectrolytes of the Moringa oleifera whole seed kernel. WIT Transactions on Ecology and the Environment. 2005;**80**:251-258

[38] Subramanium S, Vikashni N, Matakite M, Kanayathu K. Moringa oleifera and other local seeds in water purification in developing countries. Research Journal of Chemistry and Environment. 2011;**15**:135-138

[39] Vikashni N, Matakite M, Kanayathu K, Subramanium S. Water purification using Moringa oleifera and other locally available seeds in Fiji for heavy metal removal. International Journal of Applied Science and Technology. 2012;**5**:125-129

[40] Ravikumar K, Sheeja A. Heavy metal removal from water using Moringa oleifera seed coagulant and double filtration. International Journal of Scientific and Engineering Research. 2013;**4**(5):10-13

[41] Mataka L, Sajidu S, Masamba W, Mwatseteza J. Cadmium sorption by Moringa stemopetala and Moringa oleifera seed powders: Batch, time, temperature, pH and adsorption isotherm studies. International Journal of Water Resources and Environmental Engineering. 2010;**2**(3):50-59